Instructor Solutions Manual

Jay R. Schaffer

University of Northern Colorado

FOURTH EDITION

Elementary Statistics

Picturing *the* World

Larson | Farber

PEARSON

Prentice Hall

Upper Saddle River, NJ 07458

Editor-in-Chief, Mathematics & Statistics: Deirdre Lynch
Print Supplement Editor: Joanne Wendelken
Senior Managing Editor: Linda Mihatov Behrens
Project Manager, Production: Kristy S. Mosch
Art Director: Heather Scott
Supplement Cover Manager: Paul Gourhan
Supplement Cover Designer: Victoria Colotta
Operations Specialist: Ilene Kahn

The author and publisher of this book have used their best efforts in preparing this book. These efforts include the development, research, and testing of the theories and programs to determine their effectiveness. The author and publisher make no warranty of any kind, expressed or implied, with regard to these programs or the documentation contained in this book. The author and publisher shall not be liable in any event for incidental or consequential damages in connection with, or arising out of, the furnishing, performance, or use of these programs.

Printed in the United States of America

10 9 8 7 6 5 4

ISBN-13: 978-0-13-206291-6

ISBN-10:   0-13-206291-7

Pearson Education Ltd., *London*
Pearson Education Australia Pty. Ltd., *Sydney*
Pearson Education Singapore, Pte. Ltd.
Pearson Education North Asia Ltd., *Hong Kong*
Pearson Education Canada, Inc., *Toronto*
Pearson Educación de Mexico, S.A. de C.V.
Pearson Education—Japan, *Tokyo*
Pearson Education Malaysia, Pte. Ltd.

CONTENTS

CONTENTS

Introduction to Statistics

1.1 Try It Yourself Solutions

1a. The population consists of the prices per gallon of regular gasoline at all gasoline stations in the United States.

b. The sample consists of the prices per gallon of regular gasoline at the 800 surveyed stations.

c. The data set consists of the 800 prices.

2a. Because the numerical measure of $2,326,706,685 is based on the entire collection of player's salaries, it is from a population.

b. Because the numerical measure is a characteristic of a population, it is a parameter.

3a. Descriptive statistics involve the statement "76% of women and 60% of men had a physical examination within the previous year."

b. An inference drawn from the study is that a higher percentage of women had a physical examination within the previous year.

1.1 EXERCISE SOLUTIONS

1. A sample is a subset of a population.

2. It is usually impractical (too expensive and time consuming) to obtain all the population data.

3. A parameter is a numerical description of a population characteristic. A statistic is a numerical description of a sample characteristic.

4. Descriptive statistics and inferential statistics.

5. False. A statistic is a numerical measure that describes a sample characteristic.

6. True

7. True

8. False. Inferential statistics involves using a sample to draw conclusions about a population.

9. False. A population is the collection of *all* outcomes, responses, measurements, or counts that are of interest.

10. True

11. The data set is a population because it is a collection of the ages of all the members of the House of Representatives.

12. The data set is a sample because only every fourth person is measured.

13. The data set is a sample because the collection of the 500 spectators is a subset within the population of the stadium's 42,000 spectators.

14. The data set is a population because it is a collection of the annual salaries of all lawyers at a firm.

15. Sample, because the collection of the 20 patients is a subset within the population.

16. The data set is a population since it is a collection of the number of televisions in all U.S. households.

17. Population: Party of registered voters in Warren County.

    Sample: Party of Warren County voters responding to phone survey.

18. Population: Major of college students at Caldwell College.

    Sample: Major of college students at Caldwell College who take statistics.

19. Population: Ages of adults in the United States who own computers.

    Sample: Ages of adults in the United States who own Dell computers.

20. Population: Income of all homeowners in Texas.

    Sample: Income of homeowners in Texas with mortgages.

21. Population: All adults in the United States that take vacations.

    Sample: Collection of 1000 adults surveyed that take vacations.

22. Population: Collection of all infants in Italy.

    Sample: Collection of the 33,043 infants in the study.

23. Population: Collection of all households in the U.S.

    Sample: Collection of 1906 households surveyed.

24. Population: Collection of all computer users.

    Sample: Collection of 1000 computer users surveyed.

25. Population: Collection of all registered voters.

    Sample: Collection of 1045 registered voters surveyed.

26. Population: Collection of all students at a college.

    Sample: Collection of 496 college students surveyed.

27. Population: Collection of all women in the U.S.

    Sample: Collection of the 546 U.S. women surveyed.

28. Population: Collection of all U.S. vacationers.

    Sample: Collection of the 791 U.S. vacationers surveyed.

29. Statistic. The value $68,000 is a numerical description of a sample of annual salaries.

30. Statistic. 43% is a numerical description of a sample of high school students.

31. Parameter. The 62 surviving passengers out of 97 total passengers is a numerical description of all of the passengers of the Hindenburg that survived.

32. Parameter. 44% is a numerical description of the total number of governors.

33. Statistic. 8% is a numerical description of a sample of computer users.

34. Parameter. 12% is a numerical description of all new magazines.

35. Statistic. 53% is a numerical description of a sample of people in the United States.

36. Parameter. 21.1 is a numerical description of ACT scores for all graduates.

37. The statement "56% are the primary investor in their household" is an application of descriptive statistics.

   An inference drawn from the sample is that an association exists between U.S. women and being the primary investor in their household.

38. The statement "spending at least $2000 for their next vacation" is an application of descriptive statistics.

   An inference drawn from the sample is that U.S. vacationers are associated with spending more than $2000 for their next vacation.

39. Answers will vary.

40. (a) The volunteers in the study represent the sample.

   (b) The population is the collection of all individuals who completed the math test.

   (c) The statement "three times more likely to answer correctly" is an application of descriptive statistics.

   (d) An inference drawn from the sample is that individuals who are not sleep deprived will be three times more likely to answer math questions correctly than individuals who are sleep deprived.

41. (a) An inference drawn from the sample is that senior citizens who live in Florida have better memory than senior citizens who do not live in Florida.

   (b) It implies that if you live in Florida, you will have better memory.

42. (a) An inference drawn from the sample is that the obesity rate among boys ages 2 to 19 is increasing.

   (b) It implies the same trend will continue in future years.

43. Answers will vary.

1.2 DATA CLASSIFICATION

1.2 Try It Yourself Solutions

1a. One data set contains names of cities and the other contains city populations.

   b. City: Nonnumerical
   Population: Numerical

   c. City: Qualitative
   Population: Quantitative

2a. (1) The final standings represent a ranking of basketball teams.

   (2) The collection of phone numbers represents labels. No mathematical computations can be made.

   b. (1) Ordinal, because the data can be put in order.

   (2) Nominal, because you cannot make calculations on the data.

3a. (1) The data set is the collection of body temperatures.

   (2) The data set is the collection of heart rates.

   b. (1) Interval, because the data can be ordered and meaningful differences can be calculated, but it does not make sense writing a ratio using the temperatures.

   (2) Ratio, because the data can be ordered, can be written as a ratio, you can calculate meaningful differences, and the data set contains an inherent zero.

1.2 EXERCISE SOLUTIONS

1. Nominal and ordinal   **2.** Ordinal, Interval, and Ratio

3. False. Data at the ordinal level can be qualitative or quantitative.

4. False. For data at the interval level, you can calculate meaningful differences between data entries. You cannot calculate meaningful differences at the nominal or ordinal level.

5. False. More types of calculations can be performed with data at the interval level than with data at the nominal level.

6. False. Data at the ratio level can be placed in a meaningful order.

7. Qualitative, because telephone numbers are merely labels.

8. Quantitative, because the daily high temperature is a numerical measure.

9. Quantitative, because the lengths of songs on an MP3 player are numerical measures.

10. Qualitative, because the player numbers are merely labels.

11. Qualitative, because the poll results are merely responses.

12. Quantitative, because the diastolic blood pressure is a numerical measure.

13. Qualitative. Ordinal. Data can be arranged in order, but differences between data entries make no sense.

14. Qualitative. Nominal. No mathematical computations can be made and data are categorized using names.

15. Qualitative. Nominal. No mathematical computations can be made and data are categorized using names.

16. Quantitative. Ratio. A ratio of two data values can be formed so one data value can be expressed as a multiple of another.

17. Qualitative. Ordinal. The data can be arranged in order, but differences between data entries are not meaningful.

18. Quantitative. Ratio. The ratio of two data values can be formed so one data value can be expressed as a multiple of another.

19. Ordinal   **20.** Ratio   **21.** Nominal   **22.** Ratio

23. (a) Interval   (b) Nominal   (c) Ratio   (d) Ordinal

24. (a) Interval   (b) Nominal   (c) Interval   (d) Ratio

25. An inherent zero is a zero that implies "none." Answers will vary.

26. Answers will vary.

1.3 EXPERIMENTAL DESIGN

1.3 Try It Yourself Solutions

1a. (1) Focus: Effect of exercise on relieving depression.

    (2) Focus: Success of graduates.

  b. (1) Population: Collection of all people with depression.

    (2) Population: Collection of all university graduates.

  c. (1) Experiment

    (2) Survey

2a. There is no way to tell why people quit smoking. They could have quit smoking either from the gum or from watching the DVD.

  b. Two experiments could be done; one using the gum and the other using the DVD.

3a. Example: start with the first digits 92630782 . . .

  b. 92|63|07|82|40|19|26

  c. 63, 7, 40, 19, 26

4a. (1) The sample was selected by only using available students.

    (2) The sample was selected by numbering each student in the school, randomly choosing a starting number, and selecting students at regular intervals from the starting number.

  b. (1) Because the students were readily available in your class, this is convenience sampling.

    (2) Because the students were ordered in a manner such that every 25th student is selected, this is systematic sampling.

1.3 EXERCISE SOLUTIONS

1. In an experiment, a treatment is applied to part of a population and responses are observed. In an observational study, a researcher measures characteristics of interest of part of a population but does not change existing conditions.

2. A census includes the entire population; a sample includes only a portion of the population.

3. Assign numbers to each member of the population and use a random number table or use a random number generator.

4. Replication is the repetition of an experiment using a large group of subjects. It is important because it gives validity to the results.

5. True

6. False. A double-blind experiment is used to decrease the placebo effect.

7. False. Using stratified sampling guarantees that members of each group within a population will be sampled.

8. False. A census is a count of an entire population.

9. False. To select a systematic sample, a population is ordered in some way and then members of the population are selected at regular intervals.

10. True

11. In this study, you want to measure the effect of a treatment (using a fat substitute) on the human digestive system. So, you would want to perform an experiment.

12. It would be nearly impossible to ask every consumer whether he or she would still buy a product with a warning label. So, you should use a survey to collect these data.

13. Because it is impractical to create this situation, you would want to use a simulation.

14. Because the U.S. Congress keeps accurate financial records of all members, you could take a census.

15. (a) The experimental units are the 30–35 year old females being given the treatment.

    (b) One treatment is used.

    (c) A problem with the design is that there may be some bias on the part of the researchers if he or she knows which patients were given the real drug. A way to eliminate this problem would be to make the study into a double-blind experiment.

    (d) The study would be a double-blind study if the researcher did not know which patients received the real drug or the placebo.

16. (a) The experimental units are the people with early signs of arthritis.

    (b) One treatment is used.

    (c) A problem with the design is that the sample size is small. The experiment could be replicated to increase validity.

    (d) In a placebo-controlled double-blind experiment, neither the subject nor the experimenter knows whether the subject is receiving a treatment or a placebo. The experimenter is informed after all the data have been collected.

    (e) The group could be randomly split into 20 males or 20 females in each treatment group.

17. Each U.S. telephone number has an equal chance of being dialed and all samples of 1599 phone numbers have an equal chance of being selected, so this is a simple random sample. Telephone sampling only samples those individuals who have telephones, are available, and are willing to respond, so this is a possible source of bias.

18. Because the persons are divided into strata (rural and urban), and a sample is selected from each stratum, this is a stratified sample.

19. Because the students were chosen due to their convenience of location (leaving the library), this is a convenience sample. Bias may enter into the sample because the students sampled may not be representative of the population of students. For example, there may be an association between time spent at the library and drinking habits.

20. Because the disaster area was divided into grids and thirty grids were then entirely selected, this is a cluster sample. Certain grids may have been much more severely damaged than others, so this is a possible source of bias.

21. Because a random sample of out-patients were selected and all samples of 1210 patients had an equal chance of being selected, this is a simple random sample.

22. Because every twentieth engine part is sampled from an assembly line, this is a systematic sample. It is possible for bias to enter into the sample if, for some reason, the assembly line performs differently on a consistent basis.

23. Because a sample is taken from each one-acre subplot (stratum), this is a stratified sample.

24. Because a sample is taken from members of a population that are readily available, this is a convenience sample. The sample may be biased if the teachers sampled are not representative of the population of teachers. For example, some teachers may frequent the lounge more often than others.

25. Because every ninth name on a list is being selected, this is a systematic sample.

26. Each telephone has an equal chance of being dialed and all samples of 1012 phone numbers have an equal chance of being selected, so this is a simple random sample. Telephone sampling only samples those individuals who have telephones, are available, and are willing to respond, so this is a possible source of bias.

27. Answers will vary.

28. Answers will vary.

29. Census, because it is relatively easy to obtain the salaries of the 50 employees.

30. Sampling, because the population of students is too large to easily record their color. Random sampling would be advised since it would be easy to randomly select students then record their favorite car color.

31. Question is biased because it already suggests that drinking fruit juice is good for you. The question might be rewritten as "How does drinking fruit juice affect your health?"

32. Question is biased because it already suggests that drivers who change lanes several times are dangerous. The question might be rewritten as "Are drivers who change lanes several times dangerous?"

33. Question is unbiased because it does not imply how many hours of sleep are good or bad.

34. Question is biased because it already suggests that the media has a negative effect on teen girls' dieting habits. The question might be rewritten as "Do you think the media has an effect on teen girls' dieting habits?"

35. The households sampled represent various locations, ethnic groups, and income brackets. Each of these variables is considered a stratum.

36. Stratified sampling ensures that each segment of the population is represented.

37. Open Question

Advantage: Allows respondent to express some depth and shades of meaning in the answer.

Disadvantage: Not easily quantified and difficult to compare surveys.

Closed Question

Advantage: Easy to analyze results.

Disadvantage: May not provide appropriate alternatives and may influence the opinion of the respondent.

38. (a) Advantage: Usually results in a savings in the survey cost.

(b) Disadvantage: There tends to be a lower response rate and this can introduce a bias into the sample.
Sampling Technique: Convenience sampling

39. Answers will vary.

40. If blinding is not used, then the placebo effect is more likely to occur.

41. The Hawthorne effect occurs when a subject changes behavior because he or she is in an experiment. However, the placebo effect occurs when a subject reacts favorably to a placebo he or she has been given.

42. Both a randomized block design and a stratified sample split their members into groups based on similar characteristics.

43. Answers will vary.

CHAPTER 1 REVIEW EXERCISE SOLUTIONS

1. Population: Collection of all U.S. adults.

Sample: Collection of the 1000 U.S. adults that were sampled.

2. Population: Collection of all nurses in San Francisco area.

Sample: Collection of 38 nurses in San Francisco area that were sampled.

3. Population: Collection of all credit cards.

Sample: Collection of 146 credit cards that were sampled.

4. Population: Collection of all physicians in the U.S.

Sample: Collection of 1205 physicians that were sampled.

5. The team payroll is a parameter since it is a numerical description of a population (entire baseball team) characteristic.

6. Since 42% is describing a characteristic of the sample, this is a statistic.

7. Since "10 students" is describing a characteristic of a population of math majors, it is a parameter.

8. Since 19% is describing a characteristic of a sample of Indiana ninth graders, this is a statistic.

9. The average late fee of $27.46 charged by credit cards is representative of the descriptive branch of statistics. An inference drawn from the sample is that all credit cards charge a late fee of $27.46.

10. 60% of all physicians surveyed consider leaving the practice of medicine because they are discouraged over the state of U.S. healthcare is representative of the descriptive branch of statistics. An inference drawn from the sample is that 60% of all physicians surveyed consider leaving the practice of medicine because they are discouraged over the state of U.S. healthcare.

11. Quantitative because monthly salaries are numerical measurements.

12. Qualitative because Social Security numbers are merely labels for employees.

13. Quantitative because ages are numerical measurements.

14. Qualitative because zip codes are merely labels for the customers.

15. Interval. It makes no sense saying that 100 degrees is twice as hot as 50 degrees.

16. Ordinal. The data are qualitative but could be arranged in order of car size.

17. Nominal. The data are qualitative and cannot be arranged in a meaningful order.

18. Ratio. The data are numerical, and it makes sense saying that one player is twice as tall as another player.

19. Because CEOs keep accurate records of charitable donations, you could take a census.

20. Because it is impractical to create this situation, you would want to perform a simulation.

21. In this study, you want to measure the effect of a treatment (fertilizer) on a soybean crop. You would want to perform an experiment.

22. Because it would be nearly impossible to ask every college student about his/her opinion on environmental pollution, you should take a survey to collect the data.

23. The subjects could be split into male and female and then be randomly assigned to each of the five treatment groups.

24. Number the volunteers and then use a random number generator to randomly assign subjects to one of the treatment groups or the control group.

25. Because random telephone numbers were generated and called, this is a simple random sample.

26. Because the student sampled a convenient group of friends, this is a convenience sample.

27. Because each community is considered a cluster and every pregnant woman in a selected community is surveyed, this is a cluster sample.

28. Because every third car is stopped, this is a systematic sample.

29. Because grade levels are considered strata and 25 students are sampled from each stratum, this is a stratified sample.

30. Because of the convenience of surveying people waiting for their baggage, this is a convenience sample.

31. Telephone sampling only samples individuals who have telephones, are available, and are willing to respond.

32. Due to the convenience sample taken, the study may be biased toward the opinions of the student's friends.

33. The selected communities may not be representative of the entire area.

32. It may be difficult for the law enforcement official to stop every third car.

CHAPTER 1 QUIZ SOLUTIONS

1. Population: Collection of all individuals with anxiety disorders.

Sample: Collection of 372 patients in study.

2. (a) Statistic. 19% is a characteristic of a sample of Internet users.

(b) Parameter. 84% is a characteristic of the entire company (population).

(c) Statistic. 40% is a characteristic of a sample of Americans.

3. (a) Qualitative, since post office box numbers are merely labels.

(b) Quantitative, since a final exam is a numerical measure.

4. (a) Nominal. Badge numbers may be ordered numerically, but there is no meaning in this order and no mathematical computations can be made.

(b) Ratio. It makes sense to say that the number of candles sold during the 1st quarter was twice as many as sold in the 2nd quarter.

(c) Interval because meaningful differences between entries can be calculated, but a zero entry is not an inherent zero.

5. (a) In this study, you want to measure the effect of a treatment (low dietary intake of vitamin C and iron) on lead levels in adults. You want to perform an experiment.

(b) Because it would be difficult to survey every individual within 500 miles of your home, sampling should be used.

6. Randomized Block Design

7. (a) Because people were chosen due to their convenience of location (on the campground), this is a convenience sample.

(b) Because every tenth part is selected from an assembly line, this is a systematic sample.

(c) Stratified sample because the population is first stratified and then a sample is collected from each stratum.

8. Convenience

Descriptive Statistics

2.1 FREQUENCY DISTRIBUTIONS AND THEIR GRAPHS

2.1 Try It Yourself Solutions

1a. The number of classes (8) is stated in the problem.

b. Min = 15   Max = 89   Class width $= \dfrac{(89-15)}{8} = 9.25 \Rightarrow 10$

c.

Lower limit	Upper limit
15	24
25	34
35	44
45	54
55	64
65	74
75	84
85	94

d. See part (e).

e.

Class	Frequency, f
15–24	16
25–34	34
35–44	30
45–54	23
55–64	13
65–74	2
75–84	0
85–94	1

2a. See part (b).

b.

Class	Frequency, f	Midpoint	Relative frequency	Cumulative frequency
15–24	16	19.5	0.13	16
25–34	34	29.5	0.29	50
35–44	30	39.5	0.25	80
45–54	23	49.5	0.19	103
55–64	13	59.5	0.11	116
65–74	2	69.5	0.02	118
75–84	0	79.5	0.00	118
85–94	1	89.5	0.01	119
	$\sum f = 119$		$\sum \dfrac{f}{n} = 1$	

c. 86% of the teams scored fewer than 55 touchdowns. 3% of the teams scored more than 65 touchdowns.

3a.

Class Boundaries
14.5–24.5
24.5–34.5
34.5–44.5
44.5–54.5
54.5–64.5
64.5–74.5
74.5–84.5
84.5–94.5

b. Use class midpoints for the horizontal scale and frequency for the vertical scale.

c.

d. 86% of the teams scored fewer than 55 touchdowns. 3% of the teams scored more than 65 touchdowns.

4a. Use class midpoints for the horizontal scale and frequency for the vertical scale.

b. See part (c).

c.

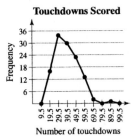

d. The number of touchdowns increases until 34.5 touchdowns, then decreases afterward.

5abc.

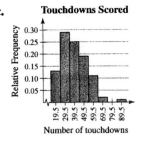

6a. Use upper class boundaries for the horizontal scale and cumulative frequency for the vertical scale.

b. See part (c).

c.

d. Approximately 80 teams scored 44 or fewer touchdowns.

e. Answers will vary.

7ab.

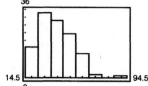

2.1 EXERCISE SOLUTIONS

1. By organizing the data into a frequency distribution, patterns within the data may become more evident.

2. Sometimes it is easier to identify patterns of a data set by looking at a graph of the frequency distribution.

3. Class limits determine which numbers can belong to that class.

   Class boundaries are the numbers that separate classes without forming gaps between them.

4. Cumulative frequency is the sum of the frequency for that class and all previous classes. Relative frequency is the proportion of entries in each class.

5. False. Class width is the difference between the lower and upper limits of consecutive classes.

6. True

7. False. An ogive is a graph that displays cumulative frequency.

8. True

9. Width $= \dfrac{\text{Max} - \text{Min}}{\text{Classes}} = \dfrac{58 - 7}{6} = 8.5 \Rightarrow 9$

   Lower class limits: 7, 16, 25, 34, 43, 52

   Upper class limits: 15, 24, 33, 42, 51, 60

10. Width $= \dfrac{\text{Max} - \text{Min}}{\text{Classes}} = \dfrac{94 - 11}{8} = 10.375 \Rightarrow 11$

    Lower class limits: 11, 22, 33, 44, 55, 66, 77, 88

    Upper class limits: 21, 32, 43, 54, 65, 76, 87, 98

11. Width $= \dfrac{\text{Max} - \text{Min}}{\text{Classes}} = \dfrac{123 - 15}{6} = 18 \Rightarrow 19$

    Lower class limits: 15, 34, 53, 72, 91, 110

    Upper class limits: 33, 52, 71, 90, 109, 128

12. Width $= \dfrac{\text{Max} - \text{Min}}{\text{Classes}} = \dfrac{171 - 24}{10} = 14.7 \Rightarrow 15$

    Lower class limits: 24, 39, 54, 69, 84, 99, 114, 129, 144, 159

    Upper class limits: 38, 53, 68, 83, 98, 113, 128, 143, 158, 173

13. (a) Class width $= 31 - 20 = 11$

(b) and (c)

Class	Frequency, f	Midpoint	Class boundaries
20–30	19	25	19.5–30.5
31–41	43	36	30.5–41.5
42–52	68	47	41.5–52.5
53–63	69	58	52.5–63.5
64–74	74	69	63.5–74.5
75–85	68	80	74.5–85.5
86–96	24	91	85.5–96.5
	$\sum f = 365$		

14a. Class width $= 10 - 0 = 10$

bc.

Class	Frequency	Midpoint	Class boundaries
0–9	188	4.5	−0.5–9.5
10–19	372	14.5	9.5–19.5
20–29	264	24.5	19.5–29.5
30–39	205	34.5	29.5–39.5
40–49	83	44.5	39.5–49.5
50–59	76	54.5	49.5–59.5
60–69	32	64.5	59.5–69.5
	$\Sigma f = 1220$		

15.

Class	Frequency, f	Midpoint	Relative frequency	Cumulative frequency
20–30	19	25	0.05	19
31–41	43	36	0.12	62
42–52	68	47	0.19	130
53–63	69	58	0.19	199
64–74	74	69	0.20	273
75–85	68	80	0.19	341
86–96	24	91	0.07	365
	$\sum f = 365$		$\sum \frac{f}{n} = 1$	

16.

Class	Frequency	Midpoint	Relative frequency	Cumulative frequency
0–9	188	4.5	0.15	188
10–19	372	14.5	0.30	560
20–29	264	24.5	0.22	824
30–39	205	34.5	0.17	1029
40–49	83	44.5	0.07	1112
50–59	76	54.5	0.06	1188
60–69	32	64.5	0.03	1220
	$\Sigma f = 1220$		$\sum \frac{f}{n} = 1$	

17. (a) Number of classes = 7   (b) Least frequency ≈ 10

(c) Greatest frequency ≈ 300 (d) Class width = 10

18. (a) Number of classes = 7   (b) Least frequency ≈ 100

(c) Greatest frequency ≈ 900 (d) Class width = 5

19. (a) 50   (b) 22.5–24.5 lbs

20. (a) 50   (b) 64–66 inches

21. (a) 24   (b) 29.5 lbs

22. (a) 44   (b) 66 inches

23. (a) Class with greatest relative frequency: 8–9 inches.
Class with least relative frequency: 17–18 inches.

(b) Greatest relative frequency $\approx$ 0.195
Least relative frequency $\approx$ 0.005

(c) Approximately 0.015

24. (a) Class with greatest relative frequency: 19–20 minutes.
Class with least relative frequency: 21–22 minutes.

(b) Greatest relative frequency $\approx$ 40%
Least relative frequency $\approx$ 2%

(c) Approximately 33%

25. Class with greatest frequency: 500–550
Class with least frequency: 250–300 and 700–750

26. Class with greatest frequency: 7.75–8.25
Class with least frequency: 6.25–6.75

27. Class width $= \dfrac{\text{Max} - \text{Min}}{\text{Number of classes}} = \dfrac{39 - 0}{5} = 7.8 \Rightarrow 8$

Class	Frequency, f	Midpoint	Relative frequency	Cumulative frequency
0–7	8	3.5	0.32	8
8–15	8	11.5	0.32	16
16–23	3	19.5	0.12	19
24–31	3	27.5	0.12	22
32–39	3	35.5	0.12	25
	$\sum f = 25$		$\sum \dfrac{f}{n} = 1$	

Class with greatest frequency: 0–7, 8–15
Class with least frequency: 16–23, 24–31, 32–39

28. Class width $= \dfrac{\text{Max} - \text{Min}}{\text{Number of classes}} = \dfrac{530 - 30}{6} = 83.3 \Rightarrow 84$

Class	Frequency	Midpoint	Relative frequency	Cumulative frequency
30–113	5	71.5	0.1724	5
114–197	7	155.5	0.2414	12
198–281	8	239.5	0.2759	20
282–365	2	323.5	0.0690	22
366–449	3	407.5	0.1034	25
450–533	4	491.5	0.1379	29
	$\sum f = 29$		$\sum \dfrac{f}{n} = 1$	

Class with greatest frequency: 198–281
Class with least frequency: 282–365

29. Class width $= \dfrac{\text{Max} - \text{Min}}{\text{Number of classes}} = \dfrac{7119 - 1000}{6} = 1019.83 \Rightarrow 1020$

Class	Frequency, f	Midpoint	Relative frequency	Cumulative frequency
1000–2019	12	1509.5	0.5455	12
2020–3039	3	2529.5	0.1364	15
3040–4059	2	3549.5	0.0909	17
4060–5079	3	4569.5	0.1364	20
5080–6099	1	5589.5	0.0455	21
6100–7119	1	6609.5	0.0455	22
	$\sum f = 22$		$\sum \dfrac{f}{n} = 1$	

July Sales for Representatives

Class with greatest frequency: 1000–2019
Class with least frequency: 5080–6099; 6100–7119

30. Class width $= \dfrac{\text{Max} - \text{Min}}{\text{Number of classes}} = \dfrac{51 - 32}{5} = 3.8 \Rightarrow 4$

Class	Frequency	Midpoint	Relative frequency	Cumulative frequency
32–35	3	33.5	0.1250	3
36–39	9	37.5	0.3750	12
40–43	8	41.5	0.3333	20
44–47	3	45.5	0.1250	23
48–51	1	49.5	0.0417	24
	$\sum f = 24$		$\sum \dfrac{f}{n} = 1$	

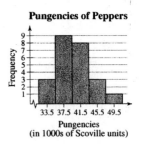

Pungencies of Peppers

Class with greatest frequency: 36–39

31. Class width $= \dfrac{\text{Max} - \text{Min}}{\text{Number of classes}} = \dfrac{514 - 291}{8} = 27.875 \Rightarrow 28$

Class	Frequency, f	Midpoint	Relative frequency	Cumulative frequency
291–318	5	304.5	0.1667	5
319–346	4	332.5	0.1333	9
347–374	3	360.5	0.1000	12
375–402	5	388.5	0.1667	17
403–430	6	416.5	0.2000	23
431–458	4	444.5	0.1333	27
459–486	1	472.5	0.0333	28
487–514	2	500.5	0.0667	30
	$\sum f = 30$		$\sum \dfrac{f}{n} = 1$	

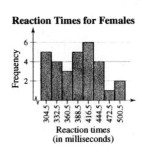

Reaction Times for Females

Class with greatest frequency: 403–430

32. Class width $= \dfrac{\text{Max} - \text{Min}}{\text{Number of classes}} = \dfrac{2888 - 2456}{5} = 86.4 \Rightarrow 87$

Class	Frequency	Midpoint	Relative frequency	Cumulative frequency
2456–2542	7	2499	0.28	7
2543–2629	3	2586	0.12	10
2630–2716	2	2673	0.08	12
2717–2803	4	2760	0.16	16
2804–2890	9	2847	0.36	25
	$\sum f = 25$		$\sum \dfrac{f}{n} = 1$	

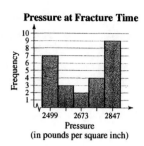

Pressure at Fracture Time

Class with greatest frequency: 2804–2890
Class with least frequency: 2630–2716

33. Class width $= \dfrac{\text{Max} - \text{Min}}{\text{Number of classes}} = \dfrac{264 - 146}{5} = 23.6 \Rightarrow 24$

Class	Frequency, f	Midpoint	Relative frequency	Cumulative frequency
146–169	6	157.5	0.2308	6
170–193	9	181.5	0.3462	15
194–217	3	205.5	0.1154	18
218–241	6	229.5	0.2308	24
242–265	2	253.5	0.0769	26
	$\sum f = 26$		$\sum \dfrac{f}{n} = 1$	

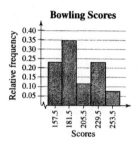

Bowling Scores

Class with greatest relative frequency: 170–193
Class with least relative frequency: 242–265

34. Class width $= \dfrac{\text{Max} - \text{Min}}{\text{Number of classes}} = \dfrac{80 - 10}{5} = 14 \Rightarrow 15$

Class	Frequency	Midpoint	Relative frequency	Cumulative frequency
10–24	11	17	0.3438	11
25–39	9	32	0.2813	20
40–54	6	47	0.1875	26
55–69	2	62	0.0625	28
70–84	4	77	0.1250	32
	$\sum f = 32$		$\sum \dfrac{f}{n} = 1$	

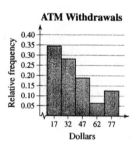

ATM Withdrawals

Class with greatest relative frequency: 10–24
Class with least relative frequency: 55–69

35. Class width $= \dfrac{\text{Max} - \text{Min}}{\text{Number of classes}} = \dfrac{52 - 33}{5} = 3.8 \Rightarrow 4$

Class	Frequency, f	Midpoint	Relative frequency	Cumulative frequency
33–36	8	34.5	0.3077	8
37–40	6	38.5	0.2308	14
41–44	5	42.5	0.1923	19
45–48	2	46.5	0.0769	21
49–52	5	50.5	0.1923	26
	$\sum f = 26$		$\sum \dfrac{f}{n} = 1$	

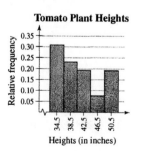

Tomato Plant Heights

Class with greatest relative frequency: 33–36
Class with least relative frequency: 45–48

36. Class width $= \dfrac{\text{Max} - \text{Min}}{\text{Number of classes}} = \dfrac{16 - 7}{5} = 1.8 \Rightarrow 2$

Class	Frequency	Midpoint	Relative frequency	Cumulative frequency
6–7	3	6.5	0.12	3
8–9	10	8.5	0.38	13
10–11	6	10.5	0.23	19
12–13	6	12.5	0.23	25
14–15	1	14.5	0.04	26
	$\sum f = 26$		$\sum \dfrac{f}{n} = 1$	

Class with greatest relative frequency: 8–9
Class with least relative frequency: 14–15

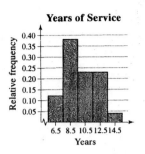

37. Class width $= \dfrac{\text{Max} - \text{Min}}{\text{Number of classes}} = \dfrac{73 - 52}{6} = 3.5 \Rightarrow 4$

Class	Frequency, f	Relative frequency	Cumulative frequency
52–55	3	0.125	3
56–59	3	0.125	6
60–63	9	0.375	15
64–67	4	0.167	19
68–71	4	0.167	23
72–75	1	0.042	24
	$\sum f = 24$	$\sum \dfrac{f}{n} \approx 1$	

Location of the greatest increase in frequency: 60–63

38. Class width $= \dfrac{\text{Max} - \text{Min}}{\text{Number of classes}} = \dfrac{57 - 16}{6} = 6.83 \Rightarrow 7$

Class	Frequency, f	Relative frequency	Cumulative frequency
16–22	2	0.10	2
23–29	3	0.15	5
30–36	8	0.40	13
37–43	5	0.25	18
44–50	0	0.00	18
51–57	2	0.10	20
	$\sum f = 20$	$\sum \dfrac{f}{n} = 1$	

Location of the greatest increase in frequency: 30–36

39. Class width $= \dfrac{\text{Max} - \text{Min}}{\text{Number of classes}} = \dfrac{18 - 2}{6} = 2.67 \Rightarrow 3$

Class	Frequency, f	Relative frequency	Cumulative frequency
2–4	9	0.3214	9
5–7	6	0.2143	15
8–10	7	0.2500	22
11–13	3	0.1071	25
14–16	2	0.0714	27
17–19	1	0.0357	28
	$\sum f = 28$	$\sum \dfrac{f}{n} \approx 1$	

Location of the greatest increase in frequency: 2–4

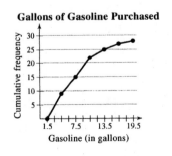

40. Class width $= \dfrac{\text{Max} - \text{Min}}{\text{Number of classes}} = \dfrac{29 - 1}{6} = 4.67 \Rightarrow 5$

Class	Frequency, f	Relative frequency	Cumulative frequency
1–5	5	0.2083	5
6–10	9	0.3750	14
11–15	3	0.1250	17
16–20	4	0.1667	21
21–25	2	0.0833	23
26–30	1	0.0417	24
	$\sum f = 24$	$\sum \dfrac{f}{n} = 1$	

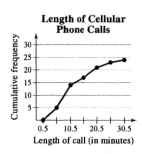

Length of Cellular Phone Calls

Location of the greatest increase in frequency: 6–10

41. Class width $= \dfrac{\text{Max} - \text{Min}}{\text{Number of classes}} = \dfrac{98 - 47}{5} = 10.2 \Rightarrow 11$

Class	Frequency, f	Midpoint	Relative frequency	Cumulative frequency
47–57	1	52	0.05	1
58–68	1	63	0.05	2
69–79	5	74	0.25	7
80–90	8	85	0.40	15
91–101	5	96	0.25	20
	$\sum f = 20$		$\sum \dfrac{f}{N} = 1$	

Exam Scores

Class with greatest frequency: 80–90
Classes with least frequency: 47–57 and 58–68

42.

Class	Frequency, f	Midpoint	Relative frequency	Cumulative frequency
0–2	16	1	0.3810	16
3–5	17	4	0.4048	33
6–8	7	7	0.1667	40
9–11	1	10	0.0238	41
12–14	0	13	0.0000	41
15–17	1	16	0.0238	42
	$\sum f = 42$		$\sum \dfrac{f}{N} = 1$	

Number of Children of First 42 Presidents

Classes with greatest frequency: 0–2
Classes with least frequency: 15–17

43. (a) Class width $= \dfrac{\text{Max} - \text{Min}}{\text{Number of classes}} = \dfrac{104 - 61}{8} = 5.375 \Rightarrow 6$

Class	Frequency, f	Midpoint	Relative frequency
61–66	1	63.5	0.0333
67–72	3	69.5	0.1000
73–78	6	75.5	0.2000
79–84	10	81.5	0.3333
85–90	5	87.5	0.1667
91–96	2	93.5	0.0667
97–102	2	99.5	0.0667
103–108	1	105.5	0.0333
	$\sum f = 30$		$\sum \dfrac{f}{N} = 1$

Daily Withdrawals

(b) 16.7%, because the sum of the relative frequencies for the last three classes is 0.167.

(c) $9600, because the sum of the relative frequencies for the last two classes is 0.10.

44. (a) Class width $= \dfrac{\text{Max} - \text{Min}}{\text{Number of classes}} = \dfrac{1359 - 410}{10} = 94.9 \Rightarrow 95$

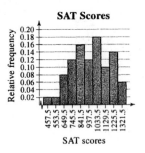

Class	Frequency	Midpoint	Relative frequency
410–505	1	457.5	0.02
506–601	1	553.5	0.02
602–697	4	649.5	0.08
698–793	6	745.5	0.12
794–889	8	841.5	0.16
890–985	6	937.5	0.12
986–1081	9	1033.5	0.18
1082–1177	5	1129.5	0.10
1178–1273	7	1225.5	0.14
1274–1369	3	1321.5	0.06
	$\sum f = 50$		$\sum \frac{f}{N} = 1$

(b) 48%, because the sum of the relative frequencies for the last four classes is 0.48.

(c) 698, because the sum of the relative frequencies for the last seven classes is 0.88.

45.

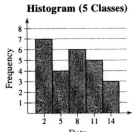

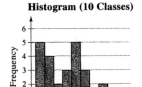

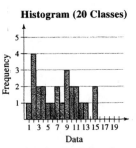

In general, a greater number of classes better preserves the actual values of the data set, but is not as helpful for observing general trends and making conclusions. When choosing the number of classes, an important consideration is the size of the data set. For instance, you would not want to use 20 classes if your data set contained 20 entries. In this particular example, as the number of classes increases, the histogram shows more fluctuation. The histograms with 10 and 20 classes have classes with zero frequencies. Not much is gained by using more than five classes. Therefore, it appears that five classes would be best.

2.2 MORE GRAPHS AND DISPLAYS

2.2 Try It Yourself Solutions

1a.
```
1
2
3
4
5
6
7
8
```

b. Key: $1|7 = 17$
```
1 | 7 5 8 8 5 5
2 | 7 6 8 9 8 9 7 9 8 7 5 3 4 6 2 5 0 1 1 2 1 4 1
3 | 9 9 7 7 8 8 6 4 7 6 5 5 5 1 4 5 9 8 2 5 2 2 2 3 3 3 2 4 1 1 0 4 2 1 2 1 0
4 | 9 8 6 8 7 8 6 4 8 5 4 6 6 7 1 1 5 4 5 3 2 2 8 3 0 4 0 5 3 0
5 | 9 4 5 4 3 5 5 9 0 2 3 5 7 0 5
6 | 8 5 1 3 3 1 1 0
7 |
8 | 9
```

c. Key: 1 | 7 = 17

```
1 | 5 5 5 7 8 8
2 | 0 1 1 1 1 2 2 3 4 4 5 5 5 6 6 7 7 7 8 8 8 9 9 9
3 | 0 0 1 1 1 1 1 2 2 2 2 2 2 2 3 3 3 4 4 4 4 5 5 5 5 5 6 6 7 7 7 8 8 8 9 9 9
4 | 0 0 0 1 1 2 2 3 3 3 3 4 4 4 4 5 5 5 5 6 6 6 6 7 7 8 8 8 8 8 9
5 | 0 0 2 3 3 4 4 5 5 5 5 7 9 9
6 | 0 1 1 1 3 3 5 8
7 |
8 | 9
```

d. It seems that most teams scored under 54 touchdowns.

2ab. Key: 1 | 7 = 17

```
1 |
1 | 5 5 5 7 8 8
2 | 0 1 1 1 1 2 2 3 4 4
2 | 5 5 6 6 7 7 7 8 8 8 9 9 9
3 | 0 0 1 1 1 1 1 2 2 2 2 2 2 2 3 3 3 4 4 4 4
3 | 5 5 5 5 5 6 6 7 7 7 8 8 8 9 9 9
4 | 0 0 0 1 1 2 2 3 3 3 4 4 4 4
4 | 5 5 5 5 6 6 6 6 7 7 8 8 8 8 8 9
5 | 0 0 2 3 3 4 4
5 | 5 5 5 5 7 9 9
6 | 0 1 1 1 3 3
6 | 5 8
7 |
7 |
8 |
8 | 9
```

3a. Use number of touchdowns for the horizontal axis.

b. **Touchdowns Scored**

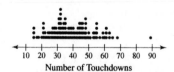

Number of Touchdowns

c. It appears that a large percentage of teams scored under 50 touchdowns.

4a.

Vehicle type	Killed (frequency)	Relative frequency	Central angle
Cars	22,423	0.64	(0.64)(360°) ≈ 230°
Trucks	10,216	0.29	(0.29)(360°) ≈ 104°
Motorcycles	2,227	0.06	(0.06)(360°) ≈ 22°
Other	425	0.01	(0.01)(360°) ≈ 4°
	$\sum f = 35{,}291$	$\sum \frac{f}{n} \approx 1$	$\sum = 360°$

b. **Motor Vehicle Occupants Killed in 1995**

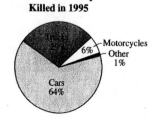

c. As a percentage of total vehicle deaths, car deaths decreased by 15%, truck deaths increased by 8%, and motorcycle deaths increased by 6%.

5a.

Cause	Frequency, f
Auto Dealers	14,668
Auto Repair	9,728
Home Furnishing	7,792
Computer Sales	5,733
Dry Cleaning	4,649

b.

c. It appears that the auto industry (dealers and repair shops) account for the largest portion of complaints filed at the BBB.

6ab.

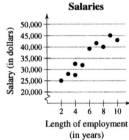

c. It appears that the longer an employee is with the company, the larger his/her salary will be.

7ab.

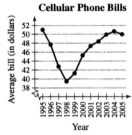

c. It appears that the average monthly bill for cellular telephone subscribers decreased significantly from 1995 to 1998, then increased from 1998 to 2004.

2.2 EXERCISE SOLUTIONS

1. Quantitative: Stem-and-Leaf Plot, Dot Plot, Histogram, Time Series Chart, Scatter Plot

Qualitative: Pie Chart, Pareto Chart

2. Unlike the histogram, the stem-and-leaf plot still contains the original data values. However, some data are difficult to organize in a stem-and-leaf plot.

3. Both the stem-and-leaf plot and the dot plot allow you to see how data are distributed, determine specific data entries, and identify unusual data values.

4. In the pareto chart, the height of each bar represents frequency or relative frequency and the bars are positioned in order of decreasing height with the tallest bar positioned to the left.

5. b   **6.** d   **7.** a   **8.** c

9. 27, 32, 41, 43, 43, 44, 47, 47, 48, 50, 51, 51, 52, 53, 53, 53, 54, 54, 54, 54, 55, 56, 56, 58, 59, 68, 68, 68, 73, 78, 78, 85

   Max: 85   Min: 27

10. 12.9, 13.3, 13.6, 13.7, 13.7, 14.1, 14.1, 14.1, 14.1, 14.3, 14.4, 14.4, 14.6, 14.9, 14.9, 15.0, 15.0, 15.0, 15.1, 15.2, 15.4, 15.6, 15.7, 15.8, 15.8, 15.8, 15.9, 16.1, 16.6, 16.7

    Max: 16.7   Min: 12.9

11. 13, 13, 14, 14, 14, 15, 15, 15, 15, 15, 16, 17, 17, 18, 19

    Max: 19   Min: 13

12. 214, 214, 214, 216, 216, 217, 218, 218, 220, 221, 223, 224, 225, 225, 227, 228, 228, 228, 228, 230, 230, 231, 235, 237, 239

    Max: 239   Min: 214

13. Anheuser-Busch is the top sports advertiser spending approximately $190 million. Honda spends the least. (Answers will vary.)

14. The value of the stock portfolio has increased fairly steadily over the past five years with the greatest increase happening between 2003 and 2006. (Answers will vary.)

15. Tailgaters irk drivers the most, while too cautious drivers irk drivers the least. (Answers will vary.)

16. The most frequent incident occurring while driving and using a cell phone is swerving. Twice as many people "sped up" than "cut off a car." (Answers will vary.)

17. Key: 6|7 = 67

    ```
    6 | 7 8
    7 | 3 5 5 6 9
    8 | 0 0 2 3 5 5 7 7 8
    9 | 0 1 1 1 2 4 5 5
    ```

    Most grades of the biology midterm were in the 80s or 90s.

18. Key: 4|0 = 40

    ```
    4 | 0 7 9 9
    5 | 0 1 2 4 6 8 9 9
    6 | 1 2 3 7
    7 | 1 3 6 8 9
    8 | 0 4 4 7
    ```

    It appears that most of the world's richest people are over 49 years old. (Answers will vary.)

19. Key: 4|3 = 4.3

    ```
    4 | 3 9
    5 | 1 8 8 8 9
    6 | 4 8 9 9 9
    7 | 0 0 2 2 2 5
    8 | 0 1
    ```

    It appears that most ice had a thickness of 5.8 centimeters to 7.2 centimeters. (Answers will vary.)

20. Key: 17|5 = 17.5

```
16 | 4 8
17 | 1 1 3 4 5 5 6 7 9
18 | 1 3 4 4 6 6 6 9
19 | 0 0 2 3 3 5 6
20 | 1 8
```

It appears that most farmers charge 17 to 19 cents per pound of apples. (Answers will vary.)

21.

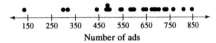

Advertisements

Number of ads

It appears that most of the 30 people from the U.S. see or hear between 450 and 750 advertisements per week. (Answers will vary.)

22.

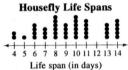

Housefly Life Spans

Life span (in days)

It appears that the lifespan of a fly tends to be between 8 and 11 days. (Answers will vary.)

23.

Category	Frequency	Relative frequency	Angle
North America	23	0.12	$(0.12)(360°) \approx 43°$
South America	12	0.06	$(0.06)(360°) \approx 23°$
Europe	43	0.22	$(0.22)(360°) \approx 81°$
Oceania	14	0.07	$(0.07)(360°) \approx 26°$
Africa	53	0.28	$(0.28)(360°) \approx 99°$
Asia	47	0.25	$(0.25)(360°) \approx 88°$
	$\sum f = 192$	$\sum \frac{f}{n} = 1$	

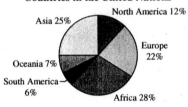

Countries in the United Nations

North America 12%
Asia 25%
Europe 22%
Oceania 7%
South America 6%
Africa 28%

Most countries in the United Nations come from Africa and the least amount come from South America. (Answers will vary.)

24.

Category	Budget frequency	Relative frequency	Angle
Science, aeronautics, and exploration	10,651	0.6343	$(0.6343)(360°) \approx 228°$
Exploration capabilities	6,108	0.3637	$(0.3637)(360°) \approx 131°$
Inspector General	34	0.0020	$(0.0020)(360°) \approx 0.7°$
	$\sum f = 16,793$	$\sum \frac{f}{N} = 1$	

2007 NASA Budget

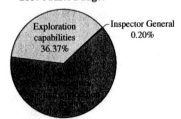

Exploration capabilities 36.37%
Inspector General 0.20%

It appears that 63.4% of NASA's budget went to science, aeronautics, and exploration. (Answers will vary.)

25.

It appears that the biggest reason for baggage delay comes from transfer baggage mishandling. (Answers will vary.)

26.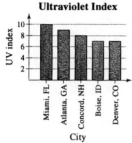

It appears that Boise, ID and Denver, CO have the same UV index. (Answers will vary.)

27.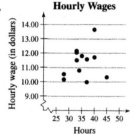

It appears that hourly wage increases as the number of hours worked increases. (Answers will vary.)

28.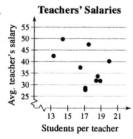

It appears that a teacher's average salary decreases as the number of students per teacher increases. (Answers will vary.)

29.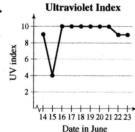

Of the period from June 14–23, 2001, in Memphis, TN, the ultraviolet index was highest from June 16–21. (Answers will vary.)

30.

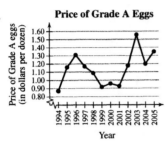

Daily High Temperatures in May

It appears that it was hottest from May 7 to May 11. (Answers will vary.)

31.

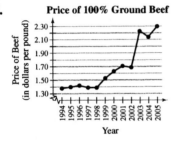

Price of Grade A Eggs

It appears the price of eggs peaked in 2003. (Answers will vary.)

32.

Price of 100% Ground Beef

It appears that the greatest increases in dollars per pound was 2002 to 2003. (Answers will vary.)

33. (a) When data are taken at regular intervals over a period of time, a time series chart should be used. (Answers will vary.)

(b)

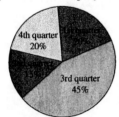

Sales for Company A

34. (a) The pie chart should be displaying all four quarters, not just the first three.

(b) Sales for Company B

35. (a) At law firm A, the lowest salary was $90,000 and the highest was $203,000; at law firm B, the lowest salary was $90,000 and the highest salary was $190,000.

   (b) There are 30 lawyers at law firm A and 32 lawyers at law firm B.

   (c) At Law Firm A, the salaries tend to be clustered at the far ends of the distribution range and at Law Firm B, the salaries tend to fall in the middle of the distribution range.

36. (a) At the 3:00 P.M. class, the youngest participant is 35 years old and the oldest participant is 85 years old. At the 8:00 P.M. class, the youngest participant is 18 years old and the oldest participant is 71 years old.

   (b) In the 3:00 P.M. class, there are 26 particpants and in the 8:00 P.M. class, there are 30 particpants.

   (c) The participants in each class are clustered at one of the ends of their distribution range. The 3:00 P.M. class mostly has particpants over 50 and the 8:00 P.M. class mostly has participants under 50. (Answers will vary.)

2.3   MEASURES OF CENTRAL TENDENCY

2.3 Try It Yourself Solutions

1a. $\Sigma x = 578$

b. $\bar{x} = \dfrac{\Sigma x}{n} = \dfrac{578}{14} = 41.3$

c. The mean age of an employee in a department is 41.3 years.

2a. 18 18, 19, 19, 19, 20, 21, 21, 21, 21, 23, 24, 24, 26, 27, 27, 29, 30, 30, 30, 33, 33, 34, 35, 38

b. median = middle entry = 24

c. The median age for the sample of fans at the concert is 24.

3a. 70, 80, 100, 130, 140, 150, 160, 200, 250, 270

b. median = mean of two middle entries {140, 150} = 145

c. The median price of the sample of MP3 players is $145.

4a. 0, 0, 1, 1, 1, 2, 3, 3, 3, 4, 5, 5, 5, 7, 9, 10, 12, 12, 13, 13, 13, 13, 13, 15, 16, 16, 17, 17, 18, 18, 18, 19, 19, 19, 20, 20, 21, 22, 23, 23, 24, 24, 25, 25, 26, 26, 26, 29, 33, 36, 37, 39, 39, 39, 39, 40, 40, 41, 41, 41, 42, 44, 44, 45, 47, 48, 49, 49, 49, 51, 53, 56, 58, 58, 59, 60, 67, 68, 68, 72

b. The age that occurs with the greatest frequency is 13 years old.

c. The mode of the ages is 13 years old.

5a. "Yes" occurs with the greatest frequency (171).

b. The mode of the responses to the survey is "Yes".

6a. $\bar{x} = \dfrac{\Sigma x}{n} = \dfrac{410}{19} \approx 21.6$

median $= 21$

mode $= 20$

b. The mean in Example 6 ($\bar{x} \approx 23.8$) was heavily influenced by the age 65. Neither the median nor the mode was affected as much by the age 65.

7ab.

Source	Score, x	Weight, w	$x \cdot w$
Test Mean	86	0.50	$(83)(0.50) = 43.0$
Midterm	96	0.15	$(96)(0.15) = 14.4$
Final	98	0.20	$(98)(0.20) = 19.6$
Computer Lab	98	0.10	$(98)(0.10) = 9.8$
Homework	100	0.05	$(100)(0.05) = 5.0$
		$\Sigma w = 1.00$	$\Sigma(x \cdot w) = 91.8$

c. $\bar{x} = \dfrac{\Sigma(x \cdot w)}{\Sigma w} = \dfrac{91.8}{1.00} = 91.8$

d. The weighted mean for the course is 91.8. So, you did get an A.

8abc.

Class	Midpoint, x	Frequency, f	$x \cdot f$
15–24	19.5	16	312
25–34	29.5	34	1003
35–44	39.5	30	1185
45–54	49.5	23	1138.5
55–64	59.5	13	773.5
65–74	69.5	2	139
75–84	79.5	0	0
85–94	89.5	1	89.5
		$N = 119$	$\Sigma(x \cdot f) = 4640.5$

d. $\mu = \dfrac{\Sigma(x \cdot f)}{N} = \dfrac{4640.5}{119} \approx 39.0$

The average number of touchdowns is approximately 39.0.

2.3 EXERCISE SOLUTIONS

1. True.

2. False. Not all data sets must have a mode.

3. False. All quantitative data sets have a median.

4. False. The mode is the only measure of central tendency that can be used for data at the nominal level of measurement.

5. False. When each data class has the same frequency, the distribution is uniform.

6. False. When the mean is greater than the median, the distrubution is skewed right.

7. Answers will vary. A data set with an outlier within it would be an example. For instance, the mean of the prices of existing home sales tends to be "inflated" due to the presence of a few very expensive homes.

8. Any data set that is symmetric has the same median and mode.

9. Skewed right because the "tail" of the distribution extends to the right.

10. Symmetric because the left and right halves of the distribution are approximately mirror images.

11. Uniform because the bars are approximately the same height.

12. Skewed left because the tail of the distribution extends to the left.

13. (11), because the distribution values range from 1 to 12 and has (approximately) equal frequencies.

14. (9), because the distribution has values in the thousands of dollars and is skewed right due to the few executives that make a much higher salary than the majority of the employees.

15. (12), because the distribution has a maximum value of 90 and is skewed left due to a few students scoring much lower than the majority of the students.

16. (10), because the distribution is rather symmetric due to the nature of the weights of seventh grade boys.

17. $\bar{x} = \dfrac{\Sigma x}{n} = \dfrac{81}{13} \approx 6.2$

    5 5 5 5 5 5 ⑥ 6 7 8 9 9

    middle value ⟹ median = 6

    mode = 5   (occurs 6 times)

18. $\bar{x} = \dfrac{\Sigma x}{n} = \dfrac{252}{10} = 25.2$

    19 20 21 22 22 23 25 30 35 35

    two middle values ⟹ median $= \dfrac{22+23}{2} = 22.5$

    mode = 22, 35   (occurs 2 times each)

19. $\bar{x} = \dfrac{\Sigma x}{n} = \dfrac{32}{7} \approx 4.57$

    3.7 4.0 4.8 ④.8 4.8 4.8 5.1

    middle value ⟹ median = 4.8

    mode = 4.8   (occurs 4 times)

20. $\bar{x} = \dfrac{\Sigma x}{n} = \dfrac{2004}{10} = 200.4$

    154 171 173 181 184 188 203 235 240 275

    two middle values ⟹ median $= \dfrac{184+188}{2} = 186$

    mode = none

    The mode cannot be found because no data points are repeated.

21. $\bar{x} = \dfrac{\Sigma x}{n} = \dfrac{661.2}{32} \approx 20.66$

10.5, 13.2, 14.9, 16.2, 16.7, 16.9, 17.6, 18.2, 18.6, 18.8, 18.8, 19.1, 19.2, 19.6, 19.8,

19.9, 20.2, 20.7, 20.9, 22.1, 22.1, 22.2, 22.9, 23.2, 23.3, 24.1, 24.9, 25.8, 26.6, 26.7, 26.7, 30.8

two middle values $\Rightarrow$ median $= \dfrac{19.9 + 20.2}{2} = 20.05$

mode $= 18.8, 22.1, 26.7$ (occurs 2 times each)

22. $\bar{x} = \dfrac{\Sigma x}{n} = \dfrac{1223}{20} = 61.2$

12 18 26 28 31 33 40 44 45 49 61 63 75 80 80 89 96 103 125 125

two middle values $\Rightarrow$ median $= \dfrac{49 + 61}{2} = 55$

mode $= 80, 125$

The modes do not represent the center of the data set because they are large values compared to the rest of the data.

23. $\bar{x} =$ not possible (nominal data)

median $=$ not possible (nominal data)

mode $=$ "Worse"

The mean and median cannot be found because the data are at the nominal level of measurement.

24. $\bar{x} =$ not possible (nominal data)

median $=$ not possible (nominal data)

mode $=$ "Watchful"

The mean and median cannot be found because the data are at the nominal level of measurement.

25. $\bar{x} = \dfrac{\Sigma x}{n} = \dfrac{1194.4}{7} \approx 170.63$

155.7, 158.1, 162.2, 169.3, 180, 181.8, 187.3

middle value $\Rightarrow$ median $= 169.3$

mode $=$ none

The mode cannot be found because no data points are repeated.

26. $\bar{x} =$ not possible (nominal data)

median $=$ not possible (nominal data)

mode $=$ "Domestic"

The mean and median cannot be found because the data are at the nominal level of measurement.

27. $\bar{x} = \dfrac{\Sigma x}{n} = \dfrac{226}{10} = 22.6$

14, 14, 15, 177, $\underbrace{18, 20}$, 22, 25, 40, 41

two middle values $\Rightarrow$ median $= \dfrac{18 + 20}{2} = 19$

mode = 14 (occurs 2 times)

28. $\bar{x} = \dfrac{\Sigma x}{n} = \dfrac{83}{5} = 16.6$

1, 10, ⑮ 25.5, 31.5

middle value $\Rightarrow$ median = 15

mode = none

The mode cannot be found because no data points are repeated.

29. $\bar{x} = \dfrac{\Sigma x}{n} = \dfrac{197.5}{14} \approx 14.11$

1.5, 2.5, 2.5, 5, 10.5, 11, $\underbrace{13, 15.5}$, 16.5, 17.5, 20, 26.5, 27, 28.5

two middle values $\Rightarrow$ median $= \dfrac{13 + 15.5}{2} = 14.25$

mode = 2.5 (occurs 2 times)

30. $\bar{x} = \dfrac{\Sigma x}{n} = \dfrac{3455}{11} \approx 314.1$

25, 35, 93, 110, 356, ③⑦④ 380, 445, 458, 480, 699

middle value $\Rightarrow$ median = 374

mode = none

The mode cannot be found because no data points are repeated.

31. $\bar{x} = \dfrac{\Sigma x}{n} = \dfrac{578}{14} = 41.3$

10, 12, 21, 24, 27, 37, $\underbrace{38, 41}$, 45, 45, 50, 57, 65, 106

two middle values $\Rightarrow$ median $= \dfrac{38 + 41}{2} = 39.5$

mode = 4.5 (occurs 2 times)

32. $\bar{x} = \dfrac{\Sigma x}{n} = \dfrac{29.9}{12} \approx 2.49$

0.8, 1.5, 1.6, 1.8, 2.1, $\underbrace{2.3, 2.4}$, 2.5, 3.0, 3.9, 4.0, 4.0

two middle values $\Rightarrow$ median $= \dfrac{2.3 + 2.4}{2} = 2.35$

mode = 4.0 (occurs 2 times)

33. $\bar{x} = \dfrac{\Sigma x}{n} = \dfrac{292}{15} \approx 19.5$

5, 8, 10, 15, 15, 15, 17, (20), 21, 22, 22, 25, 28, 32, 37

middle value $\Rightarrow$ median = 20

mode = 15 (occurs 3 times)

34. $\bar{x} = \dfrac{\Sigma x}{n} = \dfrac{2987}{14} \approx 213.4$

205, 208, 210, 212, 212, 214, 214, 214, 215, 215, 217, 217, 217, 217

two middle values $\Rightarrow$ median $= \dfrac{214 + 214}{2} = 214$

mode = 217 (occurs 4 times)

35. A = mode (data entry that occurred most often)

B = median (left of mean in skewed-right dist.)

C = mean (right of median in skewed-right dist.)

36. A = mean (left of median in skewed-left dist.)

B = median (right of mean in skewed-left dist.)

C = mode (data entry that occurred most often)

37. Mode because the data is nominal. **38.** Mean because the data are symmetric.

39. Mean because the data does not contain outliers.

40. Median because the data are skewed.

41.

Source	Score, x	Weight, w	x · w
Homework	85	0.05	(85)(0.05) = 4.25
Quiz	80	0.35	(80)(0.35) = 28
Project	100	0.20	(100)(0.20) = 20
Speech	90	0.15	(90)(0.15) = 13.5
Final Exam	93	0.25	(93)(0.25) = 23.25
		$\Sigma w = 1$	$\Sigma(x \cdot w) = 89$

$\bar{x} = \dfrac{\Sigma(x \cdot w)}{\Sigma w} = \dfrac{89}{1} = 89$

42.

Source	Score, x	Weight, w	x · w
MBAs	$45,500	8	(45,500)(8) = 364,000
Bas	$32,000	17	(32,000)(17) = 544,000
		$\Sigma w = 25$	$\Sigma(x \cdot w) = 908,000$

$\bar{x} = \dfrac{\Sigma(x \cdot w)}{\Sigma w} = \dfrac{908,000}{25} = \$36,320$

43.

Balance, x	Days, w	x · w
$523	24	(523)(24) = 12,552
$2415	2	(2415)(2) = 4830
$250	4	(250)(4) = 1000
	$\Sigma w = 30$	$\Sigma(x \cdot w) = 18,382$

$\bar{x} = \dfrac{\Sigma(x \cdot w)}{\Sigma w} = \dfrac{18,382}{30} = \$612.73$

44.

Balance, x	Days, w	$x \cdot w$
$759	15	$(759)(15) = 11,385$
$1985	5	$(1985)(5) = 9925$
$1410	5	$(1410)(5) = 7050$
$348	6	$(348)(6) = 2080$
	$\Sigma w = 31$	$\Sigma(x \cdot w) = 30,448$

$$\bar{x} = \frac{\Sigma(x \cdot w)}{\Sigma w} = \frac{30,448}{31} = \$982.19$$

45.

Grade	Points, x	Credits, w	$x \cdot w$
B	3	3	$(3)(3) = 9$
B	3	3	$(3)(3) = 9$
A	4	4	$(4)(4) = 16$
D	1	2	$(1)(2) = 2$
C	2	3	$(2)(3) = 6$
		$\Sigma w = 15$	$\Sigma(x \cdot w) = 42$

$$\bar{x} = \frac{\Sigma(x \cdot w)}{\Sigma w} = \frac{42}{15} = 2.8$$

46.

Source	Score, x	Weight, w	$x \cdot w$
Engineering	85	9	$(85)(9) = 765$
Business	81	13	$(81)(13) = 1053$
Math	90	5	$(90)(5) = 450$
		$\Sigma(x \cdot w) = 27$	$\Sigma w = 2268$

$$\bar{x} = \frac{\Sigma(x \cdot w)}{\Sigma w} = \frac{2268}{27} = 84$$

47.

Midpoint, x	Frequency, f	$x \cdot f$
61	4	$(61)(4) = 244$
64	5	$(64)(5) = 320$
67	8	$(67)(8) = 536$
70	1	$(70)(1) = 70$
	$n = 18$	$\Sigma(x \cdot f) = 1170$

$$\bar{x} = \frac{\Sigma(x \cdot f)}{n} = \frac{1170}{18} \approx 65 \text{ inches}$$

48.

Midpoint, x	Frequency, f	$x \cdot f$
64	3	$(64)(3) = 192$
67	6	$(67)(6) = 402$
70	7	$(70)(7) = 490$
73	4	$(73)(4) = 292$
76	3	$(76)(3) = 228$
	$\Sigma(x \cdot f) = 23$	$\Sigma(x \cdot f) = 1604$

$$\bar{x} = \frac{\Sigma(x \cdot f)}{n} = \frac{1604}{23} \approx 69.7 \text{ inches}$$

49.

Midpoint, x	Frequency, f	$x \cdot f$
4.5	55	$(4.5)(55) = 247.5$
14.5	70	$(14.5)(70) = 1015$
24.5	35	$(24.5)(35) = 857.5$
34.5	56	$(34.5)(56) = 1932$
44.5	74	$(44.5)(74) = 3293$
54.5	42	$(54.5)(42) = 2289$
64.5	38	$(64.5)(38) = 2451$
74.5	17	$(74.5)(17) = 1266.5$
84.5	10	$(84.5)(10) = 845$
	$n = 397$	$\Sigma(x \cdot f) = 14,196.5$

$$\bar{x} = \frac{\Sigma(x \cdot f)}{n} = \frac{14,196.5}{397} \approx 35.8 \text{ years old}$$

50.

Midpoint, x	Frequency, f	$x \cdot f$
3	12	$(3)(12) = 36$
8	26	$(8)(26) = 208$
13	20	$(13)(20) = 260$
18	7	$(18)(7) = 126$
23	11	$(23)(11) = 253$
28	7	$(28)(7) = 196$
33	4	$(33)(4) = 132$
38	4	$(38)(4) = 152$
43	1	$(43)(1) = 43$
	$n = 92$	$\Sigma(x \cdot f) = 1406$

$$\bar{x} = \frac{\Sigma(x \cdot f)}{n} = \frac{1406}{92} \approx 15.3 \text{ minutes}$$

51. Class width $= \dfrac{\text{Max} - \text{Min}}{\text{Number of classes}} = \dfrac{14 - 3}{6} = 1.83 \Rightarrow 2$

Class	Midpoint, x	Frequency, f
3–4	3.5	3
5–6	5.5	8
7–8	7.5	4
9–10	9.5	2
11–12	11.5	2
13–14	13.5	1
		$\Sigma f = 20$

Shape: Positively skewed

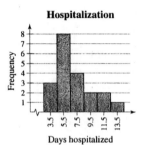

52. Class width $= \dfrac{\text{Max} - \text{Min}}{\text{Number of classes}} = \dfrac{297 - 127}{5} = 34$

Class	Midpoint, x	Frequency, f
127–161	144	9
162–196	179	8
197–231	214	3
232–266	249	3
267–301	284	1
		$\Sigma f = 24$

Shape: Positively skewed

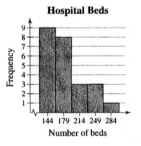

53. Class width $= \dfrac{\text{Max} - \text{Min}}{\text{Number of classes}} = \dfrac{76 - 62}{5} = 2.8 \Rightarrow 3$

Class	Midpoint, x	Frequency, f
62–64	63	3
65–67	66	7
68–70	69	9
71–73	72	8
74–76	75	3
		$\Sigma f = 30$

Shape: Symmetric

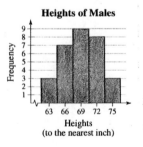

54. Class width $= \dfrac{\text{Max} - \text{Min}}{\text{Number of classes}} = \dfrac{6-1}{6} = 0.8333 \Rightarrow 1$

Class	Frequency, f
1	6
2	5
3	4
4	6
5	4
6	5
	$\Sigma f = 30$

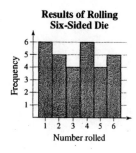

Results of Rolling
Six-Sided Die

Shape: Uniform

55. (a) $\bar{x} = \dfrac{\Sigma x}{n} = \dfrac{36.03}{6} = 6.005$

5.59, 5.99, 6, 6.02, 6.03, 6.4

two middle values $\Rightarrow$ median $= \dfrac{6 + 6.02}{2} = 6.01$

(b) $\bar{x} = \dfrac{\Sigma x}{n} = \dfrac{35.67}{6} = 5.945$

5.59, 5.99, 6, 6.02, 6.03, 6.4

two middle values $\Rightarrow$ median $= \dfrac{6 + 6.02}{2} = 6.01$

(c) mean

56. (a) $\bar{x} = \dfrac{\Sigma x}{n} = \dfrac{815.4}{19} = 42.92$

7.8, 8.2, 12.6, 12.6, 14.4, 17.8, 19.2, 21.3, 23, 24.2, 24.7, 31.1, 32.5, 41.3, 45.4, 55.2, 59.6, 134.2, 230.3

middle value $\Rightarrow$ median $= 24.2$

(b) $\bar{x} = \dfrac{\Sigma x}{n} = \dfrac{585.1}{18} \approx 32.51$

7.8, 8.2, 12.6, 12.6, 14.4, 17.8, 19.2, 21.3, 23, 24.2, 24.7, 31.1, 32.5, 41.3, 45.4, 55.5, 59.6, 134.2

two middle values $\Rightarrow$

median $= \dfrac{23 + 24.2}{2} = 23.6$

(c) mean

57. (a) $\bar{x} = \dfrac{\Sigma x}{n} = \dfrac{3222}{9} = 358$

147, 177, 336, 360, 375, 393, 408, 504, 522

middle value $\Rightarrow$ median $= 375$

(b) $\bar{x} = \dfrac{\Sigma x}{n} = \dfrac{9666}{9} = 1074$

441, 531, 1008, 1080, (1125) 1179, 1224, 1512, 1566

↑──── middle value ⟹ median = 1125

(c) $\bar{x}$ from part (b) is 3 times $\bar{x}$ from part (a). Median from part (b) is 3 times median from part (a).

(d) Multiply part (b) answers by 12.

58. (a) Mean should be used because Car A has the highest mean of the three.

(b) Median should be used because Car B has the highest median of the three.

(c) Mode should be used because Car C has the highest mode of the three.

59. Car A because it has the highest midrange of the three.

Car A: Midrange $= \dfrac{34 + 28}{2} = 32$

Car B: Midrange $= \dfrac{31 + 29}{2} = 30$

Car C: Midrange $= \dfrac{32 + 28}{2} = 30$

60. (a) $\bar{x} = 49.2$, median $= 46.5$

(b) Key: $3|6 = 36$ (c) Positively skewed

```
1 | 1 3
2 | 2 8
3 | 6 6 6 7 7 7 8  ── median
4 | 1 3 4 6 7
5 | 1 1 1 3          mean
6 | 1 2 3 4
7 | 2 2 4 6
8 | 5
9 | 0
```

61. (a) Order the data values.

11 13 22 28 36 36 36 37 37 37 38 41 43 44 46
47 51 51 51 53 61 62 63 64 72 72 74 76 85 90

Delete the lowest 10%, smallest 3 observations (11, 13, 22).

Delete the highest 10%, largest 3 observations (76, 85, 90).

Find the 10% trimmed mean using the remaining 24 observations.

10% trimmed mean = 49.2

(b) $\bar{x} = 49.2$

　median $= 46.5$

　mode $= 36,\ 37,\ 51$

　midrange $= 50.5$

(c) Using a trimmed mean eliminates potential outliers that may affect the mean of all the observations.

2.4　MEASURES OF VARIATION

2.4 Try It Yourself Solutions

1a. Min $= 23$ or $\$23,000$ and Max $= 58$ or $\$58,000$

b. Range $=$ max $-$ min $= 58 - 23 = 35$ or $\$35,000$

c. The range of the starting salaries for Corporation B is 35 or $\$35,000$ (much larger than range of Corporation A).

2a. $\mu = \dfrac{\Sigma x}{N} = \dfrac{415}{10} = 41.5$ or $\$41,500$

b.

Salary, x (1000s of dollars)	Deviation, $x - \mu$ (1000s of dollars)
23	$23 - 41.5 = -18.5$
29	$29 - 41.5 = -12.5$
32	$32 - 41.5 = -9.5$
40	$40 - 41.5 = -1.5$
41	$41 - 41.5 = -0.5$
41	$41 - 41.5 = -0.5$
49	$49 - 41.5 = 7.5$
50	$50 - 41.5 = 8.5$
52	$52 - 41.5 = 10.5$
58	$58 - 41.5 = 16.5$
$\Sigma x = 415$	$\Sigma(x - \mu) = 0$

3ab. $\mu = 41.5$ or $\$41,500$

Salary, x	$x - \mu$	$(x - \mu)^2$
23	-18.5	$(-18.5)^2 = 342.25$
29	-12.5	$(-12.5)^2 = 156.25$
32	-9.5	$(-9.5)^2 = 90.25$
40	-1.5	$(-1.5)^2 = 2.25$
41	-0.5	$(-0.5)^2 = 0.25$
41	-0.5	$(-0.5)^2 = 0.25$
49	7.5	$(7.5)^2 = 56.25$
50	8.5	$(8.5)^2 = 72.25$
52	10.5	$(10.5)^2 = 110.25$
58	16.5	$(16.5)^2 = 272.25$
$\Sigma x = 415$	$\Sigma(x - \mu) = 0$	$\Sigma(x - \mu)^2 = 1102.5$

c. $\sigma^2 = \dfrac{\Sigma(x - \mu)^2}{N} = \dfrac{1102.5}{10} \approx 110.25$

d. $\sigma = \sqrt{\sigma^2} = \sqrt{\dfrac{1102.5}{10}} = 10.5$ or $10,500

e. The population variance is 110.3 and the population standard deviation is 10.5 or $10,500.

4a.

Salary, x	$x - \bar{x}$	$(x - \bar{x})^2$
23	−18.5	342.25
29	−12.5	156.25
32	−9.5	90.25
40	−1.5	2.25
41	−0.5	0.25
41	−0.5	0.25
49	7.5	56.25
50	8.5	72.25
52	10.5	110.25
58	16.5	272.25
$\Sigma x = 415$	$\Sigma(x - \bar{x}) = 0$	$\Sigma(x - \bar{x})^2 = 1102.5$

$$SS_x = \Sigma(x - \bar{x})^2 = 1102.5$$

b. $s^2 = \dfrac{\Sigma(x - \bar{x})^2}{(n - 1)} = \dfrac{1102.5}{9} = 122.5$

c. $s = \sqrt{s^2} = \sqrt{122.5} \approx 11.1$ or $11,100

5a. (Enter data in computer or calculator)

b. $\bar{x} = 37.89,$ $s = 3.98$

6a. 7, 7, 7, 7, 7, 13, 13, 13, 13, 13

b.

x	$x - \mu$	$(x - \mu)^2$
7	$7 - 10 = -3$	$(-3)^2 = 9$
7	$7 - 10 = -3$	$(-3)^2 = 9$
7	$7 - 10 = -3$	$(-3)^2 = 9$
7	$7 - 10 = -3$	$(-3)^2 = 9$
7	$7 - 10 = -3$	$(-3)^2 = 9$
13	$13 - 10 = 3$	$(3)^2 = 9$
13	$13 - 10 = 3$	$(3)^2 = 9$
13	$13 - 10 = 3$	$(3)^2 = 9$
13	$13 - 10 = 3$	$(3)^2 = 9$
13	$13 - 10 = 3$	$(3)^2 = 9$
$\Sigma x = 100$	$\Sigma(x - \mu) = 0$	$\Sigma(x - \mu)^2 = 90$

$$\mu = \frac{\Sigma x}{N} = \frac{100}{10} = 10$$

$$\sigma = \sqrt{\frac{\Sigma(x - \mu)^2}{N}} = \sqrt{\frac{90}{10}} = \sqrt{9} = 3$$

7a. $64 - 61.25 = 2.75 = 1$ standard deviation

b. 34%

c. The estimated percent of the heights that are between 61.25 and 64 inches is 34%.

8a. $31.6 - 2(19.5) = -7.4 = > 0$

b. $31.6 + 2(19.5) = 70.6$

c. $1 - \dfrac{1}{k^2} = 1 - \dfrac{1}{(2)^2} = 1 - \dfrac{1}{4} = 0.75$

At least 75% of the data lie within 2 standard deviations of the mean. At least 75% of the population of Alaska is between 0 and 70.6 years old.

9a.

x	f	xf
0	10	$(0)(10) = 0$
1	19	$(1)(19) = 19$
2	7	$(2)(7) = 14$
3	7	$(3)(7) = 21$
4	5	$(4)(5) = 20$
5	1	$(5)(1) = 5$
6	1	$(6)(1) = 6$
	$n = 50$	$\sum xf = 85$

b. $\bar{x} = \dfrac{\sum xf}{n} = \dfrac{85}{50} = 1.7$

c.

$x - \bar{x}$	$(x - \bar{x})^2$	$(x - \bar{x})^2 \cdot f$
$0 - 1.7 = -1.70$	$(-1.70)^2 = 2.8900$	$(2.8900)(10) = 28.90$
$1 - 1.7 = -0.70$	$(-0.70)^2 = 0.4900$	$(0.4900)(19) = 9.31$
$2 - 1.7 = 0.30$	$(0.30)^2 = 0.0900$	$(0.0900)(7) = 0.63$
$3 - 1.7 = 1.30$	$(1.30)^2 = 1.6900$	$(1.6900)(7) = 11.83$
$4 - 1.7 = 2.30$	$(2.30)^2 = 5.2900$	$(5.2900)(5) = 26.45$
$5 - 1.7 = 3.30$	$(3.30)^2 = 10.9800$	$(10.9800)(1) = 10.89$
$6 - 1.7 = 4.30$	$(4.30)^2 = 18.4900$	$(18.4900)(1) = 18.49$
		$\sum (x - \bar{x})^2 f = 106.5$

d. $s = \sqrt{\dfrac{\sum (x - \bar{x})^2 f}{(n-1)}} = \sqrt{\dfrac{106.5}{49}} = \sqrt{2.17} \approx 1.5$

10a.

Class	x	f	xf
0–99	49.5	380	$(49.5)(380) = 18{,}810$
100–199	149.5	230	$(149.5)(230) = 34{,}385$
200–299	249.5	210	$(249.5)(210) = 52{,}395$
300–399	349.5	50	$(349.5)(50) = 17{,}475$
400–499	449.5	60	$(449.5)(60) = 26{,}970$
500+	650.0	70	$(650.0)(70) = 45{,}500$
		$n = 1000$	$\sum xf = 195{,}535$

b. $\bar{x} = \dfrac{\sum xf}{n} = \dfrac{195{,}535}{1000} \approx 195.5$

c.

$x - \bar{x}$	$(x - \bar{x})^2$	$(x - \bar{x})^2 \cdot f$
$49.5 - 195.5 = -146$	$(-146)^2 = 21{,}316$	$(21{,}316)(380) = 8{,}100{,}080$
$149.5 - 195.5 = -46$	$(-46)^2 = 2116$	$(2116)(230) = 486{,}680$
$249.5 - 195.5 = 54$	$(54)^2 = 2916$	$(2916)(210) = 612{,}360$
$349.5 - 195.5 = 154$	$(154)^2 = 23{,}716$	$(23{,}716)(50) = 1{,}185{,}800$
$449.5 - 195.5 = 254$	$(254)^2 = 64{,}516$	$(64{,}516)(60) = 3{,}870{,}960$
$650 - 195.5 = 454.5$	$(454.5)^2 = 206{,}570.25$	$(206{,}570.25)(70) = 14{,}459{,}917.5$
		$\sum(x - \bar{x})^2 f = 28{,}715{,}797.5$

d. $s = \sqrt{\dfrac{\sum(x - \bar{x})^2 f}{n - 1}} = \sqrt{\dfrac{28{,}715{,}797.5}{999}} = \sqrt{28{,}744.542} \approx 169.5$

2.4 EXERCISE SOLUTIONS

1. Range $=$ Max $-$ Min $= 12 - 4 = 8$

$\mu = \dfrac{\sum x}{N} = \dfrac{79}{10} = 7.9$

x	$x - \mu$	$(x - \mu)^2$
12	$12 - 7.9 = 4.1$	$(4.1)^2 = 16.81$
9	$9 - 7.9 = 1.1$	$(1.1)^2 = 1.21$
7	$7 - 7.9 = -0.9$	$(-0.9)^2 = 0.81$
5	$5 - 7.9 = -2.9$	$(-2.9)^2 = 8.41$
7	$7 - 7.9 = -0.9$	$(-0.9)^2 = 0.81$
8	$8 - 7.9 = 0.1$	$(0.1)^2 = 0.01$
10	$10 - 7.9 = 2.1$	$(2.1)^2 = 4.41$
4	$4 - 7.9 = -3.9$	$(-3.9)^2 = 15.21$
11	$11 - 7.9 = 3.1$	$(3.1)^2 = 9.61$
6	$6 - 7.9 = -1.9$	$(-1.9)^2 = 3.61$
$\sum x = 79$	$\sum(x - \mu) = 0$	$\sum(x - \mu)^2 = 60.9$

$\sigma^2 = \dfrac{\sum(x - \mu)^2}{N} = \dfrac{60.9}{10} = 6.09 \approx 6.1$

$\sigma = \sqrt{\dfrac{\sum(x - \mu)^2}{N}} = \sqrt{6.09} \approx 2.5$

2. Range $=$ Max $-$ Min $= 24 - 14 = 10$

$\mu = \dfrac{\sum x}{N} = \dfrac{264}{14} = 18.9$

x	$x - \mu$	$(x - \mu)^2$
15	$15 - 18.9 = -3.9$	$(-3.9)^2 = 15.21$
24	$24 - 18.9 = 5.1$	$(5.1)^2 = 26.01$
17	$17 - 18.9 = -1.9$	$(-1.9)^2 = 3.61$
19	$19 - 18.9 = 0.1$	$(0.1)^2 = 0.01$
20	$20 - 18.9 = 1.1$	$(1.1)^2 = 1.21$
18	$18 - 18.9 = -0.9$	$(-0.9)^2 = 0.81$
20	$20 - 18.9 = 1.1$	$(1.1)^2 = 1.21$
16	$16 - 18.9 = -2.9$	$(-2.9)^2 = 8.41$
21	$21 - 18.9 = 2.1$	$(2.1)^2 = 4.41$
23	$23 - 18.9 = 4.1$	$(4.1)^2 = 16.81$
17	$17 - 18.9 = -1.9$	$(-1.9)^2 = 3.61$
18	$18 - 18.9 = -0.9$	$(-0.9)^2 = 0.81$
22	$22 - 18.9 = 3.1$	$(3.1)2 = 9.61$
14	$14 - 18.9 = -4.9$	$(-4.9)^2 = 24.01$
$\sum x = 264$	$\sum(x - \mu) = 0$	$\sum(x - \mu)^2 = 115.74$

$$\sigma^2 = \frac{\Sigma(x - \mu)^2}{N} = \frac{115.74}{14} = 8.27 \approx 8.3$$

$$\sigma = \sqrt{\frac{\Sigma(x - \mu)^2}{N}} = \sqrt{8.27} \approx 2.9$$

3. Range = Max − Min = 18 − 6 = 12

$$\bar{x} = \frac{\Sigma x}{n} = \frac{107}{9} = 11.9$$

x	$(x - \bar{x})$	$(x - \bar{x})^2$
17	17 − 11.9 = 5.1	$(5.1)^2 = 26.01$
8	8 − 11.9 = −3.9	$(-3.9)^2 = 15.21$
13	13 − 11.9 = 1.1	$(1.1)^2 = 1.21$
18	18 − 11.9 = 6.1	$(6.1)^2 = 37.21$
15	15 − 11.9 = 3.1	$(3.1)^2 = 9.61$
9	9 − 11.9 = −2.9	$(-2.9)^2 = 8.41$
10	10 − 11.9 = −1.9	$(-1.9)^2 = 3.61$
11	11 − 11.9 = −0.9	$(-0.9)^2 = 0.81$
6	6 − 11.9 = −5.9	$(-5.9)^2 = 34.81$
$\Sigma x = 107$	$\Sigma(x - \bar{x}) = 0$	$\Sigma(x - \bar{x})^2 = 136.89$

$$s^2 = \frac{\Sigma(x - \bar{x})^2}{n - 1} = \frac{136.89}{9 - 1} = 17.1$$

$$s = \sqrt{\frac{\Sigma(x - \bar{x})^2}{n - 1}} = \sqrt{17.1} \approx 4.1$$

4. Range = Max − Min = 28 − 7 = 21

$$\bar{x} = \frac{\Sigma x}{n} = \frac{238}{13} = 18.3$$

x	$(x - \bar{x})$	$(x - \bar{x})^2$
28	28 − 18.3 = 9.7	$(9.7)^2 = 94.09$
25	25 − 18.3 = 6.7	$(6.7)^2 = 44.89$
21	21 − 18.3 = 2.7	$(2.7)^2 = 7.29$
15	15 − 18.3 = −3.3	$(-3.3)^2 = 10.89$
7	7 − 18.3 = −11.3	$(-11.3)^2 = 127.69$
14	14 − 18.3 = −4.3	$(-4.3)^2 = 18.49$
9	9 − 18.3 = −9.3	$(-9.3)^2 = 86.49$
27	27 − 18.3 = 8.7	$(8.7)^2 = 75.69$
21	21 − 18.3 = 2.7	$(2.7)^2 = 7.29$
24	24 − 18.3 = 5.7	$(5.7)^2 = 32.49$
14	14 − 18.3 = −4.3	$(-4.3)^2 = 18.49$
17	17 − 18.3 = −1.3	$(-1.3)^2 = 1.69$
16	16 − 18.3 = −2.3	$(-2.3)^2 = 5.29$
$\Sigma x = 238$	$\Sigma(x - \bar{x}) = 0$	$\Sigma(x - \bar{x})^2 = 530.77$

$$s^2 = \frac{\Sigma(x - \bar{x})^2}{n - 1} = \frac{530.77}{13 - 1} = 44.23 \approx 44.2$$

$$s = \sqrt{\frac{\Sigma(x - \bar{x})^2}{n - 1}} = \sqrt{44.23} \approx 6.7$$

5. Range = Max − Min = 96 − 23 = 73

6. Range = Max − Min = 34 − 24 = 10

7. The range is the difference between the maximum and minimum values of a data set. The advantage of the range is that it is easy to calculate. The disadvantage is that it uses only two entries from the data set.

8. The deviation, $(x - \mu)$, is the difference between an observation, x, and the mean of the data, μ. The sum of the deviations is always zero.

9. The units of variance are squared. Its units are meaningless. (Ex: dollars^2)

10. The standard deviation is the positive square root of the variance.

   Because squared deviations can never be negative, the standard deviation and variance can never be negative.

   $$\{7, 7, 7, 7, 7\} \rightarrow n = 5$$
   $$\bar{x} = 7$$
   $$s = 0$$

11. (a) Range = Max − Min = 45.6 − 21.3 = 24.3

   (b) Range = Max − Min = 65.6 − 21.3 = 44.3

   (c) The range has increased substantially.

12. $\{3, 3, 3, 7, 7, 7\} \rightarrow n = 6$
   $$\mu = 5$$
   $$s = 2$$

13. Graph (a) has a standard deviation of 24 and graph (b) has a standard deviation of 16 because graph (a) has more variability.

14. Graph (a) has a standard deviation of 2.4 and graph (b) has a standard deviation of 5. Graph (b) has more variability.

15. When calculating the population standard deviation, you divide the sum of the squared deviations by N, then take the square root of that value. When calculating the sample standard deviation, you divide the sum of the squared deviations by $n - 1$, then take the square root of that value.

16. When given a data set, one would have to determine if it represented the population or was a sample taken from a population. If the data are a population, then σ is calculated. If the data are a sample, then s is calculated.

17. Company B. Due to the larger standard deviation in salaries for company B, it would be more likely to be offered a salary of $33,000.

18. Player B. Due to the smaller standard deviation in number of strokes, player B would be the more consistent player.

19. (a) Los Angeles: range = Max − Min = 35.9 − 18.3 = 17.6

x	$(x - \bar{x})$	$(x - \bar{x})^2$
20.2	−6.06	36.67
26.1	−0.16	0.02
20.9	−5.36	28.68
32.1	5.84	34.16
35.9	9.64	93.02
23.0	−3.64	10.60
28.2	1.94	3.78
31.6	5.34	28.56
18.3	−7.96	63.29
$\Sigma x = 236.3$		$\Sigma(x - \bar{x})^2 = 298.78$

$$\bar{x} = \frac{\Sigma x}{n} = \frac{236.3}{9} = 26.26$$

$$s^2 = \frac{\Sigma(x - \bar{x})^2}{(n-1)} = \frac{298.78}{8} \approx 37.35$$

$$s = \sqrt{s^2} \approx 6.11$$

Long Beach: range = Max − Min = 26.9 − 18.2 = 8.7

x	$(x - \bar{x})$	$(x - \bar{x})^2$
20.9	−1.98	3.91
18.2	−4.68	21.88
20.8	−2.08	4.32
21.1	−1.78	3.16
26.5	3.62	13.12
26.9	4.02	16.18
24.2	1.32	1.75
25.1	2.22	4.94
22.2	−0.68	0.46
$\Sigma x = 205.9$		$\Sigma(x - \bar{x})^2 = 69.72$

$$\bar{x} = \frac{\Sigma x}{n} = \frac{205.9}{9} = 22.88$$

$$s^2 = \frac{\Sigma(x - \bar{x})^2}{(n-1)} = \frac{69.72}{8} \approx 8.71$$

$$s = \sqrt{s^2} \approx 2.95$$

(b) It appears from the data that the annual salaries in Los Angeles are more variable than the salaries in Long Beach.

20. (a) Dallas: range = Max − Min = 34.9 − 16.8 = 18.1

x	$x - \bar{x}$	$(x - \bar{x})^2$
34.9	8.92	79.61
25.7	−0.28	0.08
17.3	−8.68	75.30
16.8	−9.18	84.23
26.8	0.82	0.68
24.7	−1.28	1.63
29.4	3.42	11.71
32.7	6.72	45.19
25.5	−0.48	0.23
$\Sigma x = 233.8$		$\Sigma(x - \bar{x})^2 = 298.66$

$$\bar{x} = \frac{\Sigma x}{n} = \frac{233.8}{9} = 25.98$$

$$s^2 = \frac{\Sigma(x - \bar{x})^2}{(n-1)} = \frac{298.66}{8} \approx 37.33$$

$$s = \sqrt{s^2} \approx 6.11$$

Houston: range = Max − Min = 31.3 − 18.3 = 13

x	$x - \bar{x}$	$(x - \bar{x})^2$
25.6	−0.03	0.00
23.2	−2.43	5.92
26.7	1.07	1.14
27.7	2.07	4.27
25.4	−0.23	0.05
26.4	0.77	0.59
18.3	−7.33	53.78
26.1	0.47	0.22
31.3	5.67	32.11
$\Sigma x = 230.7$		$\Sigma(x - \bar{x})^2 = 98.08$

$$\bar{x} = \frac{\Sigma x}{n} = \frac{230.7}{9} = 25.63$$

$$s^2 = \frac{\Sigma(x - \bar{x})^2}{(n-1)} = \frac{98.08}{8} = 12.26$$

$$s = \sqrt{s^2} \approx 3.50$$

(b) It appears from the data that the annual salaries in Dallas are more variable than the salaries in Houston.

21. (a) Male: range = Max − Min = 1328 − 923 = 405

x	$(x - \bar{x})$	$(x - \bar{x})^2$
1059	−51.13	2,613.77
1328	217.8	47,469.52
1175	64.88	4,208.77
1123	12.88	165.77
923	−187.13	35,015.77
1017	−93.13	8,672.77
1214	103.88	10,790.02
1042	−68.13	4,641.02
$\Sigma x = 8881$		$\Sigma(x - \bar{x})^2 = 113,576.88$

$$\bar{x} = \frac{\Sigma x}{n} = \frac{8881}{8} = 1110.13$$

$$s^2 = \frac{\Sigma(x - \bar{x})^2}{(n - 1)} = \frac{113,576.9}{7} \approx 16,225.3$$

$$s = \sqrt{s^2} \approx 127.4$$

Female: range = Max − Min = 1393 − 841 = 552

x	$(x - \bar{x})$	$(x - \bar{x})^2$
1226	92.50	8,556.25
965	−168.50	28,392.25
841	−292.50	85,556.25
1053	−80.50	6,480.25
1056	−77.50	6,006.25
1393	259.50	67,340.25
1312	178.50	31,862.25
1222	88.50	7,832.25
$\Sigma x = 9068$		$\Sigma(x - \bar{x})^2 = 242,026.00$

$$\bar{x} = \frac{\Sigma x}{n} = \frac{9068}{8} = 1133.50$$

$$s^2 = \frac{\Sigma(x - \bar{x})^2}{(n - 1)} = \frac{242,026}{7} \approx 34,575.1$$

$$s = \sqrt{s^2} \approx 185.9$$

(b) It appears from the data, the SAT scores for females are more variable than the SAT scores for males.

22. (a) Public: range = Max − Min = 39.9 − 34.8 = 5.1

x	$(x - \bar{x})$	$(x - \bar{x})^2$
38.6	1.23	1.50
38.1	0.73	0.53
38.7	1.33	1.76
36.8	−0.58	0.33
34.8	−2.58	6.63
35.9	−1.48	2.18
39.9	2.53	6.38
36.2	−1.18	1.38
$\Sigma x = 299$		$\Sigma(x - \bar{x})^2 = 20.68$

$$\bar{x} = \frac{\Sigma x}{n} = \frac{299}{8} = 37.38$$

$$s^2 = \frac{\Sigma(x - \bar{x})^2}{(n - 1)} = \frac{20.68}{7} \approx 2.95$$

$$s = \sqrt{s^2} \approx 1.72$$

Private: range = Max − Min = 21.8 − 17.6 = 4.2

x	$(x - \bar{x})$	$(x - \bar{x})^2$
21.8	2.26	5.12
18.4	−1.14	1.29
20.3	0.76	0.58
17.6	−1.94	3.75
19.7	0.16	0.03
18.3	−1.24	1.53
19.4	−0.14	0.02
20.8	1.26	1.59
$\Sigma x = 156.3$		$\Sigma(x - \bar{x})^2 = 13.92$

$$\bar{x} = \frac{\Sigma x}{n} = \frac{156.3}{8} = 19.54$$

$$s^2 = \frac{\Sigma(x - \bar{x})^2}{(n - 1)} = \frac{13.92}{7} \approx 1.99$$

$$s = \sqrt{s^2} \approx 1.41$$

(b) It appears from the data that the annual salaries for public teachers are more variable than the salaries for private teachers.

23. (a) Greatest sample standard deviation: (ii)

Data set (ii) has more entries that are farther away from the mean.

Least sample standard deviation: (iii)

Data set (iii) has more entries that are close to the mean.

(b) The three data sets have the same mean but have different standard deviations.

24. (a) Greatest sample standard deviation: (i)

Data set (i) has more entries that are farther away from the mean.

Least sample standard deviation: (iii)

Data set (iii) has more entries that are close to the mean.

(b) The three data sets have the same mean, median, and mode, but have different standard deviations.

25. (a) Greatest sample standard deviation: (ii)

Data set (ii) has more entries that are farther away from the mean.

Least sample standard deviation: (iii)

Data set (iii) has more entries that are close to the mean.

(b) The three data sets have the same mean, median, and mode, but have different standard deviations.

26. (a) Greatest sample standard deviation: (iii)

Data set (iii) has more entries that are farther away from the mean.

Least sample standard deviation: (i)

Data set (i) has more entries that are close to the mean.

(b) The three data sets have the same mean and median but have different standard deviations.

27. Similarity: Both estimate proportions of the data contained within k standard deviations of the mean.

Difference: The Empirical Rule assumes the distribution is bell-shaped, Chebychev's Theorem makes no such assumption.

28. You must know the distribution is bell-shaped.

29. $(1300, 1700) \rightarrow (1500 - 1(200), 1500 + 1(200)) \rightarrow (\bar{x} - s, \bar{x} + s)$

68% of the farms value between \$1300 and \$1700 per acre.

30. 95% of the data falls between $\bar{x} - 2s$ and $\bar{x} + 2s$.

$\bar{x} - 2s = 2400 - 2(450) = 1500$

$\bar{x} + 2s = 2400 + 2(450) = 3300$

95% of the farm values lie between \$1500 and \$3300 per acre.

31. (a) $n = 75$

68%(75) = (0.68)(75) = 51 farm values will be between \$1300 and \$1700 per acre.

(b) $n = 25$

68%(25) = (0.68)(25) = 17 of these farm values will be between \$1300 and \$1700 per acre.

32. (a) $n = 40$

95% of the data lie within 2 standard deviations of the mean.

(95%)(40) = (0.95)(40) = 38 farm values lie between $1500 and $3300 per acre.

(b) $n = 60$

(95%)(20) = (0.95)(20) = 19 of these farm values lie between $1500 and $3300 per acre.

33. $\bar{x} = 1500$ {1000, 2000} are outliers. They are more than 2 standard deviations from the mean (1100, 1900).

$s = 200$

34. $\bar{x} = 2400$ {3325, 1490} are outliers. They are more than 2 standard deviations from the mean (1500, 3300).

$s = 450$

35. $(\bar{x} - 2s, \bar{x} + 2s) \rightarrow (1.14, 5.5)$ are 2 standard deviations from the mean.

$$1 - \frac{1}{k^2} = 1 - \frac{1}{(2)^2} = 1 - \frac{1}{4} = 0.75 \Rightarrow \text{At least 75\% of the eruption times lie between}$$
1.14 and 5.5 minutes.

If $n = 32$, at least $(0.75)(32) = 24$ eruptions will lie between 1.14 and 5.5 minutes.

36. $1 - \frac{1}{k^2} = 1 - \frac{1}{(2)^2} = 1 - \frac{1}{4} = .75 \rightarrow$ At least 75% of the 400-meter dash times lie within 2 standard deviations of mean.

$(\bar{x} - 2s, \bar{x} + 2s) \rightarrow (54.97, 59.17) \rightarrow$ At least 75% of the 400-meter dash times lie between 54.97 and 59.17 seconds.

37.

x	f	xf	$x - \bar{x}$	$(x - \bar{x})^2$	$(x - \bar{x})^2 f$
0	5	0	−2.08	4.31	21.53
1	11	11	−1.08	1.16	12.71
2	7	14	−0.08	0.01	0.04
3	10	30	0.93	0.86	8.56
4	7	28	1.93	3.71	25.94
	$n = 40$	$\sum xf = 83$			$\sum(x - \bar{x})^2 f = 68.78$

$$\bar{x} = \frac{\sum x}{n} = \frac{83}{40} \approx 2.1$$

$$s = \sqrt{\frac{\sum(x - \bar{x})^2 f}{n - 1}} = \sqrt{\frac{68.78}{39}} = \sqrt{1.76} \approx 1.3$$

38.

x	f	xf	$x - \bar{x}$	$(x - \bar{x})^2$	$(x - \bar{x})^2 f$
0	3	0	−1.74	3.03	9.08
1	15	15	−0.74	0.55	8.21
2	24	48	0.26	0.07	1.62
3	8	24	1.26	1.59	12.70
	$n = 50$	$\sum xf = 87$			$\sum(x - \bar{x})^2 f = 31.62$

$$\bar{x} = \frac{\sum xf}{n} = \frac{87}{50} \approx 1.7$$

$$s = \sqrt{\frac{\sum(x - \bar{x})^2 f}{n - 1}} \approx \sqrt{\frac{31.62}{49}} \approx \sqrt{0.645} \approx 0.8$$

39. Class width $= \dfrac{\text{Max} - \text{Min}}{5} = \dfrac{14 - 2}{5} = \dfrac{12}{5} = 2.4 \Rightarrow 3$

Class	Midpoint, x	f	xf
2–4	3	4	12
5–7	6	8	48
8–10	9	15	135
11–13	12	4	48
14–16	15	1	15
		$N = 32$	$\sum xf = 258$

$\mu = \dfrac{\Sigma xf}{N} = \dfrac{258}{32} \approx 8.1$

$x - \mu$	$(x - \mu)^2$	$(x - \mu)^2 f$
−5.1	26.01	104.04
−2.1	4.41	35.28
0.9	0.81	12.15
3.9	15.21	50.84
6.9	47.61	47.61
		$\sum(x - \mu)^2 f = 249.92$

$\sigma = \sqrt{\dfrac{\Sigma(x - \mu)^2}{N}} = \sqrt{\dfrac{249.92}{32}} \approx 2.8$

40. Class width $= \dfrac{\text{Max} - \text{Min}}{5} = \dfrac{244 - 145}{5} = 19.8 \Rightarrow 20$

Class	Midpoint, x	f	xf
145–164	154.5	8	1236.0
165–184	174.5	7	1221.5
185–204	194.5	3	583.5
205–224	214.5	1	214.5
225–244	234.5	1	234.5
		$N = 20$	$\Sigma xf = 3490.0$

$\mu = \dfrac{\Sigma xf}{N} = \dfrac{3490}{20} = 174.5$

$x - \mu$	$(x - \mu)^2$	$(x - \mu)^2 f$
−20	400	3200
0	0	0
20	400	1200
40	1600	1600
60	3600	3600
		$\Sigma(x - \mu)^2 f = 9600$

$\sigma = \sqrt{\dfrac{\Sigma(x - \mu)^2 f}{N}} = \sqrt{\dfrac{9600}{20}} = \sqrt{480} \approx 21.9$

41.

Midpoint, x	f	xf
70.5	1	70.5
92.5	12	1110.0
114.5	25	2862.5
136.5	10	1365.0
158.5	2	317.0
	$n = 50$	$\Sigma xf = 5725$

$$\bar{x} = \frac{\Sigma xf}{n} = \frac{5725}{50} = 114.5$$

$x - \bar{x}$	$(x - \bar{x})^2$	$(x - \bar{x})^2 f$
−44	1936	1936
−22	484	5808
0	0	0
22	484	4840
44	1936	3872
		$\Sigma(x - \bar{x})^2 f = 16{,}456$

$$s = \sqrt{\frac{\Sigma(x - \bar{x})^2}{n - 1}} = \sqrt{\frac{16{,}456}{49}} = \sqrt{335.83} \approx 18.33$$

42.

Class	f	xf
0	1	0
1	9	9
2	13	26
3	5	15
4	2	8
	$n = 30$	$\Sigma xf = 58$

$$\bar{x} = \frac{\Sigma xf}{n} = \frac{58}{30} \approx 1.9$$

$x - \bar{x}$	$(x - \bar{x})^2$	$(x - \bar{x})^2 f$
−1.93	3.72	3.72
−0.93	0.86	7.74
0.07	0.00	0.00
1.07	1.14	5.70
2.07	4.28	8.56
		$\Sigma(x - \bar{x})^2 f = 25.72$

$$s = \sqrt{\frac{\Sigma(x - \bar{x})^2 f}{n - 1}} = \sqrt{\frac{25.72}{29}} \approx \sqrt{0.89} \approx 0.9$$

43.

Class	Midpoint, x	f	xf
0–4	2.0	20.3	40.60
5–13	9.0	35.5	319.50
14–17	15.5	16.5	255.75
18–24	21.0	30.4	638.40
25–34	29.5	39.4	1162.30
35–44	39.5	39.0	1540.50
45–64	54.5	80.8	4403.60
65+	70.0	40.4	2828.00
		$n = 302.3$	$\Sigma xf = 11{,}188.65$

$$\bar{x} = \frac{\Sigma xf}{n} = \frac{11{,}188.65}{302.3} \approx 37.01$$

$x - \bar{x}$	$(x - \bar{x})^2$	$(x - \bar{x})^2 f$
−35.01	1225.70	24,881.77
−28.01	784.56	27,851.88
−21.51	462.68	7634.22
−16.01	256.32	7792.13
−7.51	56.40	2222.16
2.49	6.20	241.80
17.49	305.70	24,716.72
32.99	1088.34	43,968.94
		$\sum(x - \bar{x})^2 f = 139{,}309.56$

$$s = \sqrt{\frac{\Sigma(x - \bar{x})^2}{n - 1}} = \sqrt{\frac{139{,}309.56}{301.3}} = \sqrt{462.36} \approx 21.50$$

44.

Midpoint, x	f	xf
5	11.3	56.5
15	12.1	181.5
25	12.8	320.0
35	16.5	577.5
45	18.3	823.5
55	15.2	836.0
65	17.8	1157.0
75	13.4	1005.0
85	7.3	620.5
95	1.5	142.5
	$n = 126.2$	$\Sigma xf = 5720$

$$\bar{x} = \frac{\Sigma xf}{n} = \frac{5720}{126.2} \approx 45.32$$

$(x - \bar{x})$	$(x - \bar{x})^2$	$(x - \bar{x})^2 f$
−40.32	1625.70	18,370.41
−30.32	919.30	11,123.53
−20.32	412.92	5285.12
−10.32	106.50	1757.25
−0.32	0.10	1.83
9.68	93.70	1424.34
19.68	387.30	6893.94
29.68	880.90	11,804.06
39.68	1574.50	11,493.85
49.68	2468.10	3702.15
		$\sum(x - \bar{x})^2 f = 71{,}856.38$

$$s = \sqrt{\frac{\Sigma(x - \bar{x})^2 f}{n - 1}} = \sqrt{\frac{71{,}856.38}{125.2}} \approx \sqrt{573.93} \approx 23.96$$

45. $CV_{\text{heights}} = \dfrac{\sigma}{\mu} \cdot 100\% = \dfrac{3.44}{72.75} \cdot 100 \approx 4.7$

$CV_{\text{weights}} = \dfrac{\sigma}{\mu} \cdot 100\% = \dfrac{18.47}{187.83} \cdot 100 \approx 9.8$

It appears that weight is more variable than height.

46. (a)

x	x^2
1059	1,121,481
1328	1,763,584
1175	1,380,625
1123	1,261,129
923	851,929
1017	1,034,289
1214	1,473,796
1042	1,085,764
$\sum x = 8881$	$\sum x^2 = 9,972,597$

Male: $s = \sqrt{\dfrac{\sum x^2 - [(\sum x)^2/n]}{n-1}} = \sqrt{\dfrac{9,972,597 - [(8881)^2/8]}{7}}$

$= \sqrt{\dfrac{113,576.875}{7}} = \sqrt{16,225.268} \approx 127.4$

x	x^2
1226	1,503,076
965	931,225
841	707,281
1053	1,108,809
1056	1,115,136
1393	1,940,449
1312	1,721,344
1222	1,493,284
$\sum x = 9068$	$\sum x^2 = 10,520,604$

Female: $s = \sqrt{\dfrac{\sum x^2 - [(\sum x)^2/n]}{n-1}} = \sqrt{\dfrac{10,520,604 - [(9068)^2/8]}{7}}$

$= \sqrt{\dfrac{242,026}{7}} = \sqrt{34,575.143} \approx 185.9$

(b) The answers are the same as from Exercise 21.

47. (a) $\bar{x} \approx 41.5$ $s \approx 5.3$

(b) $\bar{x} \approx 43.6$ $s \approx 5.6$

(c) $\bar{x} \approx 3.5$ $s \approx 0.4$

(d) By multiplying each entry by a constant k, the new sample mean is $k \cdot \bar{x}$ and the new sample standard deviation is $k \cdot s$.

48. (a) $\bar{x} \approx 41.7$, $s \approx 6.0$

(b) $\bar{x} \approx 42.7$, $s \approx 6.0$

(c) $\bar{x} \approx 39.7$, $s \approx 6.0$

(d) By adding or subtracting a constant k to each entry, the new sample mean will be $\bar{x} + k$ with the sample standard deviation being unaffected.

49. (a) Male SAT Scores: $\bar{x} = 1110.125$ Female SAT Score: $\bar{x} = 1133.5$

x	$|x - \bar{x}|$
1059	51.125
1328	217.88
1175	64.875
1123	12.875
923	187.13
1017	93.125
1214	103.88
1042	68.125
	$\Sigma|x - \bar{x}| = 799$

x	$|x - \bar{x}|$
1226	92.5
965	168.5
841	292.5
1053	80.5
1056	77.5
1393	259.5
1312	178.5
1222	88.5
	$\Sigma|x - \bar{x}| = 1238$

$\Sigma|x - \bar{x}| = 799 \Rightarrow \dfrac{\Sigma|x - \bar{x}|}{n} = \dfrac{799}{8} = 99.9$

$s = 127.4$

$\Sigma|x - \bar{x}| = 799 \Rightarrow \dfrac{\Sigma|x - \bar{x}|}{n} = \dfrac{1238}{8} = 154.8$

$s = 185.9$

(b) Public Teachers: $\bar{x} = 37.375$ Private Teachers: $\bar{x} = 19.538$

x	$|x - \bar{x}|$
38.6	1.225
38.1	0.725
38.7	1.325
36.8	0.575
34.8	2.575
35.9	1.475
39.9	2.525
36.2	1.175
	$\Sigma|x - \bar{x}| = 11.6$

x	$|x - \bar{x}|$
21.8	2.262
18.4	1.138
20.3	0.762
17.6	1.938
19.7	0.162
18.3	1.238
19.4	0.138
20.8	1.262
	$\Sigma|x - \bar{x}| = 8.9$

$\Sigma|x - \bar{x}| = 11.6 \Rightarrow \dfrac{\Sigma|x - \bar{x}|}{n} = \dfrac{11.6}{8} = 1.45$

$s = 1.72$

$\Sigma|x - \bar{x}| = 8.9 \Rightarrow \dfrac{\Sigma|x - \bar{x}|}{n} = \dfrac{8.9}{8} = 1.11$

$s = 1.41$

50. $1 - \dfrac{1}{k^2} = 0.99 \Rightarrow 1 - 0.99 = \dfrac{1}{k^2} \Rightarrow k^2 = \dfrac{1}{0.01} \Rightarrow k = \sqrt{\dfrac{1}{0.01}} = 10$

At least 99% of the data in any data set lie within 10 standard deviations of the mean.

51. (a) $P = \dfrac{3(\bar{x} - \text{median})}{s} = \dfrac{3(17 - 19)}{2.3} \approx -2.61$; skewed left

(b) $P = \dfrac{3(\bar{x} - \text{median})}{s} = \dfrac{3(32 - 25)}{5.1} \approx 4.12$; skewed right

2.5 Try It Yourself Solutions

1a. 15, 15, 15, 17, 18, 18, 20, 21, 21, 21, 21, 22, 22, 23, 24, 24, 25, 25, 26, 26, 27, 27, 27, 28, 28, 28, 29, 29, 29, 30, 30, 31, 31, 31, 31, 31, 32, 32, 32, 32, 32, 32, 32, 33, 33, 33, 34, 34, 34, 34, 35, 35, 35, 35, 35, 36, 36, 37, 37, 37, 38, 38, 38, 39, 39, 39, 40, 40, 40, 41, 41, 42, 42, 43, 43, 43, 44, 44, 44, 44, 45, 45, 45, 45, 46, 46, 46, 46, 47, 47, 48, 48, 48, 48, 49, 50, 50, 52, 53, 53, 54, 54, 55, 55, 55, 55, 57, 59, 59, 60, 61, 61, 61, 63, 63, 65, 68, 89

b. $Q_2 = 37$

c. $Q_1 = 30$ $Q_3 = 47$

2a. (Enter the data)

b. $Q_1 = 17$ $Q_2 = 23$ $Q_3 = 28.5$

c. One quarter of the tuition costs is \$17,000 or less, one half is \$23,000 or less, and three quarters is \$28,500 or less.

3a. $Q_1 = 30$ $Q_3 = 47$

b. IQR $= Q_3 - Q_1 = 47 - 30 = 17$

c. The touchdowns in the middle half of the data set vary by 17 years.

4a. Min $= 15$ $Q_1 = 30$ $Q_2 = 37$

$Q_3 = 47$ Max $= 89$

bc.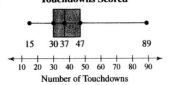

d. It appears that half of the teams scored between 30 and 47 touchdowns.

5a. 50th percentile

b. 50% of the teams scored 40 or fewer touchdowns.

6a. $\mu = 70$, $\sigma = 8$

b. $x = 60$: $z = \dfrac{x - \mu}{\sigma} = \dfrac{60 - 70}{8} = -1.25$

$x = 71$: $z = \dfrac{x - \mu}{\sigma} = \dfrac{71 - 70}{8} = 0.125$

$x = 92$: $z = \dfrac{x - \mu}{\sigma} = \dfrac{92 - 70}{8} = 2.75$

c. From the z-score, the utility bill of \$60 is 1.25 standard deviations below the mean, the bill of \$71 is 0.125 standard deviation above the mean, and the bill of \$92 is 2.75 standard deviations above the mean.

7a. Best supporting actor: $\mu = 50.1$, $\sigma = 13.9$

Best supporting actress: $\mu = 39.7$, $\sigma = 1.4$

b. Alan Arkin: $x = 72 \Rightarrow z = \dfrac{x - \mu}{\sigma} = \dfrac{72 - 50.1}{13.9} = 1.58$

Jennifer Hudson: $x = 25 \Rightarrow z = \dfrac{x - \mu}{\sigma} = \dfrac{25 - 39.7}{14} = -1.05$

c. Alan Arkin's age is 1.58 standard deviations above the mean of the best support actors. Jennifer Hudson's age is 1.05 standard deviations below the mean of the best supporting actresses. Neither actor's age is unusual.

2.5 EXERCISE SOLUTIONS

1. (a)

lower half   upper half

1 2 2 4 4 5 5 5 6 6 6 7 7 7 7 8 8 8 9 9

Q_1 Q_2 Q_3

$Q_1 = 4.5$ $Q_2 = 6$ $Q_3 = 7.5$

(b)

2. (a)

lower half   upper half

1, 1, 2, 2, 2, 2, 2, 3, 3, 3, 3, 3, 3, 4, 4, 4, 5, 5, 5, 5, 5, 6, 7, 7, 7, 7, 8, 8, 8, 9, 9, 9, 9, 9, 9

Q_1 Q_2 Q_3

$Q_1 = 3$ $Q_2 = 5$ $Q_3 = 8$

(b)

3. The soccer team scored fewer points per game than 75% of the teams in the league.

4. The salesperson sold more hardware equipment than 80% of the other sales people.

5. The student scored higher than 78% of the students who took the actuarial exam.

6. The child's IQ is higher than 93% of the other children in the same age group.

7. True

8. False. The five numbers you need to graph a box-and-whisker plot are the minimum, the maximum, Q_1, Q_3, and the median (Q_2).

9. False. The 50th percentile is equivalent to Q_2.

10. False. Any score equal to the mean will have a corresponding z-score of zero.

11. (a) Min = 10 (b) Max = 20

(c) $Q_1 = 13$ (d) $Q_2 = 15$

(e) $Q_3 = 17$ (f) IQR = $Q_3 - Q_1 = 17 - 13 = 4$

12. (a) Min = 100 (b) Max = 320

(c) $Q_1 = 130$ (d) $Q_2 = 205$

(e) $Q_3 = 270$ (f) IQR = $Q_3 - Q_1 = 270 - 130 = 140$

13. (a) Min = 900 (b) Max = 2100

(c) $Q_1 = 1250$ (d) $Q_2 = 1500$

(e) $Q_3 = 1950$ (f) IQR = $Q_3 - Q_1 = 1950 - 1250 = 700$

14. (a) Min = 25 (b) Max = 85

(c) $Q_1 = 50$ (d) $Q_2 = 65$

(e) $Q_3 = 70$ (f) IQR = $Q_3 - Q_1 = 70 - 50 = 20$

15. (a) Min = -1.9 (b) Max = 2.1

(c) $Q_1 = -0.5$ (d) $Q_2 = 0.1$

(e) $Q_3 = 0.7$ (f) IQR = $Q_3 - Q_1 = 0.7 - (-0.5) = 1.2$

16. (a) Min = -1.3 (b) Max = 2.1

(c) $Q_1 = -0.3$ (d) $Q_2 = 0.2$

(e) $Q_3 = 0.4$ (f) IQR = $Q_3 - Q_1 = 0.4 - (-0.3) = 0.7$

17. None. The data are not skewed or symmetric.

18. Skewed right. Most of the data lie to the right.

19. Skewed left. Most of the data lie to the left.

20. Symmetric

21. $Q_1 = B,$ $Q_2 = A,$ $Q_3 = C$

25% of the entries are below B, 50% are below A, and 75% are below C.

22. $P_{10} = T,$ $P_{50} = R,$ $P_{80} = S$

10% of the entries are below T, 50% are below R, and 80% are below S.

23. (a) $Q_1 = 2,$ $Q_2 = 4,$ $Q_3 = 5$

(b)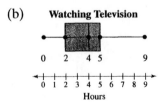

24. (a) $Q_1 = 2$, $Q_2 = 4.5$, $Q_3 = 6.5$

(b) **Vacation Days**

0 2 4.5 6.5 10

0 2 4 6 8 10
Number of days

25. (a) $Q_1 = 3$, $Q_2 = 3.85$, $Q_3 = 5.28$

(b) **Airline Distances**

1.8 3 3.85 5.28 6

0 1 2 3 4 5 6
Distances (in miles)

26. (a) $Q_1 = 15.125$, $Q_2 = 15.8$, $Q_3 = 17.65$

(b) **Railroad Equipment Manufacturers**

13.8 15.125 17.65 19.45
15.8

13.5 14.5 15.5 16.5 17.5 18.5 19.5
Hourly earnings
(in dollars)

27. (a) 5   (b) 50%   (c) 25%

28. (a) $17.65   (b) 50%   (c) 50%

29. $A \Rightarrow z = -1.43$

$B \Rightarrow z = 0$

$C \Rightarrow z = 2.14$

The z-score 2.14 is unusual because it is so large.

30. $A \rightarrow z = -1.54$

$B \rightarrow z = 0.77$

$C \rightarrow z = 1.54$

None of the z-scores are unusual.

31. (a) Statistics:

$$x = 73 \Rightarrow z = \frac{x - \mu}{\sigma} = \frac{73 - 63}{7} \approx 1.43$$

Biology:

$$x = 26 \Rightarrow z = \frac{x - \mu}{\sigma} = \frac{26 - 23}{3.9} \approx 0.77$$

(b) The student had a better score on the statistics test.

32. (a) Statistics: $x = 60 \Rightarrow z = \dfrac{x - \mu}{\sigma} = \dfrac{60 - 63}{7} \approx -0.43$

Biology: $x = 20 \Rightarrow z = \dfrac{x - \mu}{\sigma} = \dfrac{20 - 23}{3.9} \approx -0.77$

(b) The student had a better score on the statistics test.

33. (a) Statistics:

$x = 78 \Rightarrow z = \dfrac{x - \mu}{\sigma} = \dfrac{78 - 63}{7} \approx 2.14$

Biology:

$x = 29 \Rightarrow z = \dfrac{x - \mu}{\sigma} = \dfrac{29 - 23}{3.9} \approx 1.54$

(b) The student had a better score on the statistics test.

34. (a) Statistics: $x = 63 \Rightarrow z = \dfrac{x - \mu}{\sigma} = \dfrac{63 - 63}{7} = 0$

Biology: $x = 23 \Rightarrow z = \dfrac{x - \mu}{\sigma} = \dfrac{23 - 23}{3.9} = 0$

(b) The student performed equally well on the two tests.

35. (a) $x = 34{,}000 \Rightarrow z = \dfrac{x - \mu}{\sigma} = \dfrac{34{,}000 - 35{,}000}{2{,}250} \approx -0.44$

$x = 37{,}000 \Rightarrow z = \dfrac{x - \mu}{\sigma} = \dfrac{37{,}000 - 35{,}000}{2{,}250} \approx 0.89$

$x = 31{,}000 \Rightarrow z = \dfrac{x - \mu}{\sigma} = \dfrac{31{,}000 - 35{,}000}{2{,}250} \approx -1.78$

None of the selected tires have unusual life spans.

(b) $x = 30{,}500 \Rightarrow z = \dfrac{x - \mu}{\sigma} = \dfrac{30{,}500 - 35{,}000}{2{,}250} = -2 \Rightarrow$ 2.5th percentile

$x = 37{,}250 \Rightarrow z = \dfrac{x - \mu}{\sigma} = \dfrac{37{,}250 - 35{,}000}{2{,}250} = 1 \Rightarrow$ 84th percentile

$x = 35{,}000 \Rightarrow z = \dfrac{x - \mu}{\sigma} = \dfrac{35{,}000 - 35{,}000}{2{,}250} = 0 \Rightarrow$ 50th percentile

36. (a) $x = 34 \Rightarrow z = \dfrac{x - \mu}{\sigma} = \dfrac{34 - 33}{4} = 0.25$

$x = 30 \Rightarrow z = \dfrac{x - \mu}{\sigma} = \dfrac{30 - 33}{4} = -0.75$

$x = 42 \Rightarrow z = \dfrac{x - \mu}{\sigma} = \dfrac{42 - 33}{4} = 2.25$

The life span of 42 days is unusual due to a rather large z-score.

(b) $x = 29 \Rightarrow z = \dfrac{x - \mu}{\sigma} = \dfrac{29 - 33}{4} = -1 \Rightarrow$ 16th percentile

$x = 41 \Rightarrow z = \dfrac{x - \mu}{\sigma} = \dfrac{41 - 33}{4} = 2 \Rightarrow$ 97.5th percentile

$x = 25 \Rightarrow z = \dfrac{x - \mu}{\sigma} = \dfrac{25 - 33}{4} = -2 \Rightarrow$ 2.5th percentile

37. 68.5 inches

40% of the heights are below 68.5 inches.

38. 99th percentile

99% of the heights are below 76 inches.

39. $x = 74$: $z = \dfrac{x - \mu}{\sigma} = \dfrac{74 - 69.6}{3.0} \approx 1.47$

$x = 62$: $z = \dfrac{x - \mu}{\sigma} = \dfrac{62 - 69.6}{3.0} \approx -2.53$

$x = 80$: $z = \dfrac{x - \mu}{\sigma} = \dfrac{80 - 69.6}{3.0} \approx 3.47$

The height of 62 inches is unusual due to a rather small z-score. The height of 80 inches is very unusual due to a rather large z-score.

40. $x = 70$: $z = \dfrac{x - \mu}{\sigma} = \dfrac{70 - 69.6}{3.0} \approx 0.13$

$x = 66$: $z = \dfrac{x - \mu}{\sigma} = \dfrac{66 - 69.6}{3.0} \approx -1.20$

$x = 68$: $z = \dfrac{x - \mu}{\sigma} = \dfrac{68 - 69.6}{3.0} \approx -0.53$

None of the heights are unusual.

41. $x = 71.1$: $z = \dfrac{x - \mu}{\sigma} = \dfrac{71.1 - 69.6}{3.0} \approx 0.5$

Approximately the 70th percentile.

42. $x = 66.3$: $z = \dfrac{x - \mu}{\sigma} = \dfrac{66.3 - 69.6}{3.0} = -1.1$

Approximately the 12th percentile.

43. (a) 27 28 31 32 32 33 35 36 36 36 36 37 38 39 39 40 40 40 41 41

   41 42 42 42 42 42 42 43 43 43 44 44 45 45 46 47 47 47 47 47

   48 48 48 48 48 | 49 49 49 49 49 49 50 50 51 51 51 51 51 51 52

   52 52 53 53 54 | 54 54 54 54 54 | 54 54 55 56 56 56 57 57 57 59

   59 59 60 60 60 | 61 61 61 62 62 | 63 63 63 63 64 | 65 67 68 74 82

   Q_1 Q_2 Q_3

$Q_1 = 42$, $Q_2 = 49$, $Q_3 = 56$

(b) **Ages of Executives**

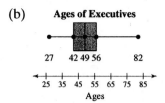

(c) Half of the ages are between 42 and 56 years.

(d) About 49 years old ($x = 49.62$ and $Q_2 = 49.00$), because half of the executives are older and half are younger.

(e) The age groups 20–29, 70–79, and 80–89 would all be considered unusual because they lie more than two standard deviations from the mean.

44. 1, 2, 3, 3, 5, 5, 7, 7, 8, 10

$$\begin{array}{ccc} \uparrow & \uparrow & \uparrow \\ Q_1 & Q_2 & Q_3 \end{array}$$

Midquartile $= \dfrac{Q_1 + Q_3}{2} = \dfrac{3 + 7}{2} = 5$

45. 22 23 24 32 33 34 36 38 39 40 41 47

$$\begin{array}{ccc} \uparrow & \uparrow & \uparrow \\ Q_1 = 28 & Q_2 & Q_3 = 39.5 \end{array}$$

Midquartile $= \dfrac{Q_1 + Q_3}{2} = \dfrac{28 + 39.5}{2} = 33.75$

46. 7.9, 8, 8.1, 9.7, 10.3, 11.2, 11.8, 12.2, 12.3, 12.7, 13.4, 15.4, 16.1

$$\begin{array}{ccc} \uparrow & \uparrow & \uparrow \\ Q_1 = 8.9 & Q_2 & Q_3 = 13.05 \end{array}$$

Midquartile $= \dfrac{Q_1 + Q_3}{2} = \dfrac{8.9 + 13.05}{2} = 10.975$

47. 13.4 15.2 15.6 16.7 17.2 18.7 19.7 19.8 19.8 20.8 21.4 22.9 28.7 30.1 31.9

$$\begin{array}{ccc} \uparrow & \uparrow & \uparrow \\ Q_1 & Q_2 & Q_3 \end{array}$$

Midquartile $= \dfrac{Q_1 + Q_3}{2} = \dfrac{16.7 + 22.9}{2} = 19.8$

48. (a) Disc 1: Symmetric

Disc 2: Skewed left

Disc 1 has less variation.

(b) Disc 2 is more likely to have outliers because its distribution is wider.

(c) Disc 1, because the distribution's typical distance from the mean is roughly 16.3.

49. **Credit Card Purchases**

Friends:
75 102.5 136 159 190

You:
28 83 115 143 215
0 25 50 75 100 125 150 175 200 225
Monthly purchases (in dollars)

You	Friends
min $= 28$	min $= 75$
$Q_1 = 83$	$Q_1 = 102.5$
$Q_2 = 115$	$Q_2 = 136$
$Q_3 = 143$	$Q_3 = 159$
max $= 215$	max $= 190$

Your distribution is symmetric and your friend's distribution is uniform.

50. Percentile $= \dfrac{\text{Number of data values less than } x}{\text{Total number of data values}} \cdot 100$

$= \dfrac{51}{80} \cdot 100 \approx$ 64th percentile

51. Percentile $= \dfrac{\text{Number of data values less than } x}{\text{Total number of data values}} \cdot 100$

$= \dfrac{75}{80} \cdot 100 \approx$ 94th percentile

CHAPTER 2 REVIEW EXERCISE SOLUTIONS

1.

Class	Midpoint	Boundaries	Frequency, f	Relative frequency	Cumulative frequency
20–23	21.5	19.5–23.5	1	0.05	1
24–27	25.5	23.5–27.5	2	0.10	3
28–31	29.5	27.5–31.5	6	0.30	9
32–35	33.5	31.5–35.5	7	0.35	16
36–39	37.5	35.5–39.5	4	0.20	20
			$\Sigma f = 20$	$\Sigma \dfrac{f}{n} = 1$	

2.

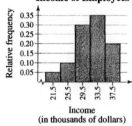

Income of Employees

Greatest relative frequency: 32–35

Least relative frequency: 20–23

3.

Liquid Volume 12-oz Cans

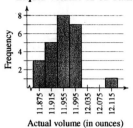

Actual volume (in ounces)

4.

Liquid Volume 12-oz Cans

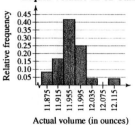

Actual volume (in ounces)

5.

Class	Midpoint, x	Frequency, f	Cumulative frequency
79–93	86	9	9
94–108	101	12	21
109–123	116	5	26
124–138	131	3	29
139–153	146	2	31
154–168	161	1	32
		$\Sigma f = 32$	

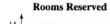
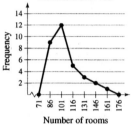

Rooms Reserved

Number of rooms

6.

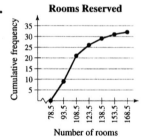

Rooms Reserved

Number of rooms

7.

```
1 | 3 7 8 9
2 | 0 1 2 3 3 3 4 4 5 5 5 7 8 8 9
3 | 1 1 2 3 4 5 7 8
4 | 3 4 7
5 | 1
```

8.

Average Daily Highs

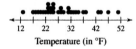

Temperature (in °F)

9.

Height of Buildings

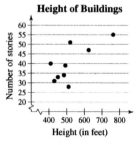

Height (in feet)

It appears as height increases, the number of stories increases.

10.

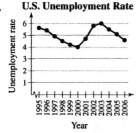

11.

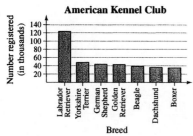

12.   American Kennel Club

Boxer
9.49%

Dachshund
9.76%

Beagle
10.57%

Golden
Retriever
11.65%

German
Shepherd
11.92%

Labrador
Retriever
33.60%

Yorkshire
Terrier
13.01%

13. $\bar{x} = 9.1$

median $= 8.5$

mode $= 7$

14. $\bar{x} = 40.6$

median $= 42$

mode $= 42$

15.

Midpoint, x	Frequency, f	xf
21.5	1	21.5
25.5	2	51.0
29.5	6	177.0
33.5	7	234.5
37.5	4	150.0
	$n = 20$	$\Sigma xf = 634$

$$\bar{x} = \frac{\Sigma xf}{n} = \frac{634}{20} = 31.7$$

16.

x	f	xf
0	13	0
1	9	9
2	19	38
3	8	24
4	5	20
5	2	10
6	4	24
	$n = 60$	$\Sigma xf = 125$

$$\bar{x} = \frac{\Sigma xf}{n} = \frac{125}{60} \approx 2.1$$

17. $\bar{x} = \dfrac{\Sigma xw}{w} = \dfrac{(78)(0.15) + (72)(0.15) + (86)(0.15) + (91)(0.15) + (87)(0.15) + (80)(0.25)}{0.15 + 0.15 + 0.15 + 0.15 + 0.15 + 0.25}$

$= \dfrac{82.1}{1} = 82.1$

18. $\bar{x} = \dfrac{\Sigma xw}{w} = \dfrac{(96)(0.20) + (85)(0.20) + (91)(0.20) + (86)(0.40)}{0.20 + 0.20 + 0.20 + 0.40}$

$\qquad = \dfrac{88.8}{1} = 88.8$

19. Skewed　　　**20.** Skewed　　　**21.** Skewed left

22. Skewed right　　**23.** Median　　**24.** Mean

25. Range $=$ Max $-$ Min $= 8.26 - 5.46 = 2.8$

26. Range $=$ Max $-$ Min $= 19.73 - 15.89 = 3.84$

27. $\mu = \dfrac{\Sigma x}{N} = \dfrac{96}{14} = 6.9$

$\qquad \sigma = \sqrt{\dfrac{\Sigma(x-\mu)^2}{N}} = \sqrt{\dfrac{(4-6.9)^2 + (2-6.9)^2 + \cdots + (3-6.9)^2 + (3-6.9)^2}{12}}$

$\qquad = \sqrt{\dfrac{295.7}{14}} \approx \sqrt{21.12} \approx 4.6$

28. $\mu = \dfrac{\Sigma x}{N} = \dfrac{602}{9} \approx 66.9$

$\qquad \sigma = \sqrt{\dfrac{\Sigma(x-\mu)^2}{N}}$

$\qquad = \sqrt{\dfrac{(52-66.9)^2 + (86-66.9)^2 + \cdots + (68-66.9)^2 + (56-66.9)^2}{9}}$

$\qquad \approx \sqrt{\dfrac{862.87}{9}} \approx \sqrt{95.87} \approx 9.8$

29. $\bar{x} = \dfrac{\Sigma x}{n} = \dfrac{36{,}801}{15} = 2453.4$

$\qquad s = \sqrt{\dfrac{\Sigma(x-\bar{x})^2}{n-1}} = \sqrt{\dfrac{(2445-2453.4)^2 + \cdots + (2.377-2453.4)^2}{14}}$

$\qquad = \sqrt{\dfrac{1{,}311{,}783.6}{14}} \approx \sqrt{93{,}698.8} \approx 306.1$

30. $\bar{x} = \dfrac{\Sigma x}{n} = \dfrac{416{,}659}{8} = 52{,}082.4$

$\qquad s = \sqrt{\dfrac{\Sigma(x-\bar{x})^2}{n-1}} = \sqrt{\dfrac{(49{,}632-52{,}082.3)^2 + \cdots + (49{,}924-52{,}082.3)^2}{7}}$

$\qquad = \sqrt{\dfrac{73{,}225{,}929.87}{7}} = \sqrt{10{,}460{,}847.12} \approx 3234.3$

31. 99.7% of the distribution lies within 3 standard deviations of the mean.

$\qquad \mu + 3\sigma = 49 + (3)(2.50) = 41.5$

$\qquad \mu - 3\sigma = 49 - (3)(2.50) = 56.5$

$\qquad$ 99.7% of the distribution lies between \$41.50 and \$56.50.

32. $(46.75, 52.25) \rightarrow (49.50 - (1)(2.75), 49.50 + (1)(2.75)) \rightarrow (\mu - \sigma, \mu + \sigma)$

68% of the cable rates lie between \$46.75 and \$52.25.

33. $n = 40 \qquad \mu = 36 \qquad \sigma = 8$

$(20, 52) \rightarrow (36 - 2(8), 36 + 2(8)) \Rightarrow (\mu - 2\sigma, \mu + 2\sigma) \Rightarrow k = 2$

$$1 = \frac{1}{k^2} = 1 - \frac{1}{(2)^2} = 1 - \frac{1}{4} = 0.75$$

At least $(40)(0.75) = 30$ customers have a mean sale between \$20 and \$52.

34. $n = 20 \qquad \mu = 7 \qquad \sigma = 2$

$(3, 11) \rightarrow (7 - 2(2), 7 + 2(2)) \rightarrow (\mu - 2\sigma, \mu + 2\sigma) \rightarrow k = 2$

$$1 - \frac{1}{k^2} = 1 - \frac{1}{(2)^2} = 1 - \frac{1}{4} = 0.75$$

At least $(20)(0.75) = 15$ shuttle flights lasted between 3 days and 11 days.

35. $\bar{x} = \dfrac{\Sigma x f}{n} = \dfrac{99}{40} \approx 2.5$

$$s = \sqrt{\frac{\Sigma (x - \bar{x})^2 f}{n - 1}} = \sqrt{\frac{(0 - 1.24)^2(1) + (1 - 1.24)^2(8) + \cdots + (5 - 1.24)^2(3)}{39}}$$

$$= \sqrt{\frac{59.975}{39}} \approx 1.2$$

36. $\bar{x} = \dfrac{\Sigma x f}{n} = \dfrac{61}{25} \approx 2.4$

$$s = \sqrt{\frac{\Sigma (x - \bar{x})^2 f}{n - 1}}$$

$$= \sqrt{\frac{(0 - 2.44)^2(4) + (1 - 2.44)^2(5) + \cdots + (6 - 2.44)^2(1)}{24}}$$

$$= \sqrt{\frac{72.16}{24}} \approx 1.7$$

37. $Q_1 = 56$ inches

38. $Q_3 = 68$ inches

39. $\text{IQR} = Q_3 - Q_1 = 68 - 56 = 12$ inches

40.

Height of Students

52 56 61   68 72

50   55   60   65   70   75
Heights

41. $\text{IQR} = Q_3 - Q_1 = 33 - 29 = 4$

42. **Weight of Football Players**

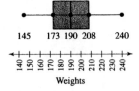

43. 23% of the students scored higher than 68.

44. $\dfrac{84}{728} \approx 0.109 \rightarrow 11\%$ have larger audiences.

The station would represent the 89th percentile, P_{89}.

45. $x = 213 \Rightarrow z = \dfrac{x - \mu}{\sigma} = \dfrac{213 - 186}{18} \approx 1.5$

This player is not unusual.

46. $x = 141 \Rightarrow z = \dfrac{x - \mu}{\sigma} = \dfrac{141 - 186}{18} = -2.5$

This is an unusually light player.

47. $x = 178 \Rightarrow z = \dfrac{x - \mu}{\sigma} = \dfrac{178 - 186}{18} = -0.44$

This player is not unusual.

48. $x = 249 \Rightarrow z = \dfrac{x - \mu}{\sigma} = \dfrac{249 - 186}{18} \approx 3.5$

This is an unusually heavy player.

CHAPTER 2 QUIZ SOLUTIONS

1. (a)

Class limits	Midpoint	Class boundaries	Frequency, f	Relative frequency	Cumulative frequency
101–112	106.5	100.5–112.5	3	0.12	3
113–124	118.5	112.5–124.5	11	0.44	14
125–136	130.5	124.5–136.5	7	0.28	21
137–148	142.5	136.5–148.5	2	0.08	23
149–160	154.5	148.5–160.5	2	0.08	25

(b) Frequency Histogram and Polygon

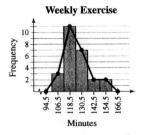

(c) Relative Frequency Histogram

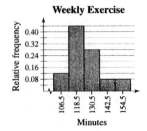

(d) Skewed

(e)
Weekly Exercise

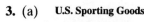
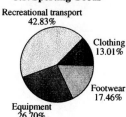

101 117.5 123 131.5   157

100 110 120 130 140 150 160
Minutes

(f)
Weekly Exercise

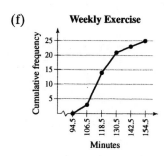

2. $\bar{x} = \dfrac{\Sigma xf}{n} = \dfrac{3130.5}{25} \approx 125.2$

$s = \sqrt{\dfrac{\Sigma(x - \bar{x})^2 f}{n - 1}} = \sqrt{\dfrac{4055.04}{24}} \approx 13.0$

3. (a) **U.S. Sporting Goods**

Recreational transport
42.83%

Clothing
13.01%

Footwear
17.46%

Equipment
26.70%

(b) **U.S. Sporting Goods**

4. (a) $\bar{x} = \dfrac{\Sigma x}{n} = 751.6$

median = 784.5

mode = (none)

The mean best describes a typical salary because there are no outliers.

(b) range = Max − Min = 575

$s^2 = \dfrac{\Sigma(x - \bar{x})^2}{n - 1} = 48{,}135.1$

$s = \sqrt{\dfrac{\Sigma(x - \bar{x})^2}{n - 1}} = 219.4$

5. $\bar{x} - 2s = 155{,}000 - 2 \cdot 15{,}000 = \$125{,}000$

$\bar{x} + 2s = 155{,}000 + 2 \cdot 15{,}000 = \$185{,}000$

95% of the new home prices fall between \$125,000 and \$185,000.

6. (a) $x = 200{,}000$ $z = \dfrac{x - \mu}{\sigma} = \dfrac{200{,}000 - 155{,}000}{15{,}000} = 3.0 \Rightarrow$ unusual price

(b) $x = 55{,}000$ $z = \dfrac{x - \mu}{\sigma} = \dfrac{55{,}000 - 155{,}000}{15{,}000} \approx -6.67 \Rightarrow$ very unusual price

(c) $x = 175{,}000$ $z = \dfrac{x - \mu}{\sigma} = \dfrac{175{,}000 - 155{,}000}{15{,}000} \approx 1.33 \Rightarrow$ not unusual

(d) $x = 122{,}000$ $z = \dfrac{x - \mu}{\sigma} = \dfrac{122{,}000 - 155{,}000}{15{,}000} = -2.2 \Rightarrow$ unusual price

7. (a) $Q_1 = 76$ $Q_2 = 80$ $Q_3 = 88$

   (b) IQR $= Q_3 - Q_1 = 88 - 76 = 12$

   (c)   **Wins for Each Team**

   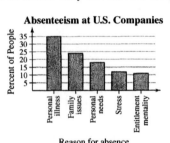

   61   76 80 88   97

   60   70   80   90   100

   Number of wins

CUMULATIVE REVIEW FOR CHAPTERS 1 AND 2

1. Systematic sampling

2. Simple Random Sampling. However, all U.S. adults may not have a telephone.

3.

Personal Illness	35%	
Family Issues	24%	
Personal Needs	18%	
Stress	12%	
Entitlement Mentality	11%	

Absenteeism at U.S. Companies

Percent of People

35
30
25
20
15
10
5

Personal illness, Family issues, Personal needs, Stress, Entitlement mentality

Reason for absence

4. $42,500 is a parameter because it is describing the average salary of all 43 employees in a company.

5. 28% is a statistic because it is describing a proportion within a sample of 1000 adults.

6. (a) $\bar{x} = 83{,}500$, $s = \$1500$

   $(80{,}500 \ \ 86{,}500) = 83{,}500 \pm 2(1500) \Rightarrow 2$ standard deviations away from the mean.

   Approximately 95% of the electrical engineers will have salaries between $(80{,}500, \ 86{,}500)$.

   (b) $40(.95) = 38$

7. Sample, because a survey of 1498 adults was taken.

8. Sample, because a study of 232,606 people was done.

9. Census, because all of the members of the Senate were included.

10. Experiment, because we want to study the effects of removing recess from schools.

11. Quantitative: Ratio

12. Qualitative: Nominal

13. min $= 0$

    $Q_1 = 2$

    $Q_2 = 12.5$

    $Q_3 = 39$

    max $= 136$

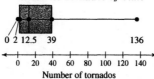

Number of Tornados by State

14. $\bar{x} = \dfrac{(0.15)(85) + (0.15)(92) + (0.15)(84) + (215)(89) + (0.40)(91)}{0.15 + 0.15 + 0.15 + 0.15 + 0.40} = 88.9$

15. (a) $\bar{x} = 5.49$

    median $= 5.4$

    mode $=$ none

    Median, because the distribution is not symmetric.

    (b) Range $= 4.1$

    $s^2 = 2.34$

    $s = 1.53$

    The standard deviation of tail lengths of alligators is 1.53 feet.

16. (a) The number of deaths due to heart disease for women is decreasing.

    (b) The study was only conducted over the past 5 years and deaths may not decrease in the next year.

17. Class width $= \dfrac{87 - 0}{8} = 10.875 \Rightarrow 11$

Class limits	Class boundaries	Class midpoint	Frequency, f	Relative frequency	Cumulative frequency
0–10	−0.5–10.5	5	8	0.296	8
11–21	10.5–21.5	16	8	0.296	16
22–32	21.5–32.5	27	1	0.037	17
33–43	32.5–43.5	38	1	0.037	18
44–54	43.5–54.5	49	1	0.037	19
55–65	54.5–65.5	60	4	0.148	23
66–76	65.5–76.5	71	0	0.000	23
77–87	76.5–87.5	82	4	0.148	27
			$n = 27$	$\sum \frac{f}{n} = 1$	

18. Skewed right

19.

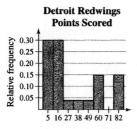

Detroit Redwings Points Scored

Least relative frequency: 66–76

Greatest relative frequency: 0–10 and 11–21

3.1 Try It Yourself Solutions

1ab. (1)

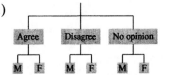

(2)

c. (1) 6 outcomes   (2) 9 outcomes

d. (1) Let A = Agree, D = Disagree, N = No Opinion, M = Male and F = Female
Sample space = {AM, AF, DM, DF, NM, NF}

(2) Let A = Agree, D = Disagree, N = No Opinion, R = Republican, De = Democrat,
O = Other
Sample space = {$AR, ADe, AO, DR, DDe, DO, NR, NDe, NO$}

2a. (1) 6 outcomes   (2) 1 outcome

b. (1) Not a simple event   (2) Simple event

3a. Manufacturer: 4

Size: 3

Color: 6

b. (4)(3)(6) = 72

c.

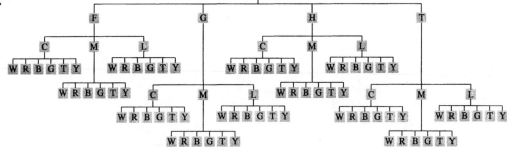

4a. (1) 26 choices for each letter

(2) 26, 25, 24, 23, 22, 21 choices

(3) 22, 26, 26, 26, 26, 26 choices

b. (1) $26 \cdot 26 \cdot 26 \cdot 26 \cdot 26 \cdot 26 = 308{,}915{,}776$

(2) $26 \cdot 25 \cdot 24 \cdot 23 \cdot 22 \cdot 21 = 165{,}765{,}600$

(3) $22 \cdot 26 \cdot 26 \cdot 26 \cdot 26 \cdot 26 = 261{,}390{,}272$

5a. (1) 52   (2) 52   (3) 52

  b. (1) 1   (2) 13   (3) 52

  c. (1) $P(7 \text{ of diamonds}) = \dfrac{1}{52} \approx 0.0192$

     (2) $P(\text{diamond}) = \dfrac{13}{52} = 0.25$

     (3) $P(\text{diamond, heart, club, or spade}) = \dfrac{52}{52} = 1$

6a. Event = the next claim processed is fraudulent (Freq = 4)

  b. Total Frequency = 100

  c. $P(\text{fraudulent claim}) = \dfrac{4}{100} = 0.04$

7a. Frequency = 54

  b. Total of the Frequencies = 1000

  c. $P(\text{age 15 to 24}) = \dfrac{54}{1000} = 0.054$

8a. Event = salmon successfully passing through a dam on the Columbia River.

  b. Estimated from the results of an experiment.

  c. Empirical probability

9a. $P(\text{age 45 to 54}) = \dfrac{180}{1000} = 0.18$

  b. $P(\text{not age 45 to 54}) = 1 - \dfrac{180}{1000} = \dfrac{820}{1000} = 0.82$

  c. $\dfrac{820}{1000}$ or 0.82

10a. 16

  b. {T1, T2, T3, T4, T5}

  c. $P(\text{tail and less than 6}) = \dfrac{5}{16} = 0.3125$

11a. $10 \cdot 10 \cdot 10 \cdot 10 \cdot 10 \cdot 10 \cdot 10 = 10{,}000{,}000$

  b. $\dfrac{1}{10{,}000{,}000}$

3.1 EXERCISE SOLUTIONS

1. (a) Yes, the probability of an event occurring must be contained in the interval [0, 1] or [0%, 100%].

    (b) Yes, the probability of an event occurring must be contained in the interval [0, 1] or [0%, 100%].

(c) No, the probability of an event occurring cannot be less than 0.

(d) Yes, the probability of an event occurring must be contained in the interval [0, 1] or [0%, 100%].

(e) Yes, the probability of an event occurring must be contained in the interval [0, 1] or [0%, 100%].

(f) No, the probability of an event cannot be greater than 1.

2. It is impossible to have more than a 100% chance of rain.

3. The fundamental counting principle counts the number of ways that two or more events can occur in sequence.

4. The law of large numbers states that as an experiment is repeated over and over, the probabilities found in an experiment will approach the actual probabilities of the event.

5. {A, B, C, D, E, F, G, H, I, J, K, L, M, N, O, P, Q, R, S, T, U, V, W, X, Y, Z}; 26

6. {HHH, HHT, HTH, HTT, THH, THT, TTH, TTT}; 8

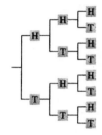

7. 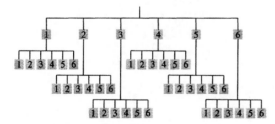 {(A, +), (B, +), (AB, +), (O, +), (A, −), (B, −), (AB, −), (O, −)} where (A, +) represents positive Rh-factor with A- blood type and (A, −) represents negative Rh-factor with A- blood type; 8.

8. {(1, 1), (1, 2), (1, 3), (1, 4), (1, 5), (1, 6), (2, 1), (2, 2), (2, 3), (2, 4), (2, 5), (2, 6), (3, 1), (3, 2), (3, 3), (3, 4), (3, 5), (3, 6), (4, 1), (4, 2), (4, 3), (4, 4), (4, 5), (4, 6), (5, 1), (5, 2), (5, 3), (5, 4), (5, 5), (5, 6), (6, 1), (6, 2), (6, 3), (6, 4), (6, 5), (6, 6)}; 36

9. Simple event because it is an event that consists of a single outcome.

10. Not a simple event because it is an event that consists of more than a single outcome.

Number less than 200 = {1, 2, 3 . . . 199}

11. Not a simple event because it is an event that consists of more than a single outcome.

   king = {king of hearts, king of spades, king of clubs, king of diamonds}

12. Simple event because it is an event that consists of a single outcome.

13. (9)(15) = 135   14. (3)(6)(4) = 72

15. (9)(10)(10)(5) = 4500   16. (2)(2)(2)(2)(2)(2) = 64

17. False. If you roll a six-sided die six times, the probability of rolling an even number at least once is approximately 0.9844.

18. False. You flip a fair coin nine times and it lands tails up each time. The probability it will land heads up on the tenth flip is 0.5.

19. False. A probability of less than 0.05 indicates an unusual event.

20. True

21. b   22. d   23. c   24. a

25. Empirical probability because company records were used to calculate the frequency of a washing machine breaking down.

26. Classical probability because each outcome is equally likely to occur.

27. $P(\text{less than } 1000) = \dfrac{999}{6296} \approx 0.159$

28. $P(\text{greater than } 1000) = \dfrac{5296}{6296} \approx 0.841$

29. $P(\text{number divisible by } 1000) = \dfrac{6}{6296} \approx 0.000953$

30. $P(\text{number not divisible by } 1000) = \dfrac{6290}{6296} \approx 0.999$

31. $P(A) = \dfrac{1}{24} \approx 0.042$ 32. $P(B) = \dfrac{3}{24} = 0.125$

33. $P(C) = \dfrac{5}{24} \approx 0.208$ 34. $P(D) = \dfrac{1}{24} \approx 0.042$

35. (a) $10 \cdot 10 \cdot 10 = 1000$ (b) $\dfrac{1}{1000} = 0.001$ (c) $\dfrac{999}{1000} = 0.999$

36. (a) $26 \cdot 9 \cdot 10 \cdot 10 \cdot 5 = 117{,}000$ (b) $\dfrac{1}{117{,}000}$ (c) $\dfrac{116{,}999}{117{,}000}$

37. {(SSS), (SSR), (SRS), (SRR), (RSS), (RSR), (RRS), (RRR)}

38. {(RRR)}

39. {(SSR), (SRS), (RSS)}

40. {(SSR), (SRS), (SRR), (RSS), (RSR), (RRS), (RRR)}

41. Let S = Sunny Day and R = Rainy Day

(a)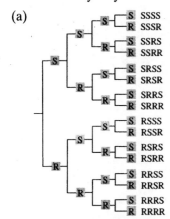

(b) {(SSSS), (SSSR), (SSRS), (SSRR), (SRSS), (SRSR)

(SRRS), (SRRR), (RSSS), (RSSR), (RSRS), (RSRR),

(RRSS), (RRSR), (RRRS), (RRRR)}

(c) {(SSSR), (SSRS), (SRSS), (RSSS)}

42.

43. $P(\text{voted in 2006 Gubernatorial election}) = \dfrac{4{,}092{,}652}{8{,}182{,}876} \approx 0.500$

44. $P(\text{did not vote Democratic}) = \dfrac{61{,}159{,}368}{114{,}413{,}842} \approx 0.535$

45. $P(\text{between 21 and 24}) = \dfrac{8.5}{142.1} \approx 0.060$

46. $P(\text{between 35 and 44}) = \dfrac{27.7}{142.1} \approx 0.195$

47. $P(\text{not between 18 and 20}) = 1 - \dfrac{5.8}{142.1} \approx 1 - 0.041 \approx 0.959$

48. $P(\text{not between 25 and 34}) = 1 - \dfrac{21.7}{142.1} \approx 1 - 0.153 \approx 0.847$

49. $P(\text{Ph.D.}) = \dfrac{8}{89} \approx 0.090$ **50.** $P(\text{Associate}) = \dfrac{18}{89} \approx 0.202$

51. $P(\text{master's}) = \dfrac{21}{89} \approx 0.236$ **52.** $P(\text{Bachelor's}) = \dfrac{33}{89} \approx 0.371$

53. (a) $P(\text{pink}) = \dfrac{2}{4} = 0.5$ (b) $P(\text{red}) = \dfrac{1}{4} = 0.25$ (c) $P(\text{white}) = \dfrac{1}{4} = 0.25$

54. $P(\text{same coloring as one of its parents}) = \dfrac{8}{16} = 0.5$

55. $P(\text{service industry}) = \dfrac{113{,}409}{144{,}428} \approx 0.785$

56. $P(\text{manufacturing industry}) = \dfrac{16{,}377}{144{,}428} \approx 0.113$

57. $P(\text{not in service industry}) = 1 - P(\text{service industry}) = 1 - 0.785 = 0.215$

58. $P(\text{not in agriculture, forestry, or fishing industry})$

$= 1 - P(\text{agriculture, forestry, or fishing industry})$

$= 1 - \left(\dfrac{2206}{144{,}428}\right) \approx 0.985$

59. (a) $P(\text{at least } 21) = \dfrac{87}{118} \approx 0.739$

(b) $P(\text{between 40 and 50 inclusive}) = \dfrac{32}{118} \approx 0.269$

(c) $P(\text{older than } 69) = \dfrac{1}{118} = 0.008$

60. (a) $P(\text{less than } \$21) = 0.25$

(b) $P(\text{between } \$21 \text{ and } \$50) = 0.50$

(c) $P(\$30 \text{ or more}) = 0.50$

61. The probability of choosing a tea drinker who does not have a college degree.

62. The probability of choosing a smoker whose mother did not smoke.

63. (a)

Sum	P(sum)	Probability
2	1/36	0.028
3	2/36	0.056
4	3/36	0.083
5	4/36	0.111
6	5/36	0.139
7	6/36	0.167
8	5/36	0.139
9	4/36	0.111
10	3/36	0.083
11	2/36	0.056
12	1/36	0.028

(b) Answers will vary.

(c) The answers in part (a) and (b) will be similar.

64. No, the odds of winning a prize are 1:6. (One winning cap and 6 losing caps) Thus, the statement should read, "one in seven game pieces win a prize."

65. (a) $P(\text{event will occur}) = \dfrac{4}{9} \approx 0.444$

(b) $P(\text{event will not occur}) = \dfrac{5}{9} \approx 0.556$

66. $13:39 = 1:3$

67. $39:13 = 3:1$

3.2 CONDITIONAL PROBABILITY AND THE MULTIPLICATION RULE

3.2 Try It Yourself Solutions

1a. (1) 30 and 102   (2) 11 and 50

b. (1) $P(\text{not have gene}) = \dfrac{30}{102} \approx 0.294$ (2) $P(\text{not have gene}|\text{normal IQ}) = \dfrac{11}{50} = 0.22$

2a. (1) Yes   (2) No

b. (1) Dependent   (2) Independent

3a. (1) Independent   (2) Dependent

b. (1) Let $A = \{\text{swimming through 1st dam}\}$

$B = \{\text{swimming through 2nd dam}\}$

$P(A \text{ and } B) = P(A) \cdot P(B|A) = (0.85) \cdot (0.85) = 0.723$

(2) Let $A = \{\text{selecting a heart}\}$

$B = \{\text{selecting a second heart}\}$

$P(A \text{ and } B) = P(A) \cdot P(B|A) = \left(\dfrac{13}{52}\right) \cdot \left(\dfrac{12}{51}\right) \approx 0.059$

4a. (1) Find probability of the event   (2) Find probability of the complement of the event

b. (1) $P(\text{3 knee surgeries successful}) = (0.90) \cdot (0.90) \cdot (0.90) = 0.729$

(2) $P(\text{at least one knee surgery successful}) = 1 - P(\text{none are successful})$

$= 1 - (0.10) \cdot (0.10) \cdot (0.10) = 0.999$

5a. (1)(2) $A = \{\text{is female}\}; B = \{\text{works in health field}\}$

b. (1) $P(A \text{ and } B) = P(A)P(B|A) = (0.65)(0.25)$

(2) $P(A \text{ and } B') = P(A)(1 - P(B|A)) = (0.65)(0.75)$

c. (1) 0.1625

(2) 0.4875

3.2 EXERCISE SOLUTIONS

1. Two events are independent if the occurrence of one of the events does not affect the probability of the occurrence of the other event.

If $P(B|A) = P(B)$ or $P(A|B) = P(A)$, then Events A and B are independent.

2. (a) Roll a die twice. The outcome of the 2nd roll is independent of the outcome of the 1st roll.

   (b) Draw two cards (without replacement) from a standard 52-card deck. The outcome of the 2nd card is dependent on the outcome of the 1st card.

3. False. If two events are independent, $P(A|B) = P(A)$.

4. False. If events A and B are independent, then $P(A \text{ and } B) = P(A) \cdot P(B)$.

5. These events are independent because the outcome of the 1st card drawn does not affect the outcome of the 2nd card drawn.

6. These events are dependent because returning a movie after its due date affects the outcome of receiving a late fee.

7. These events are dependent because the sum of the rolls depends on which numbers were rolled first and second.

8. These events are independent because the outcome of the 1st ball drawn does not affect the outcome of the 2nd ball drawn.

9. Events: depression, breathing-related sleeping disorder

   Dependent. People with depression are more likely to have a breathing-related sleeping disorder.

10. Events: stress, ulcers

    Independent. Stress only irritates already existing ulcers.

11. Events: memory loss, use of Aspartame

    Independent. The use of Aspartame does not cause memory loss.

12. Events: diabetes, obesity

    Dependent. Societies in which cases of diabetes are rare, obesity is rare.

13. Let $A = \{\text{have mutated BRCA gene}\}$ and $B = \{\text{develop breast cancer}\}$. Thus

    $$P(B) = \frac{1}{8}, P(A) = \frac{1}{600}, \text{ and } P(B|A) = \frac{8}{10}.$$

    (a) $P(B|A) = \frac{8}{10} = 0.8$

    (b) $P(A \text{ and } B) = P(A) \cdot P(B|A) = \left(\frac{1}{600}\right) \cdot \left(\frac{8}{10}\right) = 0.0013$

    (c) Dependent because $P(B|A) \neq P(B)$.

14. Let A = {drives pickup truck} and B = {drives a Ford}. Thus,

$$P(A) = \frac{1}{6}, P(B) = \frac{3}{10}, \text{ and } P(A|B) = \frac{2}{9}.$$

(a) $P(A|B) = \frac{2}{9} \approx 0.222$

(b) $P(A \text{ and } B) = P(P)P(A|B) = \left(\frac{3}{10}\right)\left(\frac{2}{9}\right) = 0.067$

(c) Dependent because $P(A|B) \neq P(A)$.

15. Let A = {own a computer} and B = {summer vacation this year}.

(a) $P(B') = \frac{45}{146} = 0.308$

(b) $P(A) = \frac{57}{146} = 0.390$

(c) $P(B|A) = \frac{46}{57} \approx 0.807$

(d) $P(A \text{ and } B) = P(A)P(B|A) = \left(\frac{57}{146}\right)\left(\frac{46}{57}\right) = 0.315$

(e) Dependent, because $P(B|A) \neq P(B)$. The probability that the family takes a summer vacation depends on whether or not they own a computer.

16. Let A = {male} and B = {Nursing major}.

(a) $P(B) = \frac{795}{3537} = 0.225$ (b) $P(A) = \frac{1110}{3537} = 0.314$

(c) $P(B|A) = \frac{95}{1110} = 0.086$

(d) $P(A \text{ and } B) = P(A)P(B|A) = \left(\frac{1110}{3537}\right)\left(\frac{95}{1110}\right) = 0.027$

(e) Dependent because $P(B|A) \neq P(B)$.

17. Let A = {pregnant} and B = {multiple births}. Thus $P(A) = 0.35$ and $P(B|A) = 0.28$.

(a) $P(A \text{ and } B) = P(A) \cdot P(B|A) = (0.35) \cdot (0.28) = 0.098$

(b) $P(B'|A) = 1 - P(B|A) = 1 - 0.28 = 0.72$

(c) It is not unusual because the probability of a pregnancy and multiple births is 0.098.

18. Let A = {has the opinion that race relations have improved} and B = {has the opinion that the rate of civil rights progress is too slow}. Thus, $P(A) = 0.6$ and $P(B|A) = 0.4$.

(a) $P(A \text{ and } B) = P(A) \cdot P(B|A) = (0.6) \cdot (0.4) = 0.24$

(b) $P(B'|A) = 1 - P(B|A) = 1 - 0.4 = 0.6$

(c) It is not unusual because the probability of an adult who thinks race relations have improved and says civil rights progress is too slow is 0.24.

19. Let A = {household in U.S. has a computer} and B = {has Internet access}.

$P(A \text{ and } B) = P(A)P(B|A) = (0.62)(0.88) = 0.546$

20. Let A = {survives bypass surgery} and B = {heart damage will heal}.

$P(A \text{ and } B) = P(A)P(B|A) = (0.60)(0.50) = 0.30$

21. Let A = {1st person is left-handed} and B = {2nd person is left-handed}.

(a) $P(A \text{ and } B) = P(A) \cdot P(A|B) = \left(\dfrac{120}{1000}\right) \cdot \left(\dfrac{119}{999}\right) \approx 0.014$

(b) $P(A' \text{ and } B') = P(A') \cdot P(B'|A') = \left(\dfrac{880}{1000}\right) \cdot \left(\dfrac{879}{999}\right) \approx 0.774$

(c) $P(\text{at least one is left-handed}) = 1 - P(A' \text{ and } B') = 1 - 0.774 = 0.226$

22. Let A = {1st bulb fails} and B = {2nd bulb fails}.

(a) $P(A \text{ and } B) = P(A) \cdot P(B|A) = \left(\dfrac{3}{12}\right)\left(\dfrac{2}{11}\right) \approx 0.045$

(b) $P(A' \text{ and } B') = P(A') \cdot P(B'|A') = \left(\dfrac{9}{12}\right)\left(\dfrac{8}{11}\right) \approx 0.545$

(c) $P(\text{at least one bulb failed}) = 1 - P(\text{none failed})$

$$= 1 - P(A' \text{ and } B')$$

$$= 1 - 0.545$$

$$= 0.455$$

23. Let A = {have one month's income or more} and B = {man}.

(a) $P(A) = \dfrac{138}{287} \approx 0.481$

(b) $P(A'|B) = \dfrac{66}{142} \approx 0.465$

(c) $P(B'|A) = \dfrac{62}{138} \approx 0.449$

(d) Dependent because $P(A') \approx 0.519 \neq 0.465 \approx P(A'|B)$

Whether a person has at least one month's income saved depends on whether or not the person is male.

24. Let $A = \{\$100 \text{ or more}\}$ and $B = \{\text{purebred}\}$.

(a) $P(A) = \dfrac{50}{90} \approx 0.556$

(b) $P(B'|A') = \dfrac{21}{40} = 0.525$

(c) $P(A \text{ and } B') = P(A) \cdot P(B'|A) = \left(\dfrac{50}{90}\right) \cdot \left(\dfrac{15}{50}\right) \approx 0.167$

(d) Dependent because $P(A) \approx 0.556 \neq 0.417 \approx P(A|B')$

The probability that the owner spent $100 or more on health care depends on whether or not the dog was a mixed breed.

25. (a) $P(\text{all five have AB+}) = (0.03) \cdot (0.03) \cdot (0.03) \cdot (0.03) \cdot (0.03) = 0.0000000243$

(b) $P(\text{none have AB+}) = (0.97) \cdot (0.97) \cdot (0.97) \cdot (0.97) \cdot (0.97) \approx 0.859$

(c) $P(\text{at least one has AB+}) = 1 - P(\text{none have AB+}) = 1 - 0.859 = 0.141$

26. (a) $P(\text{all three have O+}) = (0.38) \cdot (0.38) \cdot (0.38) \approx 0.055$

(b) $P(\text{none have O+}) = (0.62) \cdot (0.62) \cdot (0.62) \approx 0.238$

(c) $P(\text{at least one has O+}) = 1 - P(\text{none have O+}) = 1 - 0.238 = 0.762$

27. (a) $P(\text{first question correct}) = 0.2$

(b) $P(\text{first two questions correct}) = (0.2) \cdot (0.2) = 0.04$

(c) $P(\text{first three questions correct}) = (0.2)^3 = 0.008$

(d) $P(\text{none correct}) = (0.8)^3 = 0.512$

(e) $P(\text{at least one correct}) = 1 - P(\text{none correct}) = 1 - 0.512 = 0.488$

28. (a) $P(\text{none are defective}) = (0.995)^3 \approx 0.985$

(b) $P(\text{at least one defective}) = 1 - P(\text{none are defective}) = 1 - 0.985 = 0.015$

(c) $P(\text{all are defective}) = (0.005)^3 = 0.000000125$

29. $P(\text{all three products came from the third factory}) = \dfrac{25}{110} \cdot \dfrac{24}{109} \cdot \dfrac{23}{108} \approx 0.011$

30. (a) $P(\text{all share same birthday}) = \left(\dfrac{365}{365}\right)\left(\dfrac{1}{365}\right)\left(\dfrac{1}{365}\right) \approx 0.00000751$

(b) $P(\text{none share same birthday}) = \left(\dfrac{365}{365}\right)\left(\dfrac{364}{365}\right)\left(\dfrac{363}{365}\right) = 0.992$

31. $P(A|B) = \dfrac{P(A) \cdot P(B|A)}{P(A) \cdot P(B|A) + P(A') \cdot P(B|A')}$

$= \dfrac{\left(\frac{2}{3}\right) \cdot \left(\frac{1}{5}\right)}{\left(\frac{2}{3}\right) \cdot \left(\frac{1}{5}\right) + \left(\frac{1}{3}\right) \cdot \left(\frac{1}{2}\right)} = \dfrac{0.133}{0.133 + 0.167} = 0.444$

32. $P(A|B) = \dfrac{P(A) \cdot P(B|A)}{P(A) \cdot P(B|A) + P(A') \cdot P(B|A')}$

$$= \dfrac{\left(\dfrac{3}{8}\right) \cdot \left(\dfrac{2}{3}\right)}{\left(\dfrac{3}{8}\right) \cdot \left(\dfrac{2}{3}\right) + \left(\dfrac{5}{8}\right) \cdot \left(\dfrac{3}{5}\right)} = \dfrac{0.25}{0.25 + 0.375} = 0.4$$

33. $P(A|B) = \dfrac{P(A)P(B|A)}{(A)P(B|A) + P(A')P(B|A')}$

$$= \dfrac{(0.25)(0.3)}{(0.25)(0.3) + (0.75)(0.5)} = \dfrac{0.075}{0.075 + 0.375} = 0.167$$

34. $P(A|B) = \dfrac{P(A)P(A|B)}{P(A)P(A|B) + P(A')P(B|A')}$

$$= \dfrac{(0.62)(0.41)}{(0.62)(0.41) + (0.38)(0.17)} = \dfrac{0.254}{0.254 + 0.065} = 0.797$$

35. $P(A) = \dfrac{1}{200} = 0.005$

$P(B|A) = 0.80$

$P(B|A') = 0.05$

(a) $P(A|B) = \dfrac{P(A) \cdot P(B|A)}{P(A) \cdot P(B|A) + P(A') \cdot P(B|A')}$

$$= \dfrac{(0.005) \cdot (0.8)}{(0.005) \cdot (0.8) + (0.995) \cdot (0.05)} = \dfrac{0.004}{0.004 + 0.04975} \approx 0.074$$

(b) $P(A'|B') = \dfrac{P(A') \cdot P(B'|A')}{P(A') \cdot P(B'|A') + P(A) \cdot P(B'|A)}$

$$= \dfrac{(0.995) \cdot (0.95)}{(0.995) \cdot (0.95) + (0.005) \cdot (0.2)} = \dfrac{0.94525}{0.94525 + 0.001} \approx 0.999$$

36. (a) $P(\text{different birthdays}) = \dfrac{364}{365} \cdot \dfrac{363}{365} \cdot \ldots \cdot \dfrac{342}{365} \approx 0.462$

(b) $P(\text{at least two have same birthday}) = 1 - P(\text{different birthdays}) = 1 - 0.462 = 0.538$

(c) Yes, there were 2 birthdays on the 118th day.

(d) Answers will vary.

37. Let $A = \{\text{flight departs on time}\}$ and $B = \{\text{flight arrives on time}\}$.

$$P(A|B) = \dfrac{P(A \text{ and } B)}{P(B)} = \dfrac{(0.83)}{(0.87)} \approx 0.954$$

38. Let $A = \{\text{flight departs on time}\}$ and $B = \{\text{flight arrives on time}\}$.

$$P(A|B) = \dfrac{P(A \text{ and } B)}{P(B)} = \dfrac{(0.83)}{(0.87)} \approx 0.933$$

3.3 Try It Yourself Solutions

1a. (1) None of the statements are true.

(2) None of the statements are true.

(3) All of the statements are true.

b. (1) A and B are not mutually exclusive.

(2) A and B are not mutually exclusive.

(3) A and B are mutually exclusive.

2a. (1) Mutually exclusive   (2) Not mutually exclusive

b. (1) Let $A = \{6\}$ and $B = \{odd\}$.

$$P(A) = \frac{1}{6} \text{ and } P(B) = \frac{3}{6} = \frac{1}{2}$$

(2) Let $A = \{face\ card\}$ and $B = \{heart\}$.

$$P(A) = \frac{12}{52}, P(B) = \frac{13}{52}, \text{ and } P(A \text{ and } B) = \frac{3}{52}$$

c. (1) $P(A \text{ or } B) = P(A) + P(B) = \frac{1}{6} + \frac{1}{2} \approx 0.667$

(2) $P(A \text{ or } B) = P(A) + P(B) - P(A \text{ and } B) = \frac{12}{52} + \frac{13}{52} - \frac{3}{52} \approx 0.423$

3a. $A = \{$sales between \$0 and \$24,999$\}$

$B = \{$sales between \$25,000 and \$49,000$\}$

b. A and B cannot occur at the same time $\rightarrow A$ and B are mutually exclusive

c. $P(A) = \frac{3}{36}$ and $P(B) = \frac{5}{36}$

d. $P(A \text{ or } B) = P(A) + P(B) = \frac{3}{36} + \frac{5}{36} \approx 0.222$

4a. (1) Mutually exclusive   (2) Not mutually exclusive

b. (1) $P(B \text{ or } AB) = P(B) + P(AB) = \frac{45}{409} + \frac{16}{409} \approx 0.149$

(2) $P(O \text{ or } Rh+) = P(O) + P(Rh+) - P(O \text{ and } Rh+) = \frac{184}{409} + \frac{344}{409} - \frac{156}{409} \approx 0.910$

5a. Let $A = \{linebacker\}$ and $B = \{quarterback\}$.

$$P(A \text{ or } B) = P(A) + P(B) = \frac{32}{255} + \frac{11}{255} \approx 0.169$$

b. $P(\text{not a linebacker or quarterback}) = 1 - P(A \text{ or } B) \approx 1 - 0.169 \approx 0.831$

3.3 EXERCISE SOLUTIONS

1. $P(A \text{ and } B) = 0$ because A and B cannot occur at the same time.

2. (a) Toss coin once: $A = \{\text{head}\}$ and $B = \{\text{tail}\}$

(b) Draw one card: $A = \{\text{ace}\}$ and $B = \{\text{spade}\}$

3. True

4. False, two events being independent does not imply they are mutually exclusive. Example: Toss a coin then roll a 6-sided die. Let $A = \{\text{head}\}$ and $B = \{6 \text{ on die}\}$. $P(B|A) = \frac{1}{6} = P(B)$ implies A and B are independent events. However, $P(A \text{ and } B) = \frac{1}{12}$ implies A and B are not mutually exclusive.

5. False, $P(A \text{ or } B) = P(A) + P(B) - P(A \text{ and } B)$

6. True

7. Not mutually exclusive because the two events can occur at the same time.

8. Mutually exclusive because the two events cannot occur at the same time.

9. Not mutually exclusive because the two events can occur at the same time. The worker can be female and have a college degree.

10. Not mutually exclusive because the two events can occur at the same time.

11. Mutually exclusive because the two events cannot occur at the same time. The person cannot be in both age classes.

12. Not mutually exclusive because the two events can occur at the same time.

13. (a) No, it is possible for the events {overtime} and {temporary help} to occur at the same time.

(b) $P(\text{OT or temp}) = P(\text{OT}) + P(\text{temp}) - P(\text{OT and temp}) = \dfrac{18}{52} + \dfrac{9}{52} - \dfrac{5}{52} \approx 0.423$

14. (a) Not mutually exclusive because the two events can occur at the same time.

(b) $P(W \text{ or business}) = P(W) + P(\text{business}) - P(W \text{ and business})$

$$= \dfrac{1800}{3500} + \dfrac{860}{3500} - \dfrac{425}{3500} \approx 0.639$$

15. (a) Not mutually exclusive because the two events can occur at the same time. A carton can have a puncture and a smashed corner.

(b) $P(\text{puncture or corner}) = P(\text{puncture}) + P(\text{corner}) - P(\text{puncture and corner})$

$$= 0.05 + 0.08 - 0.004 = 0.126$$

16. (a) Not mutually exclusive because the two events can occur at the same time.

(b) $P(\text{does not have puncture or does not have smashed edge})$

$$= P(\text{does not have puncture}) + P(\text{does not have smashed edge})$$

$$- P(\text{does not have puncture and does not have smashed edge})$$

$$= 0.96 + 0.93 - 0.8928 = 0.997$$

17. (a) $P(\text{diamond or 7}) = P(\text{diamond}) + P(7) \cdot P(\text{diamond and 7})$

$$= \frac{13}{52} + \frac{4}{52} - \frac{1}{52} \approx 0.308$$

(b) $P(\text{red or queen}) = P(\text{red}) + P(\text{queen}) - P(\text{red and queen})$

$$= \frac{26}{52} + \frac{4}{52} - \frac{2}{52} \approx 0.538$$

(c) $P(\text{3 or face card}) = P(3) + P(\text{face card}) - P(\text{3 and face card})$

$$= \frac{4}{52} + \frac{12}{52} - 0 \approx 0.308$$

18. (a) $P(\text{6 or greater than 4}) = P(6) + P(\text{greater than 4}) - P(\text{6 and greater than 4})$

$$= \frac{1}{6} + \frac{2}{6} - \frac{1}{6} = 0.333$$

(b) $P(\text{less than 5 or odd}) = P(\text{less than 5}) + P(\text{odd}) - P(\text{less than 5 and odd})$

$$= \frac{4}{6} + \frac{3}{6} - \frac{2}{6} = 0.833$$

(c) $P(\text{3 or even}) = P(3) + P(\text{even}) - P(\text{3 and even})$

$$= \frac{1}{6} + \frac{3}{6} - 0 = 0.667$$

19. (a) $P(\text{under 5}) = 0.068$

(b) $P(\text{not 65+}) = 1 - P(\text{65+}) = 1 - 0.147 = 0.853$

(c) $P(\text{between 18 and 34}) = P(\text{between 18 and 24 or between 25 and 34})$
$= P(\text{between 18 and 24}) + P(\text{between 25 and 34}) = 0.097 + 0.132 = 0.229$

20. (a) $P(2) = 0.298$

(b) $P(\text{2 or more}) = 1 - P(1) = 1 - 0.555 = 0.445$

(c) $P(\text{between 2 and 5}) = 0.298 + 0.076 + 0.047 + 0.014 = 0.435$

21. (a) $P(\text{not completely satisfied}) = 1 - \frac{102}{1017} = 0.900$

(b) $P(\text{somewhat satisfied or completely satisfied})$

$= P(\text{somewhat satisfied}) + P(\text{completely dissatisfied})$

$= \frac{326}{1017} + \frac{132}{1017} = \frac{458}{1017} \approx 0.450$

22. (a) $P(\text{not as good as used to be}) = \frac{171}{1005} \approx 0.170$

(b) $P(\text{too much violence or tickets too costly}) = P(\text{too much violence}) + P(\text{tickets too costly})$

$$1 - \left(\frac{322}{1005} + \frac{302}{1005}\right) = 1 - \frac{624}{1005} \approx 0.379$$

23. $A = \{\text{male}\}$; $B = \{\text{nursing major}\}$

(a) $P(A \text{ or } B) = P(A) + P(B) - P(A \text{ and } B)$

$$= \frac{1110}{3537} + \frac{795}{3537} - \frac{95}{3537} = 0.512$$

(b) $P(A' \text{ or } B') = P(A') + P(B') - P(A' \text{ and } B')$

$$= \frac{2427}{3537} + \frac{2742}{3537} - \frac{1727}{3537} = 0.973$$

(c) $P(A \text{ or } B) = 0.512$

(d) No mutually exclusive. A male can be a nursing major.

24. $A = \{\text{left handed}\}$; $B = \{\text{man}\}$

(a) $P(A \text{ or } B') = P(A) + P(B') - P(A \text{ and } B')$

$$= \frac{113}{1000} + \frac{475}{1000} - \frac{50}{1000} = 0.538$$

(b) $P(A' \text{ or } B) = P(A') + P(B) - P(A' \text{ and } B)$

$$= \frac{887}{1000} + \frac{525}{1000} - \frac{462}{1000} = 0.950$$

(c) $P(A \text{ or } B) = P(A) + P(B) - P(A \text{ and } B)$

$$= \frac{113}{1000} + \frac{525}{1000} - \frac{63}{1000} = 0.575$$

(d) $P(A' \text{ and } B') = \frac{425}{1000} = 0.425$

(e) Not mutually exclusive. A woman can be right-handed.

25. $A = \{\text{frequently}\}$; $B = \{\text{occasionally}\}$; $C = \{\text{not at all}\}$; $D = \{\text{male}\}$

(a) $P(A \text{ or } B) = \dfrac{428}{2850} + \dfrac{886}{2850} = 0.461$

(b) $P(D' \text{ or } C) = P(D') + P(C) - P(D' \text{ and } C)$

$$= \frac{1378}{2850} + \frac{1536}{2850} - \frac{741}{2850} = 0.762$$

(c) $P(D \text{ or } A) = P(D) + P(A) - P(D \text{ and } A)$

$$= \frac{1472}{2850} + \frac{428}{2850} - \frac{221}{2850} = 0.589$$

(d) $P(D' \text{ or } A') = P(D') + P(A') - P(D' \text{ and } A')$

$$= \frac{1378}{2850} + \frac{2422}{2850} - \frac{1171}{2850} = 0.922$$

(e) Not mutually exclusive. A female can be frequently involved in charity work.

26. (a) $P(\text{contacts or glasses}) = \dfrac{253}{3203} + \dfrac{1268}{3203} = 0.475$

(b) $P(\text{male or wears both}) = \dfrac{1538}{3203} + \dfrac{545}{3203} - \dfrac{177}{3203} = 0.595$

(c) $P(\text{female or neither}) = \dfrac{1665}{3203} + \dfrac{1137}{3203} - \dfrac{681}{3203} = 0.662$

(d) $P(\text{male or does not wear glasses}) = \dfrac{1538}{3203} + \dfrac{253 + 1137}{3203} - \dfrac{64 + 456}{3203} = 0.752$

(e) The events are mutually exclusive because both cannot happen at the same time.

27. Answers will vary.

Conclusion: If two events, $\{A\}$ and $\{B\}$, are independent, $P(A \text{ and } B) = P(A) \cdot P(B)$. If two events are mutually exclusive, $P(A \text{ and } B) = 0$. The only scenario when two events can be independent and mutually exclusive is if $P(A) = 0$ or $P(B) = 0$.

28. $P(A \text{ or } B \text{ or } C) = P(A) + P(B) + P(C) - P(A \text{ and } B) - P(A \text{ and } C) - P(B \text{ and } C)$

$+ P(A \text{ and } B \text{ and } C)$

$= 0.40 + 0.10 + 0.50 - 0.05 - 0.25 - 0.10 + 0.03$

$= 0.63$

29. $P(A \text{ or } B \text{ or } C) = P(A) + P(B) + P(C) - P(A \text{ and } B) - P(A \text{ and } C) - P(B \text{ and } C)$

$+ P(A \text{ and } B \text{ and } C)$

$= 0.38 + 0.26 + 0.14 - 0.12 - 0.03 - 0.09 + 0.01$

$= 0.55$

3.4 COUNTING PRINCIPLES

3.4 Try It Yourself Solutions

1a. $n = 6$ teams   **b.** $6! = 720$

2a. $_8P_3 = \dfrac{8!}{(8-3)!} = \dfrac{8!}{5!} = \dfrac{8 \cdot 7 \cdot 6 \cdot 5 \cdot 4 \cdot 3 \cdot 2 \cdot 1}{5 \cdot 4 \cdot 3 \cdot 2 \cdot 1} = 8 \cdot 7 \cdot 6 = 336$

b. There are 336 possible ways that the subject can pick a first, second, and third activity.

3a. $n = 12, r = 4$

b. $_{12}P_4 = \dfrac{12!}{(12-4)!} = \dfrac{12!}{8!} = 12 \cdot 11 \cdot 10 \cdot 9 = 11,880$

4a. $n = 20, n_1 = 6, n_2 = 9, n_3 = 5$

b. $\dfrac{n!}{n_1! \cdot n_2! \cdot n_3!} = \dfrac{20!}{6! \cdot 9! \cdot 5!} = 77,597,520$

5a. $n = 20, r = 3$

b. $_{20}C_3 = 1140$

c. There are 1140 different possible three-person committees that can be selected from 20 employees.

6a. $_{20}P_2 = 380$ **b.** $\dfrac{1}{380} \approx 0.003$

7a. 1 favorable outcome and $\dfrac{6!}{1! \cdot 2! \cdot 2! \cdot 1!} = 180$ distinguishable permutations.

b. $P(\text{Letter}) = \dfrac{1}{180} \approx 0.006$

8a. $_{15}C_5 = 3003$ **b.** $_{54}C_5 = 3{,}162{,}510$ **c.** 0.0009

9a. $(_5C_3) \cdot (_7C_0) = 10 \cdot 1 = 10$ **b.** $_{12}C_3 = 220$ **c.** $\dfrac{10}{220} \approx 0.045$

3.4 EXERCISE SOLUTIONS

1. The number of ordered arrangements of n objects taken r at a time. An exmple of a permutation is the number of seating arrangements of you and three friends.

2. A selection of r of the n objects without regard to order. An example of a combination is the number of selections of different playoff teams from a volleyball tournament.

3. False, a permutation is an ordered arrangement of objects.

4. True   **5.** True   **6.** True

7. $_7P_3 = \dfrac{7!}{(7-3)!} = \dfrac{7!}{4!} = \dfrac{5040}{24} = 210$

8. $_{14}P_2 = \dfrac{14!}{(14-2)!} = \dfrac{14!}{12!} = \dfrac{(14)(13)\ldots(3)(2)(1)}{(12)(11)\ldots(3)(2)(1)} = 14 \cdot 13 = 182$

9. $_7C_4 = \dfrac{7!}{4!(7-4)!} = \dfrac{(7)(6)(5)(4)(3)(2)(1)}{[(4)(3)(2)(1)][(3)(2)(1)]} = \dfrac{540}{(24)(6)} = 35$

10. $_8P_6 = \dfrac{8!}{(8-6)!} = \dfrac{8!}{2!} = \dfrac{40{,}320}{2} = 20{,}160$

11. $_{24}C_6 = \dfrac{24!}{6!(24-6)!} = \dfrac{24!}{6!18!} = 134{,}596$

12. $\dfrac{_8C_4}{_{12}C_6} = \dfrac{\dfrac{8!}{4!(8-4)!}}{\dfrac{12!}{6!(12-6)!}} = \dfrac{\dfrac{8!}{4!4!}}{\dfrac{12!}{6!6!}} = \dfrac{70}{924} = 0.076$

13. $\dfrac{_6P_2}{_{10}P_4} = \dfrac{\dfrac{6!}{(6-2)!}}{\dfrac{10!}{(10-4)!}} = \dfrac{\dfrac{6!}{4!}}{\dfrac{10!}{6!}} = \dfrac{\dfrac{720}{24}}{\dfrac{3{,}628{,}800}{720}} = 0.0060$

14. $\dfrac{_8C_3}{_{12}C_3} = \dfrac{\dfrac{8!}{3!(8-3)!}}{\dfrac{12!}{3!(12-3)!}} = \dfrac{\dfrac{8!}{3!\,5!}}{\dfrac{12!}{3!\,9!}} = \dfrac{8!}{3!\,5!} \cdot \dfrac{3!\,9!}{12!} = \dfrac{(8!)(9!)}{(5!)(12!)}$

$\qquad = \dfrac{[(18)(7)(6)(5)\ldots(2)(1)][(9)(8)(7)\ldots(2)(1)]}{[(5)(4)\ldots(2)(1)][(12)(11)\ldots(1)]} = \dfrac{(8)(7)(6)}{(12)(11)(10)} = 0.255$

15. Permutation, because order of the 15 people in line matters.

16. Combination, because order of the committee members does not matter.

17. Combinations, because the order of the captains does not matter.

18. Permutation, because the order of the letters matters.

19. $10 \cdot 8 \cdot {_{13}C_2} = 6240$ **20.** $8! = 40{,}320$

21. $6! = 720$ **22.** $10! = 3{,}628{,}800$

23. $_{52}C_6 = 20{,}358{,}520$ **24.** $4! = 24$

25. $\dfrac{18!}{4! \cdot 8! \cdot 6!} = 9{,}189{,}180$ **26.** $_{20}C_4 = 4845$

27. 3-S's, 3-T's, 1-A, 2-I's, 1-C

$\qquad \dfrac{10!}{3! \cdot 3! \cdot 1! \cdot 2! \cdot 1!} = 50{,}400$

28. $_{40}C_{12} = 5{,}586{,}853{,}480$

29. (a) $6! = 720$ **30.** (a) 60

(b) sample (b) event

(c) $\dfrac{1}{720} = 0.0014$ (c) $\dfrac{1}{60} = 0.017$

31. (a) 12 **32.** (a) 720

(b) tree (b) median

(c) $\dfrac{1}{12} = 0.0833$ (c) $\dfrac{1}{720} = 0.0014$

33. (a) 907,200 **34.** (a) 39,916,800

(b) population (b) distribution

(c) $\dfrac{1}{907{,}200} = 0.000001$ (c) $\dfrac{1}{39{,}916{,}800} = 0.00000003$

35. $\dfrac{1}{_{12}C_3} = \dfrac{1}{220} = 0.0045$

36. $\dfrac{1}{_9C_3} = \dfrac{1}{84} = 0.012$

37. (a) $\left(\dfrac{15}{56}\right)\left(\dfrac{14}{55}\right)\left(\dfrac{13}{54}\right) = 0.0164$ (b) $\left(\dfrac{41}{56}\right)\left(\dfrac{40}{55}\right)\left(\dfrac{39}{54}\right) = 0.385$

38. (a) $\left(\dfrac{6}{14}\right)\left(\dfrac{5}{13}\right)\left(\dfrac{4}{12}\right)\left(\dfrac{3}{11}\right) = 0.015$ (b) $\left(\dfrac{8}{14}\right)\left(\dfrac{7}{13}\right)\left(\dfrac{6}{12}\right)\left(\dfrac{5}{11}\right) = 0.070$

39. (a) $_8C_4 = 70$

(b) $2 \cdot 2 \cdot 2 \cdot 2 = 16$

(c) $_4C_2\left[\dfrac{(_2C_0)\cdot(_2C_0)\cdot(_2C_2)\cdot(_2C_2)}{_8C_4}\right] \approx 0.086$

40. (a) $26 \cdot 26 \cdot 10 \cdot 10 \cdot 10 \cdot 10 = 6{,}760{,}000$

(b) $24 \cdot 24 \cdot 10 \cdot 10 \cdot 10 \cdot 10 = 5{,}760{,}000$

(c) 0.50

41. (a) $(26)(26)(10)(10)(10)(10)(10) = 67{,}600{,}000$

(b) $(26)(25)(10)(9)(8)(7)(6) = 19{,}656{,}000$

(c) $\dfrac{1}{67{,}600{,}000} \approx 0.000000015$

42. (a) $(10)(10)(10) = 1000$

(b) $(8)(10)(10) = 800$

(c) $\dfrac{(8)(10)(5)}{(8)(10)(10)} = \dfrac{1}{2} = 0.5$

43. (a) $5! = 120$ (b) $2! \cdot 3! = 12$ (c) $3! \cdot 2! = 12$ (d) 0.4

44. (a) $(_8C_3)\cdot(_2C_0) = (56)\cdot(1) = 56$

(b) $(_8C_2)\cdot(_2C_1) = (28)\cdot(2) = 56$

(c) At least two good units = one or fewer defective units.

$56 + 56 = 112$

(d) $P(\text{at least 2 defective units}) = \dfrac{(_8C_1)\cdot(_2C_2)}{_{10}C_3} = \dfrac{(8)\cdot(1)}{120} \approx 0.067$

45. $(8\%)(1200) = (0.08)(1200) = 96$ of the 1200 rate financial shape as excellent.

$P(\text{all four rate excellent}) = \dfrac{_{96}C_4}{_{1200}C_4} = \dfrac{3{,}321{,}960}{85{,}968{,}659{,}700} \approx 0.000039$

46. $(14\%)(1200) = (0.14)(1200) = 168$ of 1200 rate financial shape as poor.

$P(\text{all 10 rate poor}) = \dfrac{_{168}C_{10}}{_{1200}C_{10}} \approx 2.29 \times 10^{-9}$

47. $(36\%)(500) = (0.36)(500) = 180$ of the 1500 rate financial shape as fair $\Rightarrow 500 - 180 = 320$ rate shape as not fair.

$P(\text{none of 80 selected rate fair}) = \dfrac{_{320}C_{80}}{_{500}C_{80}} \approx 5.03 \times 10^{-18}$

48. $(41\%)(500) = (0.41)(500) = 205$ of 500 rate financial shape as good $\Rightarrow 500 - 205 = 295$
rate as not good.

$$P(\text{none of 55 selected rate good}) = \frac{_{295}C_{55}}{_{500}C_{55}} \approx 2.52 \times 10^{-14}$$

49. (a) $_{40}C_5 = 658,008$ (b) $P(\text{win}) = \dfrac{1}{658,008} \approx 0.00000152$

50. (a) $_{200}C_{15} = 1.4629 \times 10^{22}$ (b) $_{144}C_{15} = 8.5323 \times 10^{19}$

(c) $P(\text{no minorities}) = \dfrac{\left(_{144}C_{15}\right)}{\left(_{200}C_{15}\right)} \approx 0.00583$

(d) Yes, there is a very low probability of randomly selecting 15 non-minorities.

51. (a) $\dfrac{_{13}C_1\,_4C_4\,_{12}C_1\,_4C_1}{_{52}C_5} = \dfrac{(13)(1)(12)(4)}{2,598,960} = 0.0002$

(b) $\dfrac{_{13}C_1\,_4C_3\,_{12}C_1\,_4C_2}{_{52}C_5} = \dfrac{(13)(4)(12)(6)}{2,598,960} = 0.00144$

(c) $\dfrac{_{13}C_1\,_4C_3\,_{12}C_2\,_4C_1\,_4C_1}{_{52}C_5} = \dfrac{(13)(4)(66)(4)(4)}{2,598,960} = 0.0211$

(d) $\dfrac{_{13}C_2\,_{13}C_1\,_{13}C_1\,_{13}C_1}{_{52}C_5} = \dfrac{(78)(13)(13)(13)}{2,598,960} = 0.0659$

52. (a) $\dfrac{_{24}C_8}{_{41}C_8} = \dfrac{735,471}{95,548,245} = 0.0077$ (b) $\dfrac{_{17}C_8}{_{41}C_8} = \dfrac{24,310}{95,548,245} = 0.0003$

(c) $\dfrac{_{24}C_4\,_{17}C_2}{_{41}C_8} = \dfrac{(134,596)(136)}{95,548,245} = 0.1916$ (d) $\dfrac{_{24}C_4\,_{17}C_4}{_{41}C_8} = \dfrac{(10,626)(2380)}{95,548,245} = 0.2647$

53. $_{14}C_4 = 1001$ possible 4 digit arrangements if order is not important.

Assign 1000 of the 4 digit arrangements to the 13 teams since 1 arrangement is excluded.

54. $_{14}P_4 = 24,024$

55. $P(\text{1st}) = \dfrac{250}{1000} = 0.250$ $P(\text{8th}) = \dfrac{28}{1000} = 0.028$

$P(\text{2nd}) = \dfrac{199}{1000} = 0.199$ $P(\text{9th}) = \dfrac{17}{1000} = 0.017$

$P(\text{3rd}) = \dfrac{156}{1000} = 0.156$ $P(\text{10th}) = \dfrac{11}{1000} = 0.011$

$P(\text{4th}) = \dfrac{119}{1000} = 0.119$ $P(\text{11th}) = \dfrac{8}{1000} = 0.008$

$P(\text{5th}) = \dfrac{88}{1000} = 0.088$ $P(\text{12th}) = \dfrac{7}{1000} = 0.007$

$P(\text{6th}) = \dfrac{63}{1000} = 0.063$ $P(\text{13th}) = \dfrac{6}{1000} = 0.006$

$P(\text{7th}) = \dfrac{43}{1000} = 0.043$ $P(\text{14th}) = \dfrac{5}{1000} = 0.005$

56. Let A = {team with the worst record wins second pick} and

B = {team with the best record, ranked 14th, wins first pick}.

$$P(A|B)\frac{250}{(1000-5)} \cdot \frac{250}{995} \approx 0.251$$

57. Let A = {team with the worst record wins third pick} and

B = {team with the best record, ranked 13th, wins first pick} and

C = {team ranked 2nd wins the second pick}.

$$P(A|B \text{ and } C) = \frac{250}{796} = 0.314$$

58. Let A = {neither the first- nor the second-worst teams will get the first pick} and

B = {the first- or second-worst team will get the first pick}.

$$P(A) = 1 - P(B) = 1 - \left(\frac{199}{1000} + \frac{250}{100}\right) = 1 - \left(\frac{449}{1000}\right) = 1 - 0.449 = 0.551$$

CHAPTER 3 REVIEW EXERCISE SOLUTIONS

1. Sample space:

{HHHH, HHHT, HHTH, HHTT, HTHH, HTHT, HTTH, HTTT, THHH, THHT, THTH, THTT, TTHH, TTHT, TTTH, TTTT}

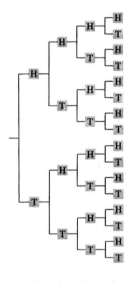

Event: Getting three heads
{HHHT, HHTH, HTHH, THHH}

2. Sample space:

{(1, 1), (1, 2), (1, 3), (1, 4), (1, 5), (1, 6), (2, 1), (2, 2), (2, 3), (2, 4), (2, 5), (2, 6), (3, 1), (3, 2), (3, 3), (3, 4), (3, 5), (3, 6), (4, 1), (4, 2), (4, 3), (4, 4), (4, 5), (4, 6), (5, 1), (5, 2), (5, 3), (5, 4), (5, 5), (5, 6), (6, 1), (6, 2), (6, 3), (6, 4), (6, 5), (6, 6)}

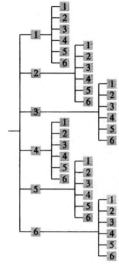

Event: sum of 4 or 5

{(1, 3), (1, 4), (2, 2), (2, 3), (3, 1), (3, 2), (4, 1)}

3. Sample space: {Jan, Feb ..., Dec}

Event: {Jan, June, July}

4. Sample space: {*GGG, GGB, GBG, GBB, BGG, BGB, BBG, BBB*}

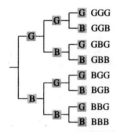

Event: The family has two boys.

{*BGB, BBG, GBB*}

5. (7)(4)(3) = 84

6. (26)(26)(26)(10)(10)(10)(10) = 175,760,000

7. Empirical probability

8. Classical probability

9. Subjective probability

10. Empirical probability

11. Classical probability

12. Empirical probability

13. P(at least 10) = 0.108 + 0.089 + 0.018 = 0.215

14. P(less than 20) = 0.608 + .177 + .108 = 0.893

15. $\dfrac{1}{(8)(10)(10)(10)(10)(10)(10)} = 1.25 \times 10^{-7}$

16. $1 - \dfrac{1}{(8)(10)(10)(10)(10)(10)(10)} = 0.999999875$

17. $P(\text{undergrad} \mid +) = 0.92$

18. $P(\text{graduate} \mid -) = 0.07$

19. Independent, the first event does not affect the outcome of the second event.

20. Dependent, the first event does affect the outcome of the second event.

21. $P(\text{correct toothpaste and correct dental rinse})$

$= P(\text{correct toothpaste}) \cdot P(\text{correct dental rinse})$

$= \left(\dfrac{1}{8}\right) \cdot \left(\dfrac{1}{5}\right) \approx 0.025$

22. $P(\text{1st pair black and 2nd pair blue or white})$

$= P(\text{1st pair black}) \cdot P(\text{2nd pair blue or white} \mid \text{1st black})$

$= \left(\dfrac{6}{18}\right) \cdot \left(\dfrac{12}{17}\right) = 0.2353$

23. Mutually exclusive because both events cannot occur at the same time.

24. Not mutually exclusive because both events can occur at the same time.

25. $P(\text{home or work}) = P(\text{home}) + P(\text{work}) - P(\text{home and work}) = 0.44 + 0.37 - 0.21 = 0.60$

26. $P(\text{silver or SUV}) = P(\text{silver}) + P(\text{SUV}) - P(\text{silver and SUV})$

$= 0.19 + 0.22 - 0.16$

$= 0.25$

27. $P(\text{4–8 or club}) = P(\text{4–8}) + P(\text{club}) - P(\text{4–8 and club}) = \dfrac{20}{52} + \dfrac{13}{52} - \dfrac{5}{52} \approx 0.538$

28. $P(\text{red or queen}) = P(\text{red}) + P(\text{queen}) - P(\text{red and queen})$

$= \dfrac{26}{52} + \dfrac{4}{52} - \dfrac{2}{52} = 0.538$

29. $P(\text{odd or less than 4}) = P(\text{odd}) + P(\text{less than 4}) - P(\text{odd and less than 4})$

$= \dfrac{6}{12} + \dfrac{3}{12} - \dfrac{2}{12} \approx 0.583$

30. $P(\text{even or greater than 6}) = P(\text{even}) + P(\text{greater than 6}) - P(\text{even and greater than 6})$

$= \dfrac{4}{8} + \dfrac{2}{8} - \dfrac{1}{8} = \dfrac{5}{8} = 0.625$

31. $P(\text{600 or more}) = P(\text{600} - \text{999}) + P(\text{1000 or more})$

$= 0.198 + 0.300 = 0.498$

32. $P(\text{300} - \text{999}) = P(\text{300} - \text{599}) + P(\text{600} - \text{999})$

$= 0.225 + 0.198 = 0.423$

33. $P(\text{poor taste or hard to find}) = P(\text{poor taste}) + P(\text{hard to find})$

$$= \frac{60}{500} + \frac{55}{500} = \frac{115}{500} = 0.23$$

34. $P(\text{no time to cook or confused about nutrition}) = P(\text{no time to cook}) + P(\text{confused about nutrition})$

$$= \frac{175}{500} + \frac{30}{500} = \frac{205}{500} = 0.41$$

35. Order is important: $_{15}P_3 = 2730$

36. $5! = 120$

37. Order is not important: $_{17}C_4 = 2380$

38. $_{13}C_2 = 78$

39. $P(\text{3 kings and 2 queens}) = \frac{_4C_3 \cdot {}_4C_2}{_{52}C_5} = \frac{4 \cdot 6}{2{,}598{,}960} \approx 0.00000923$

40. $\dfrac{1}{(23)(26)(26)(10)} = \dfrac{1}{155{,}480} \approx 0.0000064$

41. (a) $P(\text{no defectives}) = \frac{_{197}C_3}{_{200}C_3} = \frac{1{,}254{,}890}{1{,}313{,}400} \approx 0.955$

(b) $P(\text{all defective}) = \frac{_3C_3}{_{200}C_3} = \frac{1}{1{,}313{,}400} = \approx 0.000000761$

(c) $P(\text{at least one defective}) = 1 - P(\text{no defective}) = 1 - 0.955 = 0.045$

(d) $P(\text{at least one non-defective}) = 1 - P(\text{all defective}) = 1 - 0.000000761 \approx 0.999999239$

42. (a) $\frac{346}{350} \cdot \frac{345}{349} \cdot \frac{344}{348} \cdot \frac{343}{347} \approx 0.955$

(b) $\left(\frac{4}{350}\right)\left(\frac{3}{349}\right)\left(\frac{2}{348}\right)\left(\frac{1}{347}\right) = 0.000000002$

(c) $P(\text{at least one winner}) = 1 - P(\text{no winners}) = 1 - 0.955 = 0.045$

(d) $P(\text{at least one non-winner}) = 1 - P(\text{all winners})$

$$= 1 - \left(\frac{4}{350}\right)\left(\frac{3}{349}\right)\left(\frac{2}{348}\right)\left(\frac{1}{347}\right) = 0.999999998$$

CHAPTER 3 QUIZ SOLUTIONS

1. (a) $P(\text{bachelor}) = \frac{1399}{2671} \approx 0.524$

(b) $P(\text{bachelor}|F) = \frac{804}{1561} \approx 0.515$

(c) $P(\text{bachelor}|M) = \frac{595}{1110} \approx 0.536$

(d) $P(\text{associate or bachelor}) = P(\text{associate}) + P(\text{bachelor})$

$$= \frac{665}{2671} + \frac{1399}{2671} \approx 0.773$$

(e) $P(\text{doctorate}|\text{M}) = \dfrac{25}{1110} \approx 0.023$

(f) $P(\text{master or female}) = P(\text{master}) + P(\text{female}) - P(\text{master and female})$

$$= \frac{559}{2671} + \frac{1561}{2671} - \frac{329}{2671} = 0.671$$

(g) $P(\text{associate and male}) = P(\text{associate}) \cdot P(\text{male}|\text{associate})$

$$= \frac{665}{2671} \cdot \frac{260}{665} \approx 0.097$$

(h) $P(\text{F}|\text{bachelor}) = \dfrac{804}{1399} \approx 0.575$

2. Not mutually exclusive because both events can occur at the same time.

Dependent because one event can affect the occurrence of the second event.

3. (a) $_{147}C_3 = 518{,}665$

(b) $_3C_3 = 1$

(c) $_{150}C_3 - {_3}C_3 = 551{,}300 - 1 = 551{,}299$

4. (a) $\dfrac{_{147}C_3}{_{150}C_3} = \dfrac{518{,}665}{551{,}300} \approx 0.94$

(b) $\dfrac{_3C_3}{_{150}C_3} = \dfrac{1}{551{,}300} \approx 0.00000181$

(c) $\dfrac{_{150}C_3 - {_3}C_3}{_{150}C_3} = \dfrac{551{,}299}{551{,}300} \approx 0.999998$

5. $9 \cdot 10 \cdot 10 \cdot 10 \cdot 10 \cdot 5 = 450{,}000$

6. $_{30}P_4 = 657{,}720$

Discrete Probability Distributions

4.1 Try It Yourself Solutions

1a. (1) measured (2) counted

  b. (1) Random variable is continuous because x can be any amount of time needed to complete a test.

     (2) Random variable is discrete because x can be counted.

2ab.

x	f	$P(x)$
0	16	0.16
1	19	0.19
2	15	0.15
3	21	0.21
4	9	0.09
5	10	0.10
6	8	0.08
7	2	0.02
	$n = 100$	$\sum P(x) = 1$

New Employee Sales

3a. Each $P(x)$ is between 0 and 1.

  b. $\sum P(x) = 1$

  c. Because both conditions are met, the distribution is a probability distribution.

4a. (1) Yes, each outcome is between 0 and 1. (2) Yes, each outcome is between 0 and 1.

  b. (1) Yes, $\sum P(x) = 1$. (2) Yes, $\sum P(x) = 1$.

  c. (1) Is a probability distribution (2) Is a probability distribution

5ab.

x	$P(x)$	$xP(x)$
0	0.16	$(0)(0.16) = 0.00$
1	0.19	$(1)(0.19) = 0.19$
2	0.15	$(2)(0.15) = 0.30$
3	0.21	$(3)(0.21) = 0.63$
4	0.09	$(4)(0.09) = 0.36$
5	0.10	$(5)(0.10) = 0.50$
6	0.08	$(6)(0.08) = 0.48$
7	0.02	$(7)(0.02) = 0.14$
	$\sum P(x) = 1$	$\sum xP(x) = 2.60$

  c. $\mu = \sum xP(x) = 2.6$

    On average, 2.6 sales are made per day.

6ab.

x	$P(x)$	$x - \mu$	$(x - \mu)^2$	$P(x)(x - \mu)^2$
0	0.16	-2.6	6.76	$(0.16)(6.76) = 1.0816$
1	0.19	-1.6	2.56	$(0.19)(2.56) = 0.4864$
2	0.15	-0.6	0.36	$(0.15)(0.36) = 0.054$
3	0.21	0.4	0.16	$(0.21)(0.16) = 0.0336$
4	0.09	1.4	1.96	$(0.09)(1.96) = 0.1764$
5	0.10	2.4	5.76	$(0.10)(5.76) = 0.576$
6	0.08	3.4	11.56	$(0.08)(11.56) = 0.9248$
7	0.02	4.4	19.36	$(0.02)(19.36) = 0.3872$
	$\Sigma P(x) = 1$			$\Sigma P(x)(x - \mu)^2 = 3.72$

c. $\sigma = \sqrt{\sigma^2} = \sqrt{3.720} \approx 1.9$

d. A typical distance or deviation of the random variable from the mean is 1.9 sales per day.

7ab.

x	f	$P(x)$	$xP(x)$
0	25	0.111	$(0)(0.111) = 0.000$
1	48	0.213	$(1)(0.213) = 0.213$
2	60	0.267	$(2)(0.267) = 0.533$
3	45	0.200	$(3)(0.200) = 0.600$
4	20	0.089	$(4)(0.089) = 0.356$
5	10	0.044	$(5)(0.044) = 0.222$
6	8	0.036	$(6)(0.036) = 0.213$
7	5	0.022	$(7)(0.022) = 0.156$
8	3	0.013	$(8)(0.013) = 0.107$
9	1	0.004	$(9)(0.004) = 0.040$
	$n = 225$	$\Sigma P(x) \approx 1$	$\Sigma xP(x) = 2.440$

Gain, x	$P(x)$	$xP(x)$
1995	$\dfrac{1}{2000}$	$\dfrac{1995}{2000}$
995	$\dfrac{1}{2000}$	$\dfrac{995}{2000}$
495	$\dfrac{1}{2000}$	$\dfrac{495}{2000}$
245	$\dfrac{1}{2000}$	$\dfrac{245}{2000}$
95	$\dfrac{1}{2000}$	$\dfrac{95}{2000}$
-5	$\dfrac{1995}{2000}$	$-\dfrac{9975}{2000}$
	$\Sigma P(x) = 1$	$\Sigma xP(x) = -3.08$

c. $E(x) = \Sigma xP(x) = -3.08$

d. You can expect an average loss of \$3.08 for each ticket purchased.

4.1 EXERCISE SOLUTIONS

1. A random variable represents a numerical value assigned to an outcome of a probability experiment. Examples: Answers will vary.

2. A discrete probability distribution lists each possible value a random variable can assume, together with its probability.

Condition 1: $0 \leq P(x) \leq 1$

Condition 2: $\Sigma P(x) = 1$

3. An expected value of 0 represents the break even point, so the accountant will not gain or lose any money.

4. The mean of a probability distribution represents the "theoretical average" of a probability experiment.

5. False. In most applications, discrete random variables represent counted data, while continuous random variables represent measured data.

6. True

7. True

8. False. The expected value of a discrete random variable is equal to the mean of the random variable.

9. Discrete, because home attendance is a random variable that is countable.

10. Continuous, because length of time is a random variable that has an infinite number of possible outcomes and cannot be counted.

11. Continuous, because annual vehicle-miles driven is a random variable that cannot be counted.

12. Discrete, because the number of fatalities is a random variable that is countable.

13. Discrete, because the number of motorcycle accidents is a random variable that is countable.

14. Continuous, because the random variable has an infinite number of possible outcomes and cannot be counted.

15. Continuous, because the random variable has an infinite number of possible outcomes and cannot be counted.

16. Discrete, because the random variable is countable.

17. Discrete, because the random variable is countable.

18. Continuous, because the random variable has an infinite number of possible outcomes and cannot be counted.

19. Continuous, because the random variable has an infinite number of possible outcomes and cannot be counted.

20. Discrete, because the random variable is countable.

21. (a) $P(x > 2) = 0.25 + 0.10 = 0.35$

(b) $P(x < 4) = 1 - P(4) = 1 - 0.10 = 0.90$

22. (a) $P(x > 1) = 1 - P(x < 2) = 1 - (0.30 + 0.25) = 0.45$

(b) $P(x < 3) = 0.30 + 0.25 + 0.25 = 0.80$

23. $\Sigma P(x) = 1 \rightarrow P(3) = 0.22$ **24.** $\Sigma P(x) = 1 \rightarrow P(1) = 0.15$

25. Yes   **26.** No, $\Sigma P(x) \approx 1.576$

27. No, $\Sigma P(x) = 0.95$ and $P(5) < 0.$ **28.** Yes

29. (a)

x	f	$P(x)$	$xP(x)$	$(x - \mu)$	$(x - \mu)^2$	$(x - \mu)^2 P(x)$
0	1491	0.686	(0)(0.686) = 0	−0.501	0.251	(0.251)(0.686) = 0.169
1	425	0.195	(1)(0.195) = 0.195	0.499	0.249	(0.249)(0.195) = 0.049
2	168	0.077	(2)(0.077) = 0.155	1.499	2.246	(2.246)(0.077) = 0.173
3	48	0.022	(3)(0.022) = 0.066	2.499	6.244	(6.244)(0.022) = 0.138
4	29	0.013	(4)(0.013) = 0.053	3.499	12.241	(12.241)(0.013) = 0.160
5	14	0.006	(5)(0.006) = 0.030	4.499	20.238	(20.238)(0.006) = 0.122
	$n = 2175$	$\Sigma P(x) \approx 1$	$\Sigma xP(x) = 0.497$			$\Sigma(x - \mu)^2 P(x) = 0.811$

(b) $\mu = \Sigma xP(x) \approx 0.5$

(c) $\sigma^2 = \Sigma(x - \mu)^2 P(x) \approx 0.8$

(d) $\sigma = \sqrt{\sigma^2} \approx \sqrt{0.8246} \approx 0.9$

(e) A household on average has 0.5 dog with a standard deviation of 0.9 dog.

30. (a)

x	f	$P(x)$	$xP(x)$	$(x - \mu)$	$(x - \mu)^2$	$(x - \mu)^2 P(x)$
0	1941	0.7275	0	-0.5321	0.2822	$(0.2822)(0.7275) = 0.2053$
1	349	0.1308	0.1308	0.4688	0.2198	$(0.2198)(0.1308) = 0.0287$
2	203	0.0761	0.1522	1.4688	2.1574	$(2.1574)(0.0761) = 0.1642$
3	78	0.0292	0.0876	2.4688	6.0950	$(6.0950)(0.0292) = 0.1780$
4	57	0.0214	0.0856	3.4688	12.0326	$(12.0326)(0.0214) = 0.2575$
5	40	0.0150	0.075	4.4688	19.9702	$(19.9702)(0.0150) = 0.2996$
	$n = 2668$	$\sum P(x) = 1$	$\sum xP(x) = 0.531$			$\sum (x - \mu)^2 P(x) = 1.1333$

(b) $\mu = \Sigma xP(x) \approx 0.5$

(c) $\sigma^2 = \Sigma(x - \mu)^2 P(x) \approx 1.1$

(d) $\sigma = \sqrt{\sigma^2} \approx 1.1$

(e) A household has on average 0.5 cat with a standard deviation of 1.1 cats.

31. (a)

x	f	$P(x)$	$xP(x)$	$(x - \mu)$	$(x - \mu)^2$	$(x - \mu)^2 P(x)$
0	300	0.432	0.000	-0.764	0.584	$(0.584)(0.432) = 0.252$
1	280	0.403	0.403	0.236	0.056	$(0.056)(0.403) = 0.022$
2	95	0.137	0.274	1.236	1.528	$(1.528)(0.137) = 0.209$
3	20	0.029	0.087	2.236	5.000	$(5.000)(0.029) = 0.145$
	$n = 695$	$\sum P(x) \approx 1$	$\sum xP(x) = 0.764$			$\sum (x - \mu)^2 P(x) = 0.629$

(b) $\mu = \Sigma xP(x) \approx 0.8$

(c) $\sigma^2 = \Sigma(x - \mu)^2 P(x) \approx 0.6$

(d) $\sigma = \sqrt{\sigma^2} \approx 0.8$

(e) A household on average has 0.8 computer with a standard deviation of 0.8 computer.

32. (a)

x	f	$P(x)$	$xP(x)$	$(x - \mu)$	$(x - \mu)^2$	$(x - \mu)^2 P(x)$
0	93	0.250	0.000	-1.503	2.259	$(2.259)(0.250) = 0.565$
1	113	0.297	0.297	-0.503	0.253	$(0.253)(0.297) = 0.752$
2	87	0.229	0.458	0.497	0.247	$(0.247)(0.229) = 0.057$
3	64	0.168	0.505	1.497	2.241	$(2.241)(0.168) = 0.377$
4	13	0.034	0.137	2.497	6.235	$(6.235)(0.034) = 0.213$
5	8	0.021	0.105	3.497	12.229	$(12.229)(0.021) = 0.257$
	$n = 380$	$\sum P(x) = 1$	$\sum xP(x) = 1.503$			$\sum (x - \mu)^2 P(x) = 1.545$

(b) $\mu = \Sigma xP(x) \approx 1.5$

(c) $\sigma^2 = \Sigma(x - \mu)^2 P(x) \approx 1.5$

(d) $\sigma = \sqrt{\sigma^2} \approx 1.2$

(e) The average number of defects per batch is 1.5 with a standard deviation of 1.2 defects.

33. (a)

x	f	$P(x)$	$xP(x)$	$(x - \mu)$	$(x - \mu)^2$	$(x - \mu)^2 P(x)$
0	6	0.031	0.000	-3.411	11.638	$(11.638)(0.031) = 0.360$
1	12	0.063	0.063	-2.411	5.815	$(5.815)(0.063) = 0.366$
2	29	0.151	0.302	-1.411	1.992	$(1.992)(0.151) = 0.300$
3	57	0.297	0.891	-0.411	0.169	$(0.169)(0.297) = 0.050$
4	42	0.219	0.876	0.589	0.346	$(0.346)(0.219) = 0.076$
5	30	0.156	0.780	1.589	2.523	$(2.523)(0.156) = 0.394$
6	16	0.083	0.498	2.589	6.701	$(6.701)(0.083) = 0.557$
	$n = 192$	$\sum P(x) = 1$	$\sum xP(x) = 3.410$			$\sum (x - \mu)^2 P(x) = 2.103$

(b) $\mu = \Sigma xP(x) \approx 3.4$

(c) $\sigma^2 = \Sigma(x - \mu)^2 P(x) \approx 2.1$

(d) $\sigma = \sqrt{\sigma^2} \approx 1.5$

(e) An employee works an average of 3.4 overtime hours per week with a standard deviation of 1.5 hours.

34. (a)

x	f	$P(x)$	$xP(x)$	$(x - \mu)$	$(x - \mu)^2$	$(x - \mu)^2 P(x)$
0	19	0.059	0.000	-3.349	11.216	$(11.216)(0.059) = 0.662$
1	39	0.122	0.122	-2.349	5.518	$(5.518)(0.122) = 0.673$
2	52	0.163	0.326	-1.349	1.820	$(1.820)(0.163) = 0.297$
3	57	0.178	0.534	-0.349	0.122	$(0.122)(0.178) = 0.022$
4	68	0.213	0.852	0.651	0.424	$(0.424)(0.213) = 0.090$
5	41	0.128	0.640	1.651	2.726	$(2.726)(0.128) = 0.349$
6	27	0.084	0.504	2.651	7.028	$(7.028)(0.084) = 0.590$
7	17	0.053	0.371	3.651	13.330	$(13.330)(0.053) = 0.706$
	$n = 320$	$\sum P(x) = 1$	$\sum xP(x) = 3.349$			$\sum (x - \mu)^2 P(x) = 3.389$

(b) $\mu = \Sigma xP(x) \approx 3.3$

(c) $\sigma^2 = \Sigma(x - \mu)^2 P(x) \approx 3.4$

(d) $\sigma = \sqrt{\sigma^2} \approx 1.8$

(e) The average number of school related extracurricular activities per student is 3.3 with a standard deviation of 1.8 activities.

35.

x	$P(x)$	$xP(x)$	$(x - \mu)$	$(x - \mu)^2$	$(x - \mu)^2 P(x)$
0	0.02	0.00	-5.30	28.09	0.562
1	0.02	0.02	-4.30	18.49	0.372
2	0.06	0.12	-3.30	10.89	0.653
3	0.06	0.18	-2.30	5.29	0.317
4	0.08	0.32	-1.30	1.69	0.135
5	0.22	1.10	-0.30	0.09	0.020
6	0.30	1.80	0.70	0.49	0.147
7	0.16	1.12	1.70	2.89	0.462
8	0.08	0.64	2.70	7.29	0.583
	$\sum P(x) = 1$	$\sum xP(x) = 5.30$			$\sum (x - \mu)^2 P(x) = 3.250$

(a) $\mu = \Sigma xP(x) = 5.3$ (b) $\sigma^2 = \Sigma(x - \mu)^2 P(x) = 3.3$

(c) $\sigma = \sqrt{\sigma^2} = 1.8$ (d) $E[x] = \mu = \Sigma xP(x) = 5.3$

(e) The expected number of correctly answered questions is 5.3 with a standard deviation of 1.8 questions.

36.

x	$P(x)$	$xP(x)$	$(x - \mu)$	$(x - \mu)^2$	$(x - \mu)^2 P(x)$
0	0.01	0	-2.99	8.94	0.089
1	0.10	0.10	-1.99	3.96	0.396
2	0.26	0.52	-0.99	0.98	0.254
3	0.32	0.96	0.01	0.0001	0.000032
4	0.18	0.72	1.01	1.02	0.183
5	0.06	0.30	2.01	4.04	0.242
6	0.03	0.18	3.01	9.06	0.272
7	0.03	0.21	4.01	16.08	0.482
		$\sum xP(x) = 2.99$			$\sum(x - \mu)^2 P(x) = 1.921$

(a) $\mu = \Sigma xP(x) = 3.0$

(b) $\sigma^2 = \Sigma(x - \mu)^2 P(x) = 1.9$

(c) $\sigma = \sqrt{\sigma^2} = 1.4$

(d) $E[x] = \mu = \Sigma xP(x) = 3.0$

(e) The expected number of 911 calls received per hour is 3.0 with a standard deviation of 1.4 calls.

37. (a) $\mu = \Sigma xP(x) \approx 2.0$

(b) $\sigma^2 = \Sigma(x - \mu)^2 P(x) \approx 1.0$

(c) $\sigma = \sqrt{\sigma^2} \approx 1.0$

(d) $E[x] = \mu = \Sigma xP(x) = 2.0$

(e) The expected number of hurricanes that hit the U.S. is 2.0 with a standard deviation of 1.0.

38. (a) $\mu = \Sigma xP(x) = 1.7$

(b) $\sigma^2 = \Sigma(x - \mu)^2 P(x) \approx 1.0$

(c) $\sigma = \sqrt{\sigma^2} \approx 1.0$

(d) $E[x] = \mu = \Sigma xP(x) = 1.7$

(e) The expected car occupancy is 1.7 with a standard deviation of 1.0.

39. (a) $\mu = \Sigma xP(x) \approx 2.5$

(b) $\sigma^2 = \Sigma(x - \mu)^2 P(x) \approx 1.9$

(c) $\sigma = \sqrt{\sigma^2} \approx 1.4$

(d) $E[x] = \mu = \Sigma xP(x) = 2.5$

(e) The expected household size is 2.5 persons with a standard deviation of 1.4 persons.

40. (a) $\mu = \Sigma xP(x) = 1.6$

(b) $\sigma^2 = \Sigma(x - \mu)^2 P(x) \approx 1.9$

(c) $\sigma = \sqrt{\sigma^2} \approx 1.4$

(d) $E[x] = \mu = \Sigma xP(x) = 1.6$

(e) The expected car occupancy is 1.6 with a standard deviation of 1.4.

41. (a) $P(x < 2) = 0.686 + 0.195 = 0.881$

(b) $P(x \geq 1) = 1 - P(x = 0) = 1 - 0.686 = 0.314$

(c) $P(1 \leq x \leq 3) = 0.195 + 0.077 + 0.022 = 0.294$

42. (a) $P(0) = 0.432$

(b) $P(x \geq 1) = 1 - P(0) = 1 - 0.432 = 0.568$

(c) $P(0 \leq x \leq 2) = 1 - P(x = 3) = 1 - 0.029 = 0.971$

43. A household with three dogs is unusual because the probability is only 0.022.

44. A household with no computers is not unusual because the probability is 0.432.

45. $E(x) = \mu = \Sigma xP(x) = (-1) \cdot \left(\dfrac{37}{38}\right) + (35) \cdot \left(\dfrac{1}{38}\right) \approx -\$0.05$

46. $E(x) = \mu = \Sigma x P(x)$

$$= (3146) \cdot \left(\frac{1}{5000}\right) + (446) \cdot \left(\frac{1}{5000}\right) + (21) \cdot \left(\frac{15}{5000}\right) + (-4) \cdot \left(\frac{4983}{5000}\right) \approx -\$3.21$$

47. $\mu_4 = a + b\mu_x = 1000 + 1.05(36{,}000) = \$38{,}800$

48. $\sigma_y = |b|\sigma_x = (1.04)(3899) = 4054.96$

49. $\mu_{x+y} = \mu_x + \mu_y = 1532 + 1506 = 3038$

$\mu_{x-y} = \mu_x - \mu_y = 1532 - 1506 = 26$

50. $\sigma_{x-y}^2 = \sigma_x^2 + \sigma_y^2 = (312)^2 + (304)^2 = 189{,}760 \Rightarrow \sigma_{x-y} = \sqrt{\sigma_{x-y}^2} = 435.61$

4.2 BINOMIAL DISTRIBUTIONS

4.2 Try It Yourself Solutions

1a. Trial: answering a question (10 trials)
Success: question answered correctly

b. Yes, the experiment satisfies the four conditions of a binomial experiment.

c. It is a binomial experiment.

$n = 10, p = 0.25, q = 0.75, x = 0, 1, 2, \ldots, 9, 10$

2a. Trial: drawing a card with replacement (5 trials)
Success: card drawn is a club
Failure: card drawn is not a club

b. $n = 5, p = 0.25, q = 0.75, x = 3$

c. $P(3) = \dfrac{5!}{2!3!}(0.25)^3(0.75)^2 \approx 0.088$

3a. Trial: selecting a worker and asking a question (7 trials)
Success: Selecting a worker who will rely on pension
Failure: Selecting a worker who will not rely on pension

b. $n = 7, p = 0.26, q = 0.74, x = 0, 1, 2, \ldots, 6, 7$

c. $P(0) = {}_7C_0(0.26)^0(0.74)^7 = 0.1215$
$P(1) = {}_7C_1(0.26)^1(0.74)^6 = 0.2989$
$P(2) = {}_7C_2(0.26)^2(0.74)^5 = 0.3150$
$P(3) = {}_7C_3(0.26)^3(0.74)^4 = 0.1845$
$P(4) = {}_7C_4(0.26)^4(0.74)^3 = 0.0648$
$P(5) = {}_7C_5(0.26)^5(0.74)^2 = 0.0137$
$P(6) = {}_7C_6(0.26)^6(0.74)^1 = 0.0016$
$P(7) = {}_7C_7(0.26)^7(0.74)^0 = 0.0001$

d.

x	$P(x)$
0	0.1215
1	0.2989
2	0.3150
3	0.1845
4	0.0648
5	0.0137
6	0.0016
7	0.0001

4a. $n = 250, p = 0.71, x = 178$

b. $P(178) \approx 0.056$

c. The probability that exactly 178 people from a random sample of 250 people in the United States will use more than one topping on their hot dog is about 0.056.

5a. (1) $x = 2$ (2) $x = 2, 3, 4,$ or 5 (3) $x = 0$ or 1

b. (1) $P(2) \approx 0.217$

(2) $P(0) = {}_5C_0(0.21)^0(0.79)^5 = 0.308$

$P(1) = {}_5C_1(0.21)^1(0.79)^4 = 0.409$

$P(x \geq 2) = 1 - P(0) - P(1) = 1 - 0.308 - 0.409 = 0.283$

or

$P(x \geq 2) = P(2) + P(3) + P(4) + P(5)$

$= 0.217 + 0.058 + 0.008 + 0.0004$

$= 0.283$

(3) $P(x < 2) = P(0) + P(1) = 0.308 + 0.409 = 0.717$

c. (1) The probability that exactly two of the five men consider fishing their favorite leisure-time activity is about 0.217.

(2) The probability that at least two of the five men consider fishing their favorite leisure-time activity is about 0.283.

(3) The probability that fewer than two of the five men consider fishing their favorite leisure-time activity is about 0.717.

6a. Trial: selecting a business and asking if it has a Web site. (10 trials)

Success: Selecting a business with a Web site

Failure: Selecting a business without a site

b. $n = 10, p = 0.45, x = 4$

c. $P(4) \approx 0.238$

d. The probability of randomly selecting 10 small businesses and finding exactly 4 that have a website is 0.238.

7a. $P(0) = {}_6C_0(0.62)^0(0.38)^6 = 1(0.62)^0(0.38)^6 = 0.003$

$P(1) = {}_6C_1(0.62)^1(0.38)^5 = 6(0.62)^1(0.38)^5 = 0.029$

$P(2) = {}_6C_2(0.62)^0(0.38)^4 = 15(0.62)^2(0.38)^4 = 0.120$

$P(3) = {}_6C_3(0.62)^3(0.38)^3 = 20(0.62)^3(0.38)^3 = 0.262$

$P(4) = {}_6C_4(0.62)^4(0.38)^2 = 15(0.62)^4(0.38)^2 = 0.320$

$P(5) = {}_6C_5(0.62)^5(0.38)^1 = 6(0.62)^5(0.38)^1 = 0.209$

$P(6) = {}_6C_6(0.62)^6(0.38)^0 = 1(0.62)^6(0.38)^0 = 0.057$

b.

x	$P(x)$
0	0.003
1	0.029
2	0.120
3	0.262
4	0.320
5	0.209
6	0.057

c.

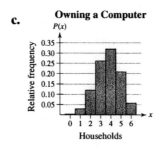

Owning a Computer

8a. Success: Selecting a clear day

$n = 31, p = 0.44, q = 0.56$

b. $\mu = np = (31)(0.44) = 13.6$

c. $\sigma^2 = npq = (31)(0.44)(0.56) = 7.6$

d. $\sigma = \sqrt{\sigma^2} = 2.8$

e. On average, there are about 14 clear days during the month of May. The standard deviation is about 3 days.

4.2 EXERCISE SOLUTIONS

1. (a) $p = 0.50$ (graph is symmetric)

(b) $p = 0.20$ (graph is skewed right $\rightarrow p < 0.5$)

(c) $p = 0.80$ (graph is skewed left $\rightarrow p > 0.5$)

2. (a) $p = 0.75$ (graph is skewed left $\rightarrow p > 0.5$)

(b) $p = 0.50$ (graph is symmetric)

(c) $p = 0.25$ (graph is skewed right $\rightarrow p < 0.5$)

3. (a) $n = 12, (x = 0, 1, 2, \ldots, 12)$

(b) $n = 4, (x = 0, 1, 2, 3, 4)$

(c) $n = 8, (x = 0, 1, 2, \ldots, 8)$

As n increases, the probability distribution becomes more symmetric.

4. (a) $n = 10, (x = 0, 1, 2, \ldots, 10)$

(b) $n = 15, (x = 0, 1, 2, \ldots, 15)$

(c) $n = 5, (x = 0, 1, 2, 3, 4, 5)$

As n increases the probability distribution becomes more symmetric.

5. (a) $0, 1$ (b) $0, 5$ (c) $4, 5$

6. (a) $0, 1, 2, 3, 4$ (b) $0, 1, 2, 3, 4, 5, 6, 7, 8, 15$ (c) $0, 1$

7. Is a binomial experiment.

Success: baby recovers

$n = 5, p = 0.80, q = 0.20, x = 0, 1, 2, \ldots, 5$

8. Is a binomial experiment.

Success: person does not make a purchase

$n = 18$, $p = 0.74$, $q = 0.26$, $x = 0, 1, 2, \ldots, 18$

9. Is a binomial experiment.

Success: Person said taxcuts hurt the economy.

$n = 15, p = 0.21, q = 0.79; x = 0, 1, 2, \ldots, 15$

10. Not a binomial experiment because the probability of a success is not the same for each trial.

11. $\mu = np = (80)(0.3) = 24$

$\sigma^2 = npq = (80)(0.3)(0.7) = 16.8$

$\sigma = \sqrt{\sigma^2} = 4.1$

12. $\mu = np = (64)(0.85) = 54.4$

$\sigma^2 = npq = (64)(0.85)(0.15) \approx 8.2$

$\sigma = \sqrt{\sigma^2} \approx 2.9$

13. $\mu = np = (124)(0.26) = 32.24$

$\sigma^2 = npq = (124)(0.26)(0.74) = 23.858$

$\sigma = \sqrt{\sigma^2} = 4.884$

14. $\mu = np = (316)(0.72) \approx 227.5$

$\sigma^2 = npq = (316)(0.72)(0.28) \approx 63.7$

$\sigma = \sqrt{\sigma^2} \approx 8.0$

15. $n = 5, p = .25$

(a) $P(3) \approx 0.088$

(b) $P(x \geq 3) = P(3) + P(4) + P(5) = 0.088 + 0.015 + .001 = 0.104$

(c) $P(x < 3) = 1 - P(x \geq 3) = 1 - 0.104 = 0.896$

16. $n = 7, p = 0.70$

(a) $P(5) = 0.318$

(b) $P(x \geq 5) = P(5) + P(6) + P(7) = 0.318 + 0.247 + 0.082 = 0.647$

(c) $P(x < 5) = 1 - P(x \geq 5) = 1 - 0.647 = 0.353$

17. $n = 10, p = 0.59$ (using binomial formula)

(a) $P(8) = 0.111$

(b) $P(x \geq 8) = P(8) + P(9) + P(10) = 0.111 + 0.036 + 0.005 = 0.152$

(c) $P(x < 8) = 1 - P(x \geq 8) = 1 - 0.152 = 0.848$

18. $n = 12, p = 0.10$

(a) $P(4) \approx 0.021$

(b) $P(x \geq 4) = 1 - P(x < 4) = 1 - P(0) - P(1) - P(2) - P(3)$
$= 1 - 0.282 - 0.377 - 0.230 - 0.085 = 0.026$

(c) $P(x < 4) = 1 - P(x \geq 4) = 1 - 0.026 = 0.974$

19. $n = 10, p = 0.21$ (using binomial formula)

   (a) $P(3) \approx 0.213$

   (b) $P(x > 3) = 1 - P(0) - P(1) - P(2) - P(3)$

$$= 1 - 0.095 - 0.252 - 3.01 - 0.213 = 0.139$$

   (c) $P(x \le 3) = 1 - P(x > 3) = 1 - 0.139 = 0.861$

20. $n = 20, \ p = 0.70$ (use binomial formula)

   (a) $P(1) \approx 1.627 \times 10^{-9}$

   (b) $P(x > 1) = P(0) - P(1) \approx 1 - 3.487 \times 10^{-11} - 1.627 \times 10^{-9} = 0.9999999983 \approx 1$

   (c) $P(x \le 1) = 1 - P(x > 1) \approx 1 - 0.9999999983 \approx 1.662 \times 10^{-9}$

21. (a) $P(3) = 0.028$

   (b) $P(x \ge 4) = 1 - P(x \le 3)$

$$= 1 - (P(0) + P(1) + P(2) + P(3))$$

$$= 1 - (0.000 + 0.001 + 0.007 + 0.028)$$

$$= 0.964$$

   (c) $P(x \le 2) = P(0) + P(1) + P(2) = 0.000 + 0.001 + 0.007 = 0.008$

22. (a) $P(2) = 0.264$

   (b) $P(x > 6) = P(7) + P(8) + P(9) + P(10)$

$$= (0.000 + 0.000 + 0.000 + 0.000)$$

$$= 0.000$$

   (c) $P(x \le 5) = P(0) + P(1) + P(2) + P(3)$

$$= 0.221 + 0.360 + 0.264 + 0.115 + 0.033 + 0.006$$

$$= 0.999$$

23. (a) $P(2) = 0.255$

   (b) $P(x > 2) = 1 - P(x \le 2)1 - (P(0) + P(1) + P(2))$

$$= 1 - (0.037 + 0.146 + 0.255)$$

$$= 0.562$$

   (c) $P(2 \le x \le 5) = P(2) + P(3) + P(4) + P(5)$

$$= 0.255 + 0.264 + 0.180 + 0.084$$

$$= 0.783$$

24. (a) $P(4) = 0.183$

(b) $P(x > 4) = 1 - P(x \leq 4)$

$$= 1 - (P(0) + P(1) + P(2) + P(3) + P(4))$$

$$= 1 - (0.037 + 0.141 + 0.244 + 0.257 + 0.183)$$

$$= 0.138$$

(c) $P(4 \leq x \leq 8) = P(4) + P(5) + P(6) + P(7) + P(8)$

$$= 0.182 + 0.092 + 0.034 + 0.009 + 0.002$$

$$= 0.320$$

25. (a) $n = 6, p = 0.37$ (b)   (c) Skewed right

x	P(x)
0	0.063
1	0.220
2	0.323
3	0.253
4	0.112
5	0.026
6	0.003

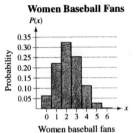

Women Baseball Fans

(d) $\mu = np = (6)(0.37) = 2.2$

(e) $\sigma^2 = npq = (6)(0.37)(0.63) \approx 1.4$

(f) $\sigma = \sqrt{\sigma^2} \approx 1.2$

(g) On average, 2.2 out of 6 women would consider themselves basketball fans. The standard deviation is 1.2 women.

$x = 0, 5,$ or 6 would be unusual due to their low probabilities.

26. (a) $n = 5, p = 0.25$ (b)   (c) Skewed right

x	P(x)
0	0.237
1	0.396
2	0.264
3	0.088
4	0.015
5	0.001

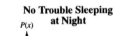
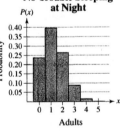

No Trouble Sleeping at Night

(d) $\mu = np = (5)(0.25) = 1.3$

(e) $\sigma^2 = npq = (5)(0.25)(0.75) \approx 0.9$

(f) $\sigma = \sqrt{\sigma^2} \approx 1.0$

(g) On average, 1.3 adults out of every 5 have no trouble sleeping at night. The standard deviation is 1.0 adults. Four or five would be uncommon due to their low probabilities.

27. (a) $n = 4, p = 0.05$ (b)

x	$P(x)$
0	0.814506
1	0.171475
2	0.013537
3	0.000475
4	0.000006

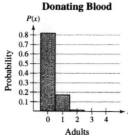

Donating Blood

(c) Skewed right

(d) $\mu = np = (4)(0.05) = 0.2$

(e) $\sigma^2 = npq = (4)(0.05)(0.95) = 0.2$

(f) $\sigma = \sqrt{\sigma^2} \approx 0.4$

(g) On average, 0.2 eligible adult out of every 4 give blood. The standard deviation is 0.4 adult.

     $x = 2, 3,$ or 4 would be uncommon due to their low probabilities.

28. (a) $n = 5, p = 0.38$ (b)

x	$P(x)$
0	0.092
1	0.281
2	0.344
3	0.211
4	0.065
5	0.008

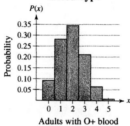

Blood Type

(c) Skewed right   (d) $\mu = np = (5)(0.38) = 1.9$

(e) $\sigma^2 = npq = (5)(0.38)(0.62) = 1.2$ (f) $\sigma = \sqrt{\sigma^2} \approx 1.1$

(g) On average, 1.9 adults out of every 5 have O+ blood. The standard deviation is 1.1 adults. Five would be uncommon due to its low probabilities.

29. (a) $n = 6, p = 0.29$ (b) $P(2) = 0.321$

x	$P(x)$
0	0.128
1	0.314
2	0.321
3	0.175
4	0.053
5	0.009
6	0.001

(c) $P(\text{at least } 5) = P(5) + P(6) = 0.009 + 0.001 = 0.010$

30. (a) $n = 5, p = 0.48$ (b) $P(2) = 0.325$

x	$P(x)$
0	0.038
1	0.175
2	0.324
3	0.299
4	0.138
5	0.025

(c) $P(\text{less than } 4) = 1 - P(4) - P(5)$

$$= 1 - 0.138 - 0.025$$

$$= 0.837$$

31. $\mu = np = (6)(0.29) = 1.7$

$\sigma^2 = npq = (6)(0.29)(0.71) = 1.2$

$\sigma = \sqrt{\sigma^2} = 1.1$

If 6 drivers are randomly selected, on average 1.7 drivers will name talking on cell phones as the most annoying habit of other drivers.

Five of the six or all of the six randomly selected drivers naming talking on cell phones as the most annoying habit of other drivers is rare because their probabilities are less than 0.05.

32. $\mu = np = (5)(0.48) = 2.4$

$\sigma^2 = npq = (5)(0.48)(0.52) = 1.2$

$\sigma = \sqrt{\sigma^2} = 1.1$

If 5 employees are randomly selected, on average 2.4 employees would claim lack of time was a barrier. It would be unusual for no employees or all 5 employees to claim lack of time was a barrier because their probabilities are less than 0.05.

33. $P(5, 2, 2, 1) = \dfrac{10!}{5!2!2!1!}\left(\dfrac{9}{16}\right)^5\left(\dfrac{3}{16}\right)^2\left(\dfrac{3}{16}\right)^2\left(\dfrac{1}{16}\right)^1 \approx 0.033$

34. $P(5, 2, 2, 1) = \dfrac{10!}{5!2!2!1!}\left(\dfrac{5}{16}\right)^5\left(\dfrac{4}{16}\right)^2\left(\dfrac{1}{16}\right)^2\left(\dfrac{6}{16}\right)^1 \approx 0.002$

4.3 MORE DISCRETE PROBABILITY DISTRIBUTIONS

4.3 Try It Yourself Solutions

1a. $P(1) = (0.23)(0.77)^0 = 0.23$

$P(2) = (0.23)(0.77)^1 = 0.177$

$P(3) = (0.23)(0.77)^2 = 0.136$

b. $P(x < 4) = P(1) + P(2) + P(3) = 0.543$

c. The probability that your first sale will occur before your fourth sales call is 0.543.

2a. $P(0) = \dfrac{3^0(2.71828)^{-3}}{0!} \approx 0.050 \qquad P(1) = \dfrac{3^1(2.71828)^{-3}}{1!} \approx 0.149$

$P(2) = \dfrac{3^2(2.71828)^{-3}}{2!} \approx 0.224 \qquad P(3) = \dfrac{3^3(2.71828)^{-3}}{3!} \approx 0.224$

$P(4) = \dfrac{3^4(2.71828)^{-3}}{4!} \approx 0.168$

b. $P(0) + P(1) + P(2) + P(3) + P(4) \approx 0.050 + 0.149 + 0.224 + 0.224 + 0.168 \approx 0.815$

c. $1 - 0.815 \approx 0.185$

d. The probability that more than four accidents will occur in any given month at the intersection is 0.185.

3a. $\mu = \dfrac{2000}{20,000} = 0.10$

  b. $\mu = 0.10, x = 3$

  c. $P(3) = 0.0002$

  d. The probability of finding three brown trout in any given cubic meter of the lake is 0.0002.

4.3 EXERCISE SOLUTIONS

1. $P(2) = (0.60)(0.4)^1 = 0.24$

2. $P(1) = (0.25)(0.75)^0 = 0.25$

3. $P(6) = (0.09)(0.91)^5 = 0.056$

4. $P(5) = (0.38)(0.62)^4 = 0.056$

5. $P(3) = \dfrac{(4)^3(e^{-4})}{3!} = 0.195$

6. $P(5) = \dfrac{(6)^5(e^{-6})}{5!} = 0.161$

7. $P(2) = \dfrac{(1.5)^2(e^{-1.5})}{2!} = 0.251$

8. $P(4) = \dfrac{(8.1)^4(e^{-8.1})}{4!} = 0.054$

9. The binomial distribution counts the number of successes in n trials. The geometric counts the number of trials until the first success is obtained.

10. The binomial distribution counts the number of successes in n trials. The Poisson distribution counts the number of occurrences that take place within a given unit of time.

11. Geometric. You are interested in counting the number of trials until the first success.

12. Poisson. You are interested in counting the number of occurrences that takes place within a given unit of time.

13. Poisson. You are interested in counting the number of occurrences that take place within a given unit of space.

14. Binomial. You are interested in counting the number of successes out of n trials.

15. Binomial. You are interested in counting the number of successes out of n trials.

16. Geometric. You are interested in counting the number of trials until the first success.

17. $p = 0.19$

  (a) $P(5) = (0.19)(0.81)^4 \approx 0.082$

  (b) P(sale on 1st, 2nd, or 3rd call)

   $= P(1) + P(2) + P(3) = (0.19)(0.81)^0 + (0.19)(0.81)^1 + (0.19)(0.81)^2 \approx 0.469$

  (c) $P(x > 3) = 1 - P(x \leq 3) = 1 - 0.469 = 0.531$

18. $p = 0.526$

  (a) $P(2) = (0.526)(0.474)^1 \approx 0.249$

  (b) P(makes 1st or 2nd shot) $= P(1) + P(2) = (0.526)(0.474)^0 + (0.526)(0.474)^1 \approx 0.775$

  (c) (Binomial: $n = 2, p = 0.526$)

   $P(0) = \dfrac{2!}{0!2!}(0.526)^0(0.474)^2 \approx 0.225$

19. $p = 0.002$

    (a) $P(10) = (0.002)(0.998)^9 \approx 0.002$

    (b) P(1st, 2nd, or 3rd part is defective) $= P(1) + P(2) + P(3)$

$$= (0.002)(0.998)^0 + (0.002)(0.998)^1 + (0.002)(0.998)^2 \approx 0.006$$

    (c) $P(x > 10) = 1 - P(x \leq 10) = 1 - [P(1) + P(2) + \cdots + P(10)] = 1 - [0.020] \approx 0.980$

20. $p = 0.25$

    (a) $P(4) = (0.25)(0.75)^3 \approx 0.105$

    (b) P(prize with 1st, 2nd, or 3rd purchase) $= P(1) + P(2) + P(3)$

$$= (0.25)(0.75)^0 + (0.25)(0.75)^1 + (0.25)(0.75)^2 \approx 0.578$$

    (c) $P(x > 4) = 1 - P(x \leq 4) = 1 - [P(1) + P(2) + P(3) + P(4)] = 1 - [0.684] \approx 0.316$

21. $\mu = 3$

    (a) $P(5) = \dfrac{3^5 e^{-3}}{5!} \approx 0.101$

    (b) $P(x \geq 5) = 1 - (P(0) + P(1) + P(2) + P(3) + P(4))$

$$\approx 1 - (0.050 + 0.149 + 0.224 + 0.224 + 0.168)$$

$$= 0.185$$

    (c) $P(x > 5) = 1 - (P(0) + P(1) + P(2) + P(3) + P(4) + P(5))$

$$\approx 1 - (0.050 + 0.149 + 0.224 + 0.224 + 0.168 + 0.101)$$

$$= 0.084$$

22. $\mu = 4$

    (a) $P(3) = 0.195$

    (b) $P(x \leq 3) = P(0) + P(1) + P(2) + P(3) = 0.0183 + 0.0733 + 0.1465 + 0.1954 = 0.433$

    (c) $P(x > 3) = 1 - P(x \leq 3) = 1 - 0.4335 = 0.567$

23. $\mu = 0.6$

    (a) $P(1) = 0.329$

    (b) $P(x \leq 1) = P(0) + P(1) = 0.549 + 0.329 = 0.878$

    (c) $P(x > 1) = 1 - P(x \leq 1) = 1 - 0.878 = 0.122$

24. $\mu = 8.7$

    (a) $P(9) = \dfrac{8.7^9 e^{-8.7}}{9!} \approx 0.131$

    (b) $P(x \leq 9) = 0.627$

    (c) $P(x > 9) = 1 - P(x \leq 9) = 1 - 0.627 = 0.373$

25. (a) $n = 6000$, $p = 0.0004$

$$P(4) = \frac{6000!}{5996!4!}(0.0004)^4(0.9996)^{5996} \approx 0.1254235482$$

(b) $\mu = \frac{6000}{2500} = 2.4$ warped glass items per 2500.

$P(4) = 0.1254084986$

The results are approximately the same.

26. (a) $P(0) = \frac{_2C_0 \, _{13}C_3}{_{15}C_3} = \frac{(1)(286)}{(455)} \approx 0.629$

(b) $P(1) = \frac{_2C_1 \, _{13}C_2}{_{15}C_3} = \frac{(2)(78)}{(455)} \approx 0.343$

(c) $P(2) = \frac{_2C_2 \, _{13}C_1}{_{15}C_3} = \frac{(1)(13)}{(455)} \approx 0.029$

27. $p = 0.001$

(a) $\mu = \frac{1}{p} = \frac{1}{0.001} = 1000$

$\sigma^2 = \frac{q}{p^2} = \frac{0.999}{(0.001)^2} = 999,000$

$\sigma = \sqrt{\sigma^2} \approx 999.5$

On average you would have to play 1000 times until you won the lottery. The standard deviation is 999.5 times.

(b) 1000 times

Lose money. On average you would win $500 every 1000 times you play the lottery. So, the net gain would be $-\$500$.

28. $p = 0.005$

(a) $\mu = \frac{1}{p} = \frac{1}{0.005} = 200$

$\sigma^2 = \frac{q}{p^2} = \frac{0.995}{(0.005)^2} = 39,800$

$\sigma = \sqrt{\sigma^2} \approx 199.5$

(b) 200

On average 200 records will be examined before finding one that has been miscalculated. The standard deviation is 199.5 records.

29. $\mu = 3.8$

(a) $\sigma^2 = 3.8$

$\sigma = \sqrt{\sigma^2} \approx 1.9$

The standard deviation is 1.9 strokes.

(b) $P(x > 72) = 1 - P(x \le 72) = 1 - 0.695 = 0.305$

30. $\mu = 7.6$

(a) $\sigma^2 = 7.6$

$\sigma = \sqrt{\sigma^2} \approx 2.8$

The standard deviation is 2.8 inches.

(b) $P(X > 12) = 1 - P(X \le 12) \approx 1 - 0.954 \approx 0.046$

CHAPTER 4 REVIEW EXERCISE SOLUTIONS

1. Discrete

2. Continuous

3. Continuous

4. Discrete

5. No, $\Sigma P(x) \ne 1$.

6. Yes

7. Yes

8. No, $P(5) > 1$ and $\Sigma P(x) \ne 1$.

9. Yes

10. No, $\Sigma P(x) \ne 1$.

11. (a)

x	Frequency	P(x)	xP(x)	x − μ	(x − μ)²	(x − μ)²P(x)
2	3	0.005	0.009	−4.371	19.104	0.088
3	12	0.018	0.055	−3.371	11.362	0.210
4	72	0.111	0.443	−2.371	5.621	0.623
5	115	0.177	0.885	−1.371	1.879	0.332
6	169	0.260	1.560	−0.371	0.137	0.036
7	120	0.185	1.292	0.629	0.396	0.073
8	83	0.128	1.022	1.629	2.654	0.339
9	48	0.074	0.665	2.629	6.913	0.510
10	22	0.034	0.338	3.629	13.171	0.446
11	6	0.009	0.102	4.629	21.430	0.198
	n = 650	$\sum P(x) = 1$	$\sum xP(x) = 6.371$			$\sum (x-\mu)^2 P(x) = 2.855$

(b)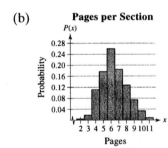

Pages per Section

(c) $\mu = \Sigma xP(x) \approx 6.4$

$\sigma^2 = \Sigma(x - \mu)^2 P(x) \approx 2.9$

$\sigma = \sqrt{\sigma^2} \approx 1.7$

12. (a)

x	Frequency	P(x)	xP(x)	x − μ	(x − μ)²	(x − μ)²P(x)
0	29	0.207	0.000	−1.293	1.672	0.346
1	62	0.443	0.443	−0.293	0.086	0.038
2	33	0.236	0.471	0.707	0.500	0.118
3	12	0.086	0.257	1.707	2.914	0.250
4	3	0.021	0.086	2.707	7.328	0.157
5	1	0.007	0.036	3.707	13.742	0.098
	n = 140	$\sum P(x) = 1$	$\sum xP(x) = 1.293$			$\sum (x-\mu)^2 P(x) = 1.007$

(b) **Hits Per Game**

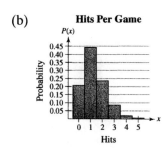

(c) $\mu = \Sigma x P(x) \approx 1.3$

$\sigma^2 = \Sigma(x - \mu)^2 P(x) \approx 1.0$

$\sigma = \sqrt{\sigma^2} \approx 1.0$

13. (a)

x	Frequency	$P(x)$	$xP(x)$	$x - \mu$	$(x - \mu)^2$	$(x - \mu)^2 P(x)$
0	3	0.015	0.000	−2.315	5.359	0.080
1	38	0.190	0.190	−1.315	1.729	0.329
2	83	0.415	0.830	−0.315	0.099	0.041
3	52	0.260	0.780	0.685	0.469	0.122
4	18	0.090	0.360	1.685	2.839	0.256
5	5	0.025	0.125	2.685	7.209	0.180
6	1	0.005	0.030	3.685	13.579	0.068
	$n = 200$	$\sum P(x) = 1$	$\sum xP(x) = 2.315$			$\sum(x - \mu)^2 P(x) = 1.076$

(b) **Televisions per Household**

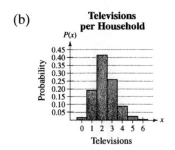

(c) $\mu = \Sigma x P(x) \approx 2.3$

$\sigma^2 = \Sigma(x - \mu)^2 P(x) \approx 1.1$

$\sigma = \sqrt{\sigma^2} \approx 1.0$

14. (a)

x	Frequency	$P(x)$	$xP(x)$	$x - \mu$	$(x - \mu)^2$	$(x - \mu)^2 P(x)$
15	76	0.134	2.014	−16.802	282.311	37.908
30	445	0.786	23.587	−1.802	3.248	2.553
60	30	0.053	3.180	28.198	795.120	42.144
90	3	0.005	0.477	58.198	3386.993	17.952
120	12	0.021	2.544	88.198	7778.866	164.923
	$n = 566$	$\sum P(x) = 1$	$\sum xP(x) = 31.802$			$\sum(x - \mu)^2 P(x) = 265.480$

(b) **Advertising Sales**

(c) $\mu = \Sigma x P(x) \approx 31.8$

$\sigma^2 = \Sigma(x - \mu)^2 P(x) \approx 265.5$

$\sigma = \sqrt{\sigma^2} \approx 16.3$

15. $E(x) = \mu = \Sigma x P(x) = 3.4$

16. $E(x) = \mu = \Sigma x P(x) = 2.5$

17. Yes, $n = 12, p = 0.24, q = 0.76, x = 0, 1, \ldots, 12.$

18. No, the experiment is not repeated for a fixed number of trials.

19. $n = 8, p = 0.25$

    (a) $P(3) = 0.208$

    (b) $P(x \geq 3) = 1 - P(x < 3) = 1 - [P(0) + P(1) + P(2)] \approx 1 - [0.100 + 0.267 + 0.311] = 0.322$

    (c) $P(x > 3) = 1 - P(x \leq 3) = 1 - [P(0) + P(1) + P(2) + P(3)]$

$$= 1 - [0.100 + 0.267 + 0.311 + 0.208] = 0.114$$

20. $n = 12, \ p = 0.25$

    (a) $P(2) = 0.232$

    (b) $P(x \geq 2) = 1 - P(0) - P(1) = 1 - 0.0317 - 0.1267 = 0.842$

    (c) $P(x > 2) = 1 - P(0) - P(1) - P(2)$

$$= 1 - 0.032 - 0.127 - 0.232 = 0.609$$

21. $n = 7, p = 0.43$ (use binomial formula)

    (a) $P(3) = 0.294$

    (b) $P(x \geq 3) = 1 - P(x < 3) = 1 - [P(0) + P(1) + P(2)] \approx 1 - [0.0195 + 0.1032 + 0.2336] = 0.644$

    (c) $P(x > 3) = 1 - P(x \leq 3) \approx 1 - [P(0) + P(1) + P(2) + P(3)]$

$$\approx 1 - (0.0195 + 0.1032 + 0.2336 + 0.2937) = 0.350$$

22. $n = 5, \ p = 0.31$

    (a) $P(2) = 0.316$

    (b) $P(x \geq 2) = P(2) + P(3) + P(4) + P(5) = 0.316 + 0.142 + 0.032 + 0.003 \approx 0.492$

    (c) $P(x > 2) = P(3) + P(4) + P(5) \approx 0.142 + 0.032 + 0.003 \approx 0.177$

23. (a)

x	$P(x)$
0	0.007
1	0.059
2	0.201
3	0.342
4	0.291
5	0.099

(b)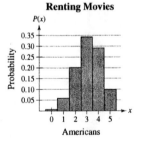

Renting Movies

(c) $\mu = np = (5)(0.63) \approx 3.2$

$\sigma^2 = npq = (5)(0.63)(0.37) = 1.2$

$\sigma = \sqrt{\sigma^2} \approx 1.1$

24. (a)

x	$P(x)$
0	0.001
1	0.014
2	0.073
3	0.206
4	0.328
5	0.279
6	0.099

(b)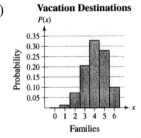

Vacation Destinations

(c) $\mu = np = (6)(0.68) = 4.1$

$\sigma^2 = npq = (6)(0.68)(0.32) = 1.3$

$\sigma = \sqrt{\sigma^2} \approx 1.1$

25. (a)

x	P(x)
0	0.130
1	0.346
2	0.346
3	0.154
4	0.026

(b)

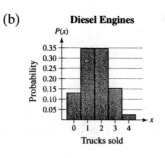

Diesel Engines

(c) $\mu = np = (4)(0.40) \approx 1.6$

$\sigma^2 = npq = (4)(0.40)(0.60) = 1.0$

$\sigma = \sqrt{\sigma^2} \approx 1.0$

26. $n = 5, p = 0.15$

(a)

x	P(x)
0	0.444
1	0.392
2	0.138
3	0.024
4	0.002
5	0.0001

(b)

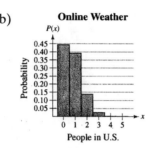

Online Weather

(c) $\mu = np = (5)(0.15) = 0.8$

$\sigma^2 = npq = (5)(0.15)(0.85) = 0.6$

$\sigma = \sqrt{\sigma^2} \approx 0.8$

27. $p = 0.167$

(a) $P(4) \approx 0.096$

(b) $P(x \le 4) = P(1) + P(2) + P(3) + P(4) \approx 0.518$

(c) $P(x > 3) = 1 - P(x \le 3) = 1 - [P(1) + P(2) + P(3)] \approx 0.579$

28. $p = \dfrac{73}{153} = 0.477$

(a) $P(1) \approx 0.477$ (b) $P(2) \approx 0.249$

(c) $P(1 \text{ or } 2) = P(1) + P(2) \approx 0.477 + 0.249 = 0.726$

(d) $P(X \le 3) = P(1) + P(2) + P(3) = 0.477 + 0.249 + 0.130 = 0.856$

29. $\mu = \dfrac{2457}{36} \approx 68.25$ lightning deaths/year $\rightarrow \mu = \dfrac{68.25}{365} \approx 0.1869$ deaths/day

(a) $P(0) = \dfrac{0.1869^0 e^{0.1869}}{0!} \approx 0.830$ (b) $P(1) = \dfrac{0.1869^1 e^{1869}}{1!} \approx 0.155$

(c) $P(x > 1) = 1 - [P(0) + P(1)] = 1 - [0.830 + 0.155] = 0.015$

30. (a) $\mu = 10$

$P(x \ge 3) = 1 - P(x < 3)$

$\qquad = 1 - [P(0) + P(1) + P(2)] = 1 - [0.0000 + 0.0005 + 0.0023] \approx 0.997$

(b) $\mu = 5$

$P(x \ge 3) = 1 - P(x < 3)$

$\qquad = 1 - [P(0) + P(1) + P(2)] = 1 - [0.0067 + 0.0337 + 0.0842] \approx 0.875$

(c) $\mu = 15$

$P(x \ge 3) = 1 - P(x < 3) = 1 - [P(0) + P(1) + P(2)] \approx 1 - [0.0000 + 0.0000 + 0.0000] = 1$

CHAPTER 4 QUIZ SOLUTIONS

1. (a) Discrete because the random variable is countable.

   (b) Continuous because the random variable has an infinite number of possible outcomes and cannot be counted.

2. (a)

x	Frequency	$P(x)$	$xP(x)$	$x - \mu$	$(x - \mu)^2$	$(x - \mu)^2P(x)$
1	70	0.398	0.398	−1.08	1.166	0.464
2	41	0.233	0.466	−0.08	0.006	0.001
3	49	0.278	0.835	0.92	0.846	0.236
4	13	0.074	0.295	1.92	3.686	0.272
5	3	0.017	0.085	2.92	8.526	0.145
	$n = 176$	$\sum P(x) = 1$	$\sum xP(x) = 2.080$			$\sum(x - \mu)^2P(x) = 1.119$

   (b)

Hurricane Intensity

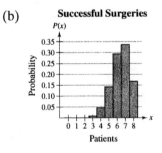

   (c) $\mu = \Sigma xP(x) \approx 2.1$

   $\sigma^2 = \Sigma(x - \mu)^2P(x) \approx 1.1$

   $\sigma = \sqrt{\sigma^2} \approx 1.1$

   On average the intensity of a hurricane will be 2.1. The standard deviation is 1.1.

   (d) $P(x \geq 4) = P(4) + P(5) = 0.074 + 0.017 = 0.091$

3. $n = 8, p = 0.80$

   (a)

x	$P(x)$
0	0.000003
1	0.000082
2	0.001147
3	0.009175
4	0.045875
5	0.146801
6	0.293601
7	0.335544
8	0.167772

   (b) **Successful Surgeries**

   (c) $\mu = np = (8)(0.80) = 6.4$

   $\sigma^2 = npq = (8)(0.80)(0.20) = 1.3$

   $\sigma = \sqrt{\sigma^2} \approx 1.1$

   (d) $P(2) = 0.001$ (e) $P(x < 2) = P(0) + P(1) = 0.000003 + 0.000082 = 0.000085$

4. $\mu = 5$

   (a) $P(5) = 0.176$

   (b) $P(x < 5) = P(0) + P(1) + P(2) + P(3) + P(4)$

   $= 0.00674 + 0.03369 + 0.08422 + 0.14037 + 0.17547 \approx 0.440$

   (c) $P(0) \approx 0.007$

Normal Probability Distributions

5.1 Try It Yourself Solutions

1a. *A*: 45, *B*: 60, *C*: 45 (*B* has the greatest mean.)

b. Curve *C* is more spread out, so curve *C* has the greatest standard deviation.

2a. Mean = 3.5 feet

b. Inflection points: 3.3 and 3.7
Standard deviation = 0.2 foot

3a. (1) 0.0143 (2) 0.9850

4a.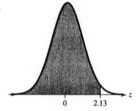

5a.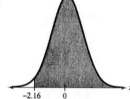

b. 0.9834

b. 0.0154

c. Area = 1 − 0.0154 = 0.9846

6a. 0.0885 **b.** 0.0154

c. Area = 0.0885 − 0.0154 = 0.0731

5.1 EXERCISE SOLUTIONS

1. Answers will vary. **2.** 1

3. Answers will vary.
Similarities: Both curves will have the same line of symmetry.
Differences: One curve will be more spread out than the other.

4. Answers will vary.
Similarities: Both curves will have the same shape (i.e., equal standard deviations)
Differences: The two curves will have different lines of symmetry.

5. $\mu = 0$, $\sigma = 1$

6. Transform each data value *x* into a *z*-score. This is done by subtracting the mean from *x* and dividing by the standard deviation. In symbols,

$$z = \frac{x - \mu}{\sigma}.$$

7. "The" standard normal distribution is used to describe one specific normal distribution ($\mu = 0$, $\sigma = 1$). "A" normal distribution is used to describe a normal distribution with any mean and standard deviation.

8. (c) is true because a z-score equal to zero indicates that the corresponding x-value is equal to the mean. (a) and (b) are not true because it is possible to have a z-score equal to zero and the mean is not zero or the corresponding x-value is not zero.

9. No, the graph crosses the x-axis.

10. No, the graph is not symmetric.

11. Yes, the graph fulfills the properties of the normal distribution.

12. No, the graph is skewed left.

13. No, the graph is skewed to the right.

14. No, the graph is not bell-shaped.

15. The histogram represents data from a normal distribution because it's bell-shaped.

16. The histogram does not represent data from a normal distribution because it's skewed right.

17. (Area left of $z = 1.2$) − (Area left of $z = 0$) = 0.8849 − 0.5 = 0.3849

18. (Area left of $z = 0$) − (Area left of $z = -2.25$) = 0.5 − 0.0122 = 0.4878

19. (Area left of $z = 1.5$) − (Area left of $z = -0.5$) = 0.9332 − 0.3085 = 0.6247

20. (Area right of $z = 2$) = 1 − (Area left of $z = 2$) = 1 − 0.9772 = 0.0228

21. 0.9131   22. 0.5319   23. 0.975   24. 0.8997

25. 1 − 0.2578 = 0.7422   26. 1 − 0.0256 = 0.9744   27. 1 − 0.8997 = 0.1003

28. 1 − 0.9994 = 0.0006   29. 0.005   30. 0.0008

31. 1 − 0.9469 = 0.0531   32. 1 − 0.9940 = 0.006   33. 0.9382 − 0.5 = 0.4382

34. 0.9979 − 0.5 = 0.4979   35. 0.5 − 0.0630 = 0.437   36. 0.5 − 0.3050 = 0.195

37. 0.9750 − 0.0250 = 0.95   38. 0.9901 − 0.0099 = 0.9802

39. 0.1003 + 0.1003 = 0.2006   40. 0.0250 + 0.0250 = 0.05

41. (a)

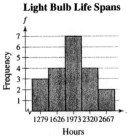

Light Bulb Life Spans

It is reasonable to assume that the life span is normally distributed because the histogram is nearly symmetric and bell-shaped.

(b) $\bar{x} = 1941.35$ $s \approx 432.385$

(c) The sample mean of 1941.35 hours is less than the claimed mean, so on the average the bulbs in the sample lasted for a shorter time. The sample standard deviation of 432 hours is greater than the claimed standard deviation, so the bulbs in the sample had a greater variation in life span than the manufacturer's claim.

42. (a)
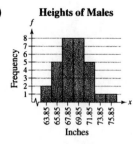
Heights of Males

It is reasonable to assume that the heights are normally distributed because the histogram is nearly symmetric and bell-shaped.

(b) $\bar{x} \approx 68.75$, $s \approx 2.85$

(c) The mean of your sample is 0.85 inch less than that of the previous study so the average height from the sample is less than in the previous study. The standard deviation is about 0.15 less than that of the previous study, so the heights are slightly less spread out than in the previous study.

43. (a) A = 92.994   B = 93.004   C = 93.014   D = 93.018

(b) $x = 93.014 \Rightarrow z = \dfrac{x - \mu}{\sigma} = \dfrac{93.014 - 93.01}{0.005} = 0.8$

$x = 93.018 \Rightarrow z = \dfrac{x - \mu}{\sigma} = \dfrac{93.018 - 93.01}{0.005} = 1.6$

$x = 93.004 \Rightarrow z = \dfrac{x - \mu}{\sigma} = \dfrac{93.004 - 93.01}{0.005} = -1.2$

$x = 92.994 \Rightarrow z = \dfrac{x - \mu}{\sigma} = \dfrac{92.994 - 93.01}{0.005} = -3.2$

(c) $x = 92.994$ is unusual due to a relatively small z-score (-3.2).

44. (a) A = 328   B = 330   C = 338   D = 341

(b) $x = 328 \Rightarrow z = \dfrac{x - \mu}{\sigma} = \dfrac{328 - 336}{3.5} = -2.29$

$x = 338 \Rightarrow z = \dfrac{x - \mu}{\sigma} = \dfrac{338 - 336}{3.5} = 0.57$

$x = 330 \Rightarrow z = \dfrac{x - \mu}{\sigma} = \dfrac{330 - 336}{3.5} = -1.71$

$x = 341 \Rightarrow z = \dfrac{x - \mu}{\sigma} = \dfrac{341 - 336}{3.5} = 1.43$

(c) $x = 328$ is unusual due to a relatively small z-score (-2.29).

45. (a) A = 1186   B = 1406   C = 1848   D = 2177

(b) $x = 1406 \Rightarrow z = \dfrac{x - \mu}{\sigma} = \dfrac{1406 - 1518}{308} = -0.36$

$x = 1848 \Rightarrow z = \dfrac{x - \mu}{\sigma} = \dfrac{1848 - 1518}{308} = 1.07$

$x = 2177 \Rightarrow z = \dfrac{x - \mu}{\sigma} = \dfrac{2177 - 1518}{308} = 2.14$

$x = 1186 \Rightarrow z = \dfrac{x - \mu}{\sigma} = \dfrac{1186 - 1518}{308} = -1.08$

(c) $x = 2177$ is unusual due to a relatively large z-score (2.14).

46. (a) A = 14   B = 18   C = 25   D = 32

(b) $x = 18 \Rightarrow z = \dfrac{x - \mu}{\sigma} = \dfrac{18 - 21}{4.8} = -0.63$

$x = 32 \Rightarrow z = \dfrac{x - \mu}{\sigma} = \dfrac{32 - 21}{4.8} = 2.29$

$x = 14 \Rightarrow z = \dfrac{x - \mu}{\sigma} = \dfrac{14 - 21}{4.8} = -1.46$

$x = 25 \Rightarrow z = \dfrac{x - \mu}{\sigma} = \dfrac{25 - 21}{4.8} = 0.83$

(c) $x = 32$ is unusual due to a relatively large z-score (2.29).

47. 0.6915

48. 0.1587

49. $1 - 0.95 = 0.05$

50. $1 - 0.1003 = 0.8997$

51. $0.8413 - 0.3085 = 0.5328$

52. $0.9772 - 0.6915 = 0.2857$

53. $P(z < 1.45) = 0.9265$

54. $P(z < 0.45) = 0.6736$

55. $P(z > -0.95) = 1 - P(z < -0.95) = 1 - 0.1711 = 0.8289$

56. $P(z > -1.85) = 1 - P(z < -1.85) = 1 - 0.0322 = 0.9678$

57. $P(-0.89 < z < 0) = 0.5 - 0.1867 = 0.3133$

58. $P(-2.08 < z < 0) = 0.5 - 0.0188 = 0.4812$

59. $P(-1.65 < z < 1.65) = 0.9505 - 0.0495 = 0.901$

60. $P(-1.54 < z < 1.54) = 0.9382 - 0.0618 = 0.8764$

61. $P(z < -2.58 \text{ or } z > 2.58) = 2(0.0049) = 0.0098$

62. $P(z < -1.54 \text{ or } z > 1.54) = 2(0.618) = 0.1236$

63.

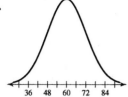

The normal distribution curve is centered at its mean (60) and has 2 points of inflection (48 and 72) representing $\mu \pm \sigma$.

64.

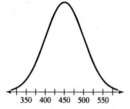

The normal distribution curve is centered at its mean (450) and has 2 points of inflection (400 and 500) representing $\mu \pm \sigma$.

65. (a) Area under curve = area of rectangle = (base)(height) = (1)(1) = 1

(b) $P(0.25 < x < 0.5) = $ (base)(height) = (0.25)(1) = 0.25

(c) $P(0.3 < x < 0.7) = $ (base)(height) = (0.4)(1) = 0.4

66. (a)

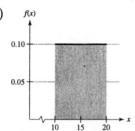

Area under curve = Area of rectangle

= (base)(height)

= $(20 - 10) \cdot (0.10)$

= 1

(b) $P(12 < x < 15) = $ (base)(height) = (3)(0.1) = 0.3

(c) $P(13 < x < 18) = $ (base)(height) = (5)(0.1) = 0.5

5.2 NORMAL DISTRIBUTIONS: FINDIING PROBABILITIES

5.2 Try It Yourself Solutions

1a.

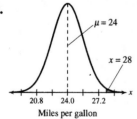

b. $z = \dfrac{x - \mu}{\sigma} = \dfrac{28 - 24}{1.6} = 2.50$

c. $P(z < 2.50) = 0.9938$

$P(z > 2.50) = 1 - 0.9938 = 0.0062$

d. The probability that a randomly selected manual transmission Focus will get more than 28 mpg in city driving is 0.0062.

2a.

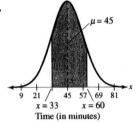

$\mu = 45$

9   21   45   57   69   81 x
$x = 33$ $x = 60$
Time (in minutes)

b. $z = \dfrac{x - \mu}{\sigma} = \dfrac{33 - 45}{12} = -1$

$z = \dfrac{x - \mu}{\sigma} = \dfrac{60 - 45}{12} = 1.25$

c. $P(z < -1) = 0.1587$

$P(z > 1.25) = 0.8944$

$0.8944 - 0.1587 = 0.7357$

d. If 150 shoppers enter the store, then you would expect $150(0.7357) \approx 110$ shoppers to be in the store between 33 and 60 minutes.

3a. Read user's guide for the technology tool.

b. Enter the data.

c. $P(190 < x < 225) = P(-1 < z < 0.4) = 0.4968$

The probability that a randomly selected U.S. man's cholesterol is between 190 and 225 is about 0.4968.

5.2 EXERCISE SOLUTIONS

1. $P(x < 80) = P(z < -1.2) = 0.1151$

2. $P(x < 100) = P(z < 2.8) = 0.9974$

3. $P(x > 92) = P(z > 1.2) = 1 - 0.8849 = 0.1151$

4. $P(x > 75) = P(z > -2.2) = 1 - 0.0139 = 0.9861$

5. $P(70 < x < 80) = P(-3.2 < z < -1.2) = 0.1151 - 0.0007 = 0.1144$

6. $P(85 < x < 95) = P(-0.2 < z < 1.8) = 0.9641 - 0.4207 = 0.5434$

7. $P(200 < x < 450) = P(-2.68 < z < -0.47) = 0.3192 - 0.0037 = 0.3155$

8. $P(670 < x < 800) = P(1.32 < z < 2.45) = 0.9929 - 0.9066 = 0.0863$

9. $P(200 < x < 239) = P(0.39 < z < 1.48) = 0.9306 - 0.6517 = 0.2789$

10. $P(200 < x < 239) = P(-0.46 < z < 0.48) = 0.6844 - 0.3228 = 0.3616$

11. $P(141 < x < 151) = P(0.84 < z < 2.94) = 0.9984 - 0.7995 = 0.1989$

12. $P(140 < x < 149) = P(-1.74 < z < 0) = 0.5 - 0.0409 = 0.4591$

13. (a) $P(x < 66) = P(z < -1.2) = 0.1151$

  (b) $P(66 < x < 72) = P(-1.2 < z < 0.8) = 0.7881 - 0.1151 = 0.6730$

  (c) $P(x > 72) = P(z > 0.8) = 1 - P(z < 0.8) = 1 - 0.7881 = 0.2119$

14. (a) $P(x < 7) = P(z < -1.5) = 0.0668$

  (b) $P(7 < x < 15) = P(-1.5 < z < 2.5) = 0.9938 - 0.0668 = 0.9270$

  (c) $P(x > 15) = P(z > 2.5) = 1 - P(z < 2.5) = 1 - 0.9938 = 0.0062$

15. (a) $P(x < 17) = P(z < -1.67) = 0.0475$

  (b) $P(20 < x < 29) = P(-0.98 < z < 1.12) = 0.8686 - 0.1635 = 0.7051$

  (c) $P(x > 32) = P(z > 1.81) = 1 - P(z < 1.81) = 1 - 0.9649 = 0.0351$

16. (a) $P(x < 23) = P(z < -0.67) = 0.2514$

  (b) $P(23 < x < 25) = P(-0.67 < z < 0) = 0.5 - 0.2514 = 0.2486$

  (c) $P(x > 27) = P(z > 0.67) = 1 - P(z < 0.67) = 1 - 0.7486 = 0.2514$

17. (a) $P(x < 5) = P(z < -2) = 0.0228$

  (b) $P(5.5 < x < 9.5) = P(-1.5 < z < 2.5) = 0.9938 - 0.0668 = 0.927$

  (c) $P(x > 10) = P(z > 3) = 1 - P(z < 3) = 1 - 0.9987 = 0.0013$

18. (a) $P(x < 70) = P(z < -2.5) = 0.0062$

  (b) $P(90 < x < 120) = P(-0.83 < z < 1.67) = 0.9525 - 0.2033 = 0.7492$

  (c) $P(x > 140) = P(z > 3.33) = 1 - P(z < 3.33) = 1 - 0.9996 = 0.0004$

19. (a) $P(x < 4) = (z < -2.44) = 0.0073$

  (b) $P(5 < x < 7) = P(-1.33 < z < 0.89) = 0.8133 - 0.0918 = 0.7215$

  (c) $P(x > 8) = P(z > 2) = 1 - 0.9772 = 0.0228$

20. (a) $P(x < 17) = P(z < -0.6) = 0.2743$

  (b) $P(20 < x < 28) = P(0 < z < 1.6) = 0.9452 - 0.5 = 0.4452$

  (c) $P(x > 30) = P(z > 2) = 1 - 0.9772 = 0.0228$

21. (a) $P(x < 600) = P(z < 0.86) = 0.8051 \Longrightarrow 80.51\%$

  (b) $P(x > 550) = P(z > 0.42) = 1 - P(z < 0.42) = 1 - 0.6628 = 0.3372$
    $(1000)(0.3372) = 337.2 \Longrightarrow 337$ scores

22. (a) $P(x < 500) = P(z < -0.16) = 0.4364 \Longrightarrow 43.64\%$

  (b) $P(x > 600) = P(z > 0.71) = 1 - P(z < 0.71) = 1 - 0.7611 = 0.2389$
    $(1500)(0.2389) = 358.35 \Longrightarrow 358$ scores

23. (a) $P(x < 200) = P(z < 0.39) = 0.6517 \Rightarrow 65.17\%$

(b) $P(x > 240) = P(z > 1.51) = 1 - P(z < 1.51) = 1 - 0.9345 = 0.0655$

$(250)(0.0655) = 16.375 \Rightarrow 16$ women

24. (a) $P(x < 239) = P(z < 0.48) = 0.6844$

(b) $P(x > 200) = P(z > -0.46) = 1 - P(z < -0.46) = 1 - 0.3228 = 0.6772$

$(200)(0.6672) = 135.44 \Rightarrow 135$ women

25. (a) $P(x > 11) = P(z > 0.5) = 1 - P(z < 0.5) = 1 - 0.6915 = 0.3085 \Rightarrow 30.85\%$

(b) $P(x < 8) = P(z < -1) = 0.1587$

$(200)(0.1587) = 31.74 \Rightarrow 32$ fish

26. (a) $P(x > 30) = P(z > 1.67) = 1 - P(z < 1.67) = 1 - 0.9525 = 0.0475 \Rightarrow 4.75\%$

(b) $P(x < 22) = P(z < -1) = 0.1587$

$(50)(0.1587) = 7.935 \Rightarrow 8$ beagles

27. (a) $P(x > 4) = P(z > -3) = 1 - P(z < -3) = 1 - 0.0013 = 0.9987 \Rightarrow 99.87\%$

(b) $P(x < 5) = P(z < -2) = 0.0228$

$(35)(0.0228) = 0.798 \Rightarrow 1$ adult

28. (a) $P(x > 125) = P(z > 2.08) = 1 - P(z < 2.08) = 1 - 0.9812 = 0.0188 \rightarrow 1.88\%$

(b) $P(x < 90) = P(z < -0.83) = 0.2033$

$(300)(0.2033) = 60.99 \Rightarrow 61$ bills

29. $P(x > 2065) = P(z > 2.17) = 1 - P(z < 2.17) = 1 - 0.9850 = 0.0150 \Rightarrow 1.5\%$

It is unusual for a battery to have a life span that is more than 2065 hours because of the relatively large z-score (2.17).

30. $P(x < 3.1) = P(z < -1.56) = 0.0594 \Rightarrow 5.94\%$

It is not unusual for a person to consume less than 3.1 pounds of peanuts because the z-score is within 2 standard deviations of the mean.

31. Out of control, because the 10th observation plotted beyond 3 standard deviations.

32. Out of control, because two out of three consecutive points lie more than 2 standard deviations from the mean. (8th and 10th observations.)

33. Out of control, because the first nine observations lie below the mean and since two out of three consecutive points lie more than 2 standard deviations from the mean.

34. In control, because none of the three warning signals detected a change.

5.3 NORMAL DISTRIBUTIONS: FINDING VALUES

5.3 Try It Yourself Solutions

1ab. (1)

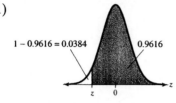

$1 - 0.9616 = 0.0384$ 0.9616

(2)

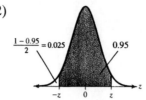

$\dfrac{1 - 0.95}{2} = 0.025$ 0.95

  c. (1) $z = -1.77$ (2) $z = \pm 1.96$

2a. (1)

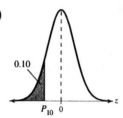

0.10

P_{10} 0

(2)

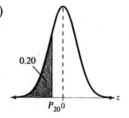

0.20

P_{20} 0

(3)

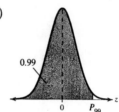

0.99

0 P_{99}

  b. (1) use area $= 0.1003$ (2) use area $= 0.2005$ (3) use area $= 0.9901$

  c. (1) $z = -1.28$ (2) $z = -0.84$ (3) $z = 2.33$

3a. $\mu = 70, \sigma = 8$

  b. $z = -0.75 \Rightarrow x = \mu + z\sigma = 70 + (-0.75)(8) = 64$

     $z = 4.29 \Rightarrow x = \mu + z\sigma = 70 + (4.29)(8) = 104.32$

     $z = -1.82 \Rightarrow x = \mu + z\sigma = 70 + (-1.82)(8) = 55.44$

  c. 64 and 55.44 are below the mean. 104.32 is above the mean.

4a.

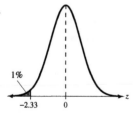

1%

-2.33 0

  b. $z = -2.33$

  c. $x = \mu + z\sigma = 142 + (-2.33)(6.51) \approx 126.83$

  d. So, the longest braking distance a Honda Accord could have and still be in the top 1% is 127 feet.

5a.

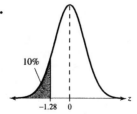

10%

-1.28 0

b. $z = -1.28$

c. $x = \mu + z\sigma = 11.2 + (-1.28)(2.1) = 8.512$

d. So, the maximum length of time an employee could have worked and still be laid off is 8 years.

5.3 EXERCISE SOLUTIONS

1. $z = -0.70$ **2.** $z = -0.81$ **3.** $z = 0.34$

4. $z = -1.33$ **5.** $z = -0.16$ **6.** $z = -2.41$

7. $z = 2.39$ **8.** $z = 0.84$ **9.** $z = -1.645$

10. $z = 1.04$ **11.** $z = 1.555$ **12.** $z = -2.33$

13. $z = -2.33$ **14.** $z = -1.04$ **15.** $z = -0.84$

16. $z = 0.13$ **17.** $z = 1.175$ **18.** $z = 0.44$

19. $z = -0.67$ **20.** $z = 0$ **21.** $z = 0.67$

22. $z = 1.28$ **23.** $z = -0.39$ **24.** $z = 0.39$

25. $z = -0.38$ **26.** $z = 0.25$ **27.** $z = -0.58$

28. $z = 1.99$ **29.** $z = \pm 1.645$ **30.** $z = \pm 1.96$

31. 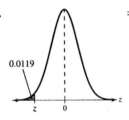 $\Rightarrow z = -1.18$ **32.** 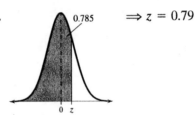 $\Rightarrow z = 0.79$

33. 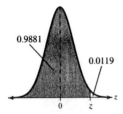 $\Rightarrow z = 1.18$ **34.** 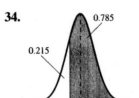 $\Rightarrow z = -0.79$

35. 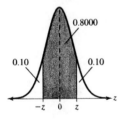 $\Rightarrow z = \pm 1.28$ **36.** 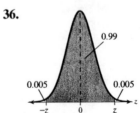 $\Rightarrow z = \pm 2.575$

37. 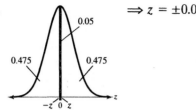 $\Rightarrow z = \pm 0.06$

38. 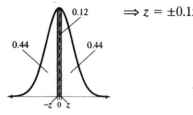 $\Rightarrow z = \pm 0.15$

39. (a) 95th percentile $\Rightarrow$ Area $= 0.95 \Rightarrow z = 1.645$

$x = \mu + z\sigma = 64.1 + (1.645)(2.71) = 68.56$ inches

(b) 1st quartile $\Rightarrow$ Area $= 0.25 \Rightarrow z = -0.67$

$x = \mu + z\sigma = 64.1 + (-0.67)(2.71) = 62.28$ inches

40. (a) 90th percentile $\Rightarrow$ Area $= 0.90 \Rightarrow z = 1.28$

$x = \mu + z\sigma = 69.6 + (1.28)(3.0) \approx 73.44$ inches

(b) 1st quartile $\Rightarrow$ Area $= 0.25 \Rightarrow z = -0.67$

$x = \mu + z\sigma = 69.6 + (-0.67)(3.0) \approx 67.59$ inches

41. (a) 10th percentile $\Rightarrow$ Area $= 0.10 \Rightarrow z = -1.28$

$x = \mu + z\sigma = 17.1 + (-1.28)(4) = 11.98$ pounds

(b) 3rd quartile $\Rightarrow$ Area $= 0.75 \Rightarrow z = 0.67$

$x = \mu + z\sigma = 17.1 + (0.67)(4) = 19.78$ pounds

42. (a) 5th percentile $\Rightarrow$ Area $= 0.05 \Rightarrow z = -1.645$

$x = \mu + z\sigma = 11.4 + (-1.645)(3) = 6.47$ pounds

(b) 3rd quartile $\Rightarrow$ Area $= 0.75 \Rightarrow z = 0.67$

$x = \mu + z\sigma = 11.4 + (0.67)(3) = 13.41$ pounds

43. (a) Top 30% $\Rightarrow$ Area $= 0.70 \Rightarrow z = 0.52$

$x = \mu + z\sigma = 127 + (0.52)(23.5) = 139.22$ days

(b) Bottom 10% $\Rightarrow$ Area $= 0.10 \Rightarrow z = -1.28$

$x = \mu + z\sigma = 127 + (-1.28)(23.5) = 96.92$ days

44. (a) Top 25% $\Rightarrow$ Area $= 0.75 \Rightarrow z = 0.67$

$x = \mu + z\sigma = 15.4 + (0.67)(2.5) = 17.08$ pounds

(b) Bottom 15% $\Rightarrow$ Area $= 0.15 \Rightarrow z = -1.04$

$x = \mu + z\sigma = 15.4 + (-1.04)(2.5) = 12.80$ pounds

45. Lower 5% $\Rightarrow$ Area $= 0.05 \Rightarrow z = -1.645$

$x = \mu + z\sigma = 20 + (-1.645)(0.07) = 19.88$

46. Upper 7.5% $\Rightarrow$ Area $= 0.955 \Rightarrow z = 1.70$

$x = \mu + z\sigma = 32 + (1.70)(0.36) = 32.61$

47. Bottom 10% $\Rightarrow$ Area $= 0.10 \Rightarrow z = -1.28$

$x = \mu + z\sigma = 30{,}000 + (-1.28)(2500) = 26{,}800$

Tires which wear out by 26,800 miles will be replaced free of charge.

48. A: Top 10% $\Rightarrow$ Area $= 0.90 \Rightarrow z = 1.28$

$x = \mu + z\sigma = 72 + (1.28)(9) = 83.52$

B: Top 30% $\Rightarrow$ Area $= 0.70 \Rightarrow z = 0.52$

$x = \mu + z\sigma = 72 + (0.52)(9) = 76.68$

C: Top 70% $\Rightarrow$ Area $= 0.30 \Rightarrow z = -0.52$

$x = \mu + z\sigma = 72 + (-0.52)(9) = 67.32$

D: Top 90% $\Rightarrow$ Area $= 0.10 \Rightarrow z = -1.28$

$x = \mu + z\sigma = 72 + (-1.28)(9) = 60.48$

49. Top 1% $\Rightarrow$ Area $= 0.99 \Rightarrow z = 2.33$

$x = \mu + z\sigma \Rightarrow 8 = \mu + (2.33)(0.03) \Rightarrow \mu = 7.930$ ounces

5.4 SAMPLING DISTRIBUTIONS AND THE CENTRAL LIMIT THEOREM

5.4 Try It Yourself Solutions

1a.

Sample	Mean	Sample	Mean	Sample	Mean	Sample	Mean
1, 1, 1	1	3, 1, 1	1.67	5, 1, 1	2.33	7, 1, 1	3
1, 1, 3	1.67	3, 1, 3	2.33	5, 1, 3	3	7, 1, 3	3.67
1, 1, 5	2.33	3, 1, 5	3	5, 1, 5	3.67	7, 1, 5	4.33
1, 1, 7	3	3, 1, 7	3.67	5, 1, 7	4.33	7, 1, 7	5
1, 3, 1	1.67	3, 3, 1	2.33	5, 3, 1	3	7, 3, 1	3.67
1, 3, 3	2.33	3, 3, 3	3	5, 3, 3	3.67	7, 3, 3	4.33
1, 3, 5	3	3, 3, 5	3.67	5, 3, 5	4.33	7, 3, 5	5
1, 3, 7	3.67	3, 3, 7	4.33	5, 3, 7	5	7, 3, 7	5.67
1, 5, 1	2.33	3, 5, 1	3	5, 5, 1	3.67	7, 5, 1	4.33
1, 5, 3	3	3, 5, 3	3.67	5, 5, 3	4.33	7, 5, 3	5
1, 5, 5	3.67	3, 5, 5	4.33	5, 5, 5	5	7, 5, 5	5.67
1, 5, 7	4.33	3, 5, 7	5	5, 5, 7	5.67	7, 5, 7	6.33
1, 7, 1	3	3, 7, 1	3.67	5, 7, 1	4.33	7, 7, 1	5
1, 7, 3	3.67	3, 7, 3	4.33	5, 7, 3	5	7, 7, 3	5.67
1, 7, 5	4.33	3, 7, 5	5	5, 7, 5	5.67	7, 7, 5	6.33
1, 7, 7	5	3, 7, 7	5.67	5, 7, 7	6.33	7, 7, 7	7

b.

$\bar{x}$	f	Probability
1	1	0.0156
1.67	3	0.0469
2.33	6	0.0938
3	10	0.1563
3.67	12	0.1875
4.33	12	0.1875
5	10	0.1563
5.67	6	0.0938
6.33	3	0.0469
7	1	0.0156

c. $\mu_{\bar{x}} = \mu = 4,$

$\sigma_{\bar{x}}^2 = \dfrac{\sigma^2}{n} = \dfrac{5}{3} = 1.667,$

$\sigma_{\bar{x}} = \dfrac{\sigma}{\sqrt{n}} = \dfrac{\sqrt{5}}{\sqrt{3}} = 1.291$

$\mu_{\bar{x}} = 4, \sigma_{\bar{x}}^2 \approx 1.667, \sigma_{\bar{x}} \approx 1.291$

2a. $\mu_{\bar{x}} = \mu = 64, \sigma_{\bar{x}} = \dfrac{\sigma}{\sqrt{n}} = \dfrac{9}{\sqrt{100}} = 0.9$

b. $n = 100$

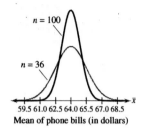

Mean of phone bills (in dollars)

c. With a larger sample size, the mean stays the same but the standard deviation decreases.

3a. $\mu_{\bar{x}} = \mu = 3.5, \sigma_{\bar{x}} = \dfrac{\sigma}{\sqrt{n}} = \dfrac{0.2}{\sqrt{16}} = 0.05$

b.

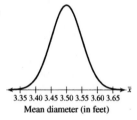

Mean diameter (in feet)

4a. $\mu_{\bar{x}} = \mu = 25, \sigma_{\bar{x}} = \dfrac{\sigma}{\sqrt{n}} = \dfrac{1.5}{\sqrt{100}} = 0.15$

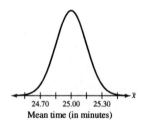

Mean time (in minutes)

b. $\bar{x} = 24.7$: $z = \dfrac{\bar{x} - \mu}{\dfrac{\sigma}{\sqrt{n}}} = \dfrac{24.7 - 25}{\dfrac{1.5}{\sqrt{100}}} = -\dfrac{0.3}{0.15} = -2$

$\bar{x} = 25.5$: $z = \dfrac{\bar{x} - \mu}{\dfrac{\sigma}{\sqrt{n}}} = \dfrac{25.5 - 25}{\dfrac{1.5}{\sqrt{100}}} = \dfrac{0.5}{0.15} = 3.33$

c. $P(z < -2) \approx 0.0228$

$P(z < 3.33) = 0.9996$

$P(8.7 < \bar{x} < 9.5) = P(-2 < z < 3.33) = 0.9996 - 0.0228 = 0.9768$

5a. $\mu_{\bar{x}} = \mu = 306{,}258, \sigma_{\bar{x}} = \dfrac{\sigma}{\sqrt{n}} = \dfrac{44{,}000}{\sqrt{12}} \approx 12{,}701.7$

280,855   306,258   331,661
Mean sales price (in dollars)

b. $\bar{x} = 280{,}000: z = \dfrac{\bar{x} - \mu}{\dfrac{\sigma}{\sqrt{n}}} = \dfrac{280{,}000 - 306{,}258}{\dfrac{44{,}000}{\sqrt{12}}} = \dfrac{-26{,}258}{12{,}701.7} = -2.07$

c. $P(\bar{x} > 200{,}000) = P(z > -2.07) = 1 - P(z < -2.07) = 1 - 0.0192 = 0.9808$

6a. $x = 700: z = \dfrac{x - \mu}{\sigma} = \dfrac{700 - 625}{150} = 0.5$

$\bar{x} = 700: z = \dfrac{\bar{x} - \mu}{\dfrac{\sigma}{\sqrt{n}}} = \dfrac{700 - 625}{\dfrac{150}{\sqrt{10}}} = \dfrac{75}{47.43} = 1.58$

b. $P(z < 0.5) = 0.6915$
$P(z < 1.58) = 0.9429$

c. There is a 69% chance an individual receiver will cost less than $700. There is a 94% chance that the mean of a sample of 10 receivers is less than $700.

5.4 EXERCISE SOLUTIONS

1. $\mu_{\bar{x}} = \mu = 100$

$\sigma_{\bar{x}} = \dfrac{\sigma}{\sqrt{n}} = \dfrac{15}{\sqrt{50}} = 2.121$

2. $\mu_{\bar{x}} = \mu = 100$

$\sigma_{\bar{x}} = \dfrac{\sigma}{\sqrt{n}} = \dfrac{15}{\sqrt{100}} = 1.5$

3. $\mu_{\bar{x}} = \mu = 100$

$\sigma_{\bar{x}} = \dfrac{\sigma}{\sqrt{n}} = \dfrac{15}{\sqrt{250}} = 0.949$

4. $\mu_{\bar{x}} = \mu = 100$

$\sigma_{\bar{x}} = \dfrac{\sigma}{\sqrt{n}} = \dfrac{15}{\sqrt{1000}} = 0.474$

5. False. As the size of the sample increases, the mean of the distribution of the sample mean does not change.

6. False. As the size of a sample increases the standard deviation of the distribution of sample means decreases.

7. False. The shape of the sampling distribution of sample means is normal for large sample sizes even if the shape of the population is non-normal.

8. True

9. $\mu_{\bar{x}} = 3.5$, $\sigma_{\bar{x}} = 1.708$

$\mu = 3.5$, $\sigma = 2.958$

The means are equal but the standard deviation of the sampling distribution is smaller.

Sample	Mean	Sample	Mean	Sample	Mean	Sample	Mean
0, 0, 0	0	2, 0, 0	0.67	4, 0, 0	1.33	8, 0, 0	2.67
0, 0, 2	0.67	2, 0, 2	1.33	4, 0, 2	2	8, 0, 2	3.33
0, 0, 4	1.33	2, 0, 4	2	4, 0, 4	2.67	8, 0, 4	4
0, 0, 8	2.67	2, 0, 8	3.33	4, 0, 8	4	8, 0, 8	5.33
0, 2, 0	0.67	2, 2, 0	1.33	4, 2, 0	2	8, 2, 0	3.33
0, 2, 2	1.33	2, 2, 2	2	4, 2, 2	2.67	8, 2, 2	4
0, 2, 4	2	2, 2, 4	2.67	4, 2, 4	3.33	8, 2, 4	4.67
0, 2, 8	3.33	2, 2, 8	4	4, 2, 8	4.67	8, 2, 8	6
0, 4, 0	1.33	2, 4, 0	2	4, 4, 0	2.67	8, 4, 0	4
0, 4, 2	2	2, 4, 2	2.67	4, 4, 2	3.33	8, 4, 2	4.67
0, 4, 4	2.67	2, 4, 4	3.33	4, 4, 4	4	8, 4, 4	5.33
0, 4, 8	4	2, 4, 8	4.67	4, 4, 8	5.33	8, 4, 8	6.67
0, 8, 0	2.67	2, 8, 0	3.33	4, 8, 0	4	8, 8, 0	5.33
0, 8, 2	3.33	2, 8, 2	4	4, 8, 2	4.67	8, 8, 2	6
0, 8, 4	4	2, 8, 4	4.67	4, 8, 4	5.33	8, 8, 4	6.67
0, 8, 8	5.33	2, 8, 8	6	4, 8, 8	6.67	8, 8, 8	8

10. {120 120, 120 140, 120 180, 120 220, 140 120, 140 140, 140 180, 140 220, 180 120, 180 140, 180 180, 180 220, 220 120, 220 140, 220 180, 220 220}

$\mu_{\bar{x}} = 165$, $\sigma_{\bar{x}} = 27.157$

$\mu = 165$, $\sigma = 38.406$

The means are equal but the standard deviation of the sampling distribution is smaller.

11. (c) Because $\mu_{\bar{x}} = 16.5$, $\sigma_{\bar{x}} = \dfrac{\sigma}{\sqrt{n}} = \dfrac{11.9}{\sqrt{100}} = 1.19$ and the graph approximates a normal curve.

12. (b) Because $\mu_{\bar{x}} = 5.8$, $\sigma_{\bar{x}} = \dfrac{\sigma}{\sqrt{n}} = \dfrac{2.3}{\sqrt{100}} \approx 0.23$ and the graph approximates a normal curve.

13. $z = \dfrac{\bar{x} - \mu}{\dfrac{\sigma}{\sqrt{n}}} = \dfrac{12.2 - 12}{\dfrac{0.95}{\sqrt{36}}} = \dfrac{0.2}{0.158} = 1.26$

$P(\bar{x} < 12.2) = P(z < 1.26) = 0.8962$

14. $z = \dfrac{\bar{x} - \mu}{\dfrac{\sigma}{\sqrt{n}}} = \dfrac{12.2 - 12}{\dfrac{0.95}{\sqrt{100}}} = \dfrac{0.2}{0.095} = 2.11$

$P(\bar{x} > 12.2) = P(z > 2.11) = 1 - P(z < 2.11) = 1 - 0.9826 = 0.0174$

The probability is unusual because it is less than 0.05.

15. $z = \dfrac{\bar{x} - \mu}{\dfrac{\sigma}{\sqrt{n}}} = \dfrac{221 - 220}{\dfrac{39}{\sqrt{75}}} = \dfrac{1}{0.450} = 2.22$

$P(\bar{x} > 221) = P(z > 2.22) = 1 - P(z < 2.22) = 1 - 0.9868 = 0.0132$

The probability is unusual because it is less than 0.05.

16. $z = \dfrac{\bar{x} - \mu}{\dfrac{\sigma}{\sqrt{n}}} = \dfrac{12{,}753 - 12{,}750}{\dfrac{1.7}{\sqrt{36}}} = \dfrac{3}{0.283} = 10.59$

$P(\bar{x} < 12{,}750 \text{ or } \bar{x} > 12{,}753) = P(z < 0 \text{ or } z > 10.59) = 0.5 + 0.000 = 0.5$

17. $\mu_{\bar{x}} = 87.5$

$\sigma_{\bar{x}} = \dfrac{\sigma}{\sqrt{n}} = \dfrac{6.25}{\sqrt{12}} \approx 1.804$

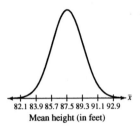

82.1 83.9 85.7 87.5 89.3 91.1 92.9
Mean height (in feet)

18. $\mu_{\bar{x}} = 800$

$\sigma_{\bar{x}} = \dfrac{\sigma}{\sqrt{n}} = \dfrac{100}{\sqrt{15}} \approx 25.820$

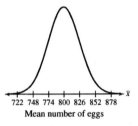

722 748 774 800 826 852 878
Mean number of eggs

19. $\mu_{\bar{x}} = 224$

$\sigma_{\bar{x}} = \dfrac{\sigma}{\sqrt{n}} = \dfrac{8}{\sqrt{40}} = 1.265$

221.5   224   226.5
Mean price (in dollars)

20. $\mu_{\bar{x}} = 47.2$

$\sigma_{\bar{x}} = \dfrac{\sigma}{\sqrt{n}} = \dfrac{3.6}{\sqrt{36}} = 0.6$

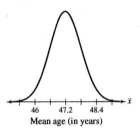

46   47.2   48.4
Mean age (in years)

21. $\mu_{\bar{x}} = 110$

$\sigma_{\bar{x}} = \dfrac{\sigma}{\sqrt{n}} = \dfrac{38.5}{\sqrt{20}} \approx 8.609$

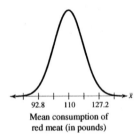

92.8   110   127.2
Mean consumption of
red meat (in pounds)

22. $\mu_{\bar{x}} = 51.5$

$\sigma_{\bar{x}} = \dfrac{\sigma}{\sqrt{n}} = \dfrac{17.1}{\sqrt{25}} \approx 3.42$

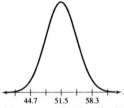

44.7   51.5   58.3
Mean consumption (in gallons)

23. $\mu_{\bar{x}} = 87.5, \sigma_{\bar{x}} = \dfrac{\sigma}{\sqrt{n}} = \dfrac{6.25}{\sqrt{24}} \approx 1.276$

$\mu_{\bar{x}} = 87.5, \sigma_{\bar{x}} = \dfrac{\sigma}{\sqrt{n}} = \dfrac{6.25}{\sqrt{36}} \approx 1.042$

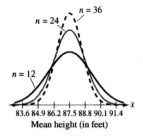

83.6 84.9 86.2 87.5 88.8 90.1 91.4
Mean height (in feet)

As the sample size increases, the standard error decreases, while the mean of the sample means remains constant.

24. $\mu_{\bar{x}} = 800, \sigma_{\bar{x}} = \dfrac{\sigma}{\sqrt{n}} = \dfrac{100}{\sqrt{30}} \approx 18.257$

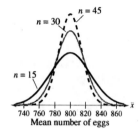

740 760 780 800 820 840 860
Mean number of eggs

$\mu_{\bar{x}} = 800, \sigma_{\bar{x}} = \dfrac{\sigma}{\sqrt{n}} = \dfrac{100}{\sqrt{45}} \approx 14.907$

As the sample size increases, the standard error decreases.

25. $z = \dfrac{\bar{x} - \mu}{\dfrac{\sigma}{\sqrt{n}}} = \dfrac{44{,}000 - 46{,}700}{\dfrac{5600}{\sqrt{42}}} = \dfrac{-2700}{864.10} \approx -3.12$

$P(\bar{x} < 44{,}000) = P(z < -3.12) = 0.0009$

26. $z = \dfrac{\bar{x} - \mu}{\dfrac{\sigma}{\sqrt{n}}} = \dfrac{55{,}000 - 59{,}100}{\dfrac{1700}{\sqrt{35}}} = \dfrac{-4100}{207.35} \approx -14.27$

$P(\bar{x} < 55{,}000) = P(z < -14.27) \approx 0$

27. $z = \dfrac{\bar{x} - \mu}{\dfrac{\sigma}{\sqrt{n}}} = \dfrac{2.768 - 2.818}{\dfrac{0.045}{\sqrt{32}}} = \dfrac{-0.05}{0.00795} \approx -6.29$

$z = \dfrac{\bar{x} - \mu}{\dfrac{\sigma}{\sqrt{n}}} = \dfrac{2.918 - 2.818}{\dfrac{0.045}{\sqrt{32}}} = \dfrac{0.1}{0.00795} \approx 12.57$

$P(2.768 < \bar{x} < 2.918) = P(-6.29 < z < 12.57) \approx 1 - 0 = 1$

28. $z = \dfrac{\bar{x} - \mu}{\dfrac{\sigma}{\sqrt{n}}} = \dfrac{3.310 - 3.305}{\dfrac{0.049}{\sqrt{38}}} = \dfrac{0.005}{0.00794} \approx 0.63$

$z = \dfrac{\bar{x} - \mu}{\dfrac{\sigma}{\sqrt{n}}} = \dfrac{3.320 - 3.305}{\dfrac{0.049}{\sqrt{38}}} = \dfrac{0.015}{0.00794} \approx 1.89$

$P(3.310 < \bar{x} < 3.320) = P(0.63 < z < 1.89) = 0.9706 - 0.7357 = 0.2349$

29. $z = \dfrac{\bar{x} - \mu}{\dfrac{\sigma}{\sqrt{n}}} = \dfrac{66 - 64.1}{\dfrac{2.71}{\sqrt{60}}} = \dfrac{1.9}{0.350} \approx 5.43$

$P(\bar{x} > 66) = P(z > 5.43) \approx 0$

30. $z = \dfrac{\bar{x} - \mu}{\dfrac{\sigma}{\sqrt{n}}} = \dfrac{70 - 69.6}{\dfrac{3.0}{\sqrt{60}}} = \dfrac{0.4}{0.387} \approx 1.03$

$P(\bar{x} > 70) = P(z > 1.03) = 1 - P(z < 1.03) = 1 - 0.8485 = 0.1515$

31. $z = \dfrac{\bar{x} - \mu}{\sigma} = \dfrac{70 - 64.1}{2.71} \approx 2.18$

$P(x < 70) = P(z < 2.18) = 0.9854$

$z = \dfrac{\bar{x} - \mu}{\dfrac{\sigma}{\sqrt{n}}} = \dfrac{70 - 64.1}{\dfrac{2.71}{\sqrt{20}}} = \dfrac{5.9}{0.606} \approx 9.73$

$P(\bar{x} < 70) = P(z < 9.73) \approx 1$

It is more likely to select a sample of 20 women with a mean height less than 70 inches, because the sample of 20 has a higher probability.

32. $z = \dfrac{x - \mu}{\sigma} = \dfrac{65 - 69.6}{3.0} \approx -1.53$

$P(x < 65) = P(z < -1.53) = 0.0630$

$z = \dfrac{\bar{x} - \mu}{\dfrac{\sigma}{\sqrt{n}}} = \dfrac{65 - 69.6}{\dfrac{3.0}{\sqrt{15}}} = \dfrac{-4.6}{0.775} \approx -5.94$

$P(\bar{x} < 65) = P(z < -5.94) \approx 0$

It is more likely to select one man with a height less than 65 inches because the probability is greater.

33. $z = \dfrac{\bar{x} - \mu}{\dfrac{\sigma}{\sqrt{n}}} = \dfrac{127.9 - 128}{\dfrac{0.20}{\sqrt{40}}} = \dfrac{-0.1}{0.032} \approx 3.16$

$P(\bar{x} < 127.9) = P(z < -3.16) = 0.0008$

Yes, it is very unlikely that we would have randomly sampled 40 cans with a mean equal to 127.9 ounces, because it is more than 2 standard deviations from the mean of the sample means.

34. $z = \dfrac{\bar{x} - \mu}{\dfrac{\sigma}{\sqrt{n}}} = \dfrac{64.05 - 64}{\dfrac{0.11}{\sqrt{40}}} = \dfrac{0.05}{0.017} \approx 2.87$

$P(\bar{x} > 64.05) = P(z > 2.87) = 1 - P(z < 2.87) = 1 - 0.9979 = 0.0021$

Yes, it is very unlikely that we would have randomly sampled 40 containers with a mean equal to 64.05 ounces, because it is more then 2 standard deviations from the mean of the sample means.

35. (a) $\mu = 96$

$\sigma = 0.5$

$z = \dfrac{\bar{x} - \mu}{\dfrac{\sigma}{\sqrt{\mu}}} = \dfrac{96.25 - 96}{\dfrac{0.5}{\sqrt{90}}} = \dfrac{0.25}{0.079} \approx 3.16$

$P(\bar{x} \geq 96.25) = P(z > 3.16) = 1 - P(z < 3.16) = 1 - 0.9992 = 0.0008$

(b) Claim is inaccurate.

(c) Assuming the distribution is normally distributed:

$$z = \frac{\bar{x} - \mu}{\sigma} = \frac{96.25 - 96}{0.5} = 0.5$$

$$P(x > 96.25) = P(z > 0.5) = 1 - P(z < 0.5) = 1 - 0.6915 = 0.3085$$

Assuming the manufacturer's claim is true, an individual board with a length of 96.25 would not be unusual. It is within 1 standard deviation of the mean for an individual board.

36. (a) $\mu = 10$

$\sigma = 0.5$

$$z = \frac{\bar{x} - \mu}{\dfrac{\sigma}{\sqrt{n}}} = \frac{10.21 - 10}{\dfrac{0.5}{\sqrt{25}}} = \frac{0.21}{0.1} = 2.1$$

$$P(\bar{x} \geq 10.21) = P(z \geq 2.1) = 1 - P(z \leq 2.1) = 1 - 0.9821 = 0.0179$$

(b) Claim is inaccurate.

(c) Assuming the distribution is normally distributed:

$$z = \frac{x - \mu}{\sigma} = \frac{10.21 - 10}{0.5} = 0.42$$

$$P(x \geq 10.21) = P(z \geq 0.42) = 1 - P(z \leq 0.42) = 1 - 0.6628 = 0.3372$$

Assuming the manufacturer's claim is true, an individual carton with a weight of 10.21 would not be unusual because it is within 1 standard deviation of the mean for an individual ice cream carton.

37. (a) $\mu = 50,000$

$\sigma = 800$

$$z = \frac{\bar{x} - \mu}{\dfrac{\sigma}{\sqrt{n}}} = \frac{49,721 - 50,000}{\dfrac{800}{\sqrt{100}}} = \frac{-279}{80} = -3.49$$

$$P(\bar{x} \leq 49,721) = P(z \leq -3.49) = 0.0002$$

(b) The manufacturer's claim is inaccurate.

(c) Assuming the distribution is normally distributed:

$$z = \frac{x - \mu}{\sigma} = \frac{49,721 - 50,000}{800} = -0.35$$

$$P(x < 49,721) = P(z < -0.35) = 0.3669$$

Assuming the manufacturer's claim is true, an individual tire with a life span of 49,721 miles is not unusual. It is within 1 standard deviation of the mean for an individual tire.

38. (a) $\mu = 38,000$

$\sigma = 1000$

$$z = \frac{\bar{x} - \mu}{\frac{\sigma}{\sqrt{n}}} = \frac{37,650 - 38,000}{\frac{1000}{\sqrt{50}}} = \frac{-350}{141.42} = -2.47$$

$P(\bar{x} \leq 37,650) = P(z \leq -2.47) = 0.0068$

(b) Claim is inaccurate.

(c) Assuming the distribution is normally distributed:

$$z = \frac{x - \mu}{\sigma} = \frac{37,650 - 38,000}{1000} = -0.35$$

$P(x < 37,650) = P(z < -0.35) = 0.3632$

Assuming the manufacturer's claim is true, an individual brake pad lasting less than 37,650 miles would not be unusual, because it is within 1 standard deviation of the mean for an individual brake pad.

39. $\mu = 518$

$\sigma = 115$

$$z = \frac{\bar{x} - \mu}{\frac{\sigma}{\sqrt{n}}} = \frac{530 - 518}{\frac{115}{\sqrt{50}}} = \frac{12}{16.26} = 0.74$$

$P(\bar{x} \geq 530) = P(z \geq 0.74) = 1 - P(z \leq 0.74) = 1 - 0.7704 = 0.2296$

The high school's claim is not justified because it is not rare to find a sample mean as large as 530.

40. $\mu = 4$

$\sigma = 0.5$

$$z = \frac{\bar{x} - \mu}{\frac{\sigma}{\sqrt{n}}} = \frac{4.2 - 4}{\frac{0.5}{\sqrt{100}}} = \frac{0.2}{0.05} = 4$$

$P(\bar{x} \geq 4.2) = P(z \geq 4) = 1 - P(z \leq 4) \approx 1 - 1 \approx 0$

It is very unlikely the machine is calibrated to produce a bolt with a mean of 4 inches.

41. Use the finite correction factor since $n = 55 > 40 = 0.05N$.

$$z = \frac{\bar{x} - \mu}{\frac{\sigma}{\sqrt{n}}\sqrt{\frac{N-n}{N-1}}} = \frac{2.871 - 2.876}{\frac{0.009}{\sqrt{55}}\sqrt{\frac{800-55}{800-1}}} = \frac{-0.005}{(0.00121)\sqrt{0.9324}} \approx -4.27$$

$P(\bar{x} < 2.871) = P(z < -4.27) \approx 0$

42. Use the finite correction factor since $n = 30 > 25 = 0.05N$.

$$z = \frac{\bar{x} - \mu}{\frac{\sigma}{\sqrt{n}}\sqrt{\frac{N-n}{N-1}}} = \frac{2.5 - 3.32}{\frac{1.09}{\sqrt{30}}\sqrt{\frac{500-30}{500-1}}} = \frac{-0.82}{(0.199)\sqrt{0.9419}} \approx -4.25$$

$$z = \frac{\bar{x} - \mu}{\frac{\sigma}{\sqrt{n}}\sqrt{\frac{N-n}{N-1}}} = \frac{4 - 3.32}{\frac{1.09}{\sqrt{30}}\sqrt{\frac{500-30}{500-1}}} = \frac{0.68}{(0.199)\sqrt{0.9419}} \approx 3.52$$

$$P(2.5 < \bar{x} < 4) = P(-4.25 < z < 3.52) \approx 1 - 0 = 1$$

43.

Sample	Number of boys	Proportion
bbb	3	1
bbg	2	$\frac{2}{3}$
bgb	2	$\frac{2}{3}$
bgg	1	$\frac{1}{3}$
gbb	2	$\frac{2}{3}$
gbg	1	$\frac{1}{3}$
ggb	1	$\frac{1}{3}$
ggg	0	0

44.

Proportion	Probability
0	$\frac{1}{8}$
$\frac{1}{3}$	$\frac{3}{8}$
$\frac{2}{3}$	$\frac{3}{8}$
1	$\frac{1}{8}$

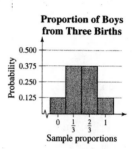

Proportion of Boys from Three Births

The spread of the histogram for each proportion is equal to the number of occurences of 0, 1, 2, and 3 boys, respectively, from the binomial distribution.

45.

Sample	Sample Mean
bbb	1
bbg	$\frac{2}{3}$
bgb	$\frac{2}{3}$
bgg	$\frac{1}{3}$
gbb	$\frac{2}{3}$
gbg	$\frac{1}{3}$
ggb	$\frac{1}{3}$
ggg	0

The sample mean is the same as the proportion of boys in each sample.

46.

Sample	Number of boys	Proportion
bbbb	4	1
bbbg	3	$\frac{3}{4}$
bbgb	3	$\frac{3}{4}$
bbgg	2	$\frac{1}{2}$
bgbb	3	$\frac{3}{4}$
bgbg	2	$\frac{1}{2}$
bggb	2	$\frac{1}{2}$
bggg	1	$\frac{1}{4}$

Sample	Number of boys	Proportion
gbbb	3	$\frac{3}{4}$
gbbg	2	$\frac{1}{2}$
gbgb	2	$\frac{1}{2}$
gbgg	1	$\frac{1}{4}$
ggbb	2	$\frac{1}{2}$
ggbg	1	$\frac{1}{4}$
gggb	1	$\frac{1}{4}$
gggg	0	0

Proportion	Probability
0	$\frac{1}{16}$
$\frac{1}{4}$	$\frac{1}{4}$
$\frac{1}{2}$	$\frac{3}{8}$
$\frac{3}{4}$	$\frac{1}{4}$
1	$\frac{1}{16}$

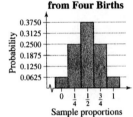

Proportion of Boys from Four Births

47. $z = \dfrac{\hat{p} - p}{\sqrt{\dfrac{pq}{n}}} = \dfrac{0.70 - 0.75}{\sqrt{\dfrac{0.75(0.25)}{90}}} = \dfrac{-0.05}{0.0456} = -1.10$

$P(p < 0.70) = P(z < -1.10) = 0.1357$

5.5 NORMAL APPROXIMATIONS TO BINOMIAL DISTRIBUTIONS

5.5 Try It Yourself Solutions

1a. $n = 70, p = 0.80, q = 0.20$

b. $np = 56, nq = 14$

c. Because $np \geq 5$ and $nq \geq 5$, the normal distribution can be used.

d. $\mu = np = (70)(0.80) = 56$

$\sigma = \sqrt{npq} = \sqrt{(70)(0.80)(0.20)} \approx 3.35$

2a. (1) $57, 58, \ldots, 83$ (2) $\ldots, 52, 53, 54$

b. (1) $56.5 < x < 83.5$ (2) $x < 54.5$

3a. $n = 70, p = 0.80$

$np = 56 \geq 5$ and $nq = 14 \geq 5$

The normal distribution can be used.

b. $\mu = np = 56$

$\sigma = \sqrt{npq} \approx 3.35$

c. $x > 50.5$

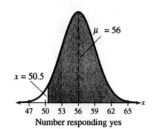

d. $z = \dfrac{x - \mu}{\sigma} = \dfrac{50.5 - 56}{3.35} \approx -1.64$

e. $P(z < -1.64) = 0.0505$

$P(x > 50.5) = P(z > -1.64) = 1 - P(z < -1.64) = 0.9495$

The probability that more than 50 respond yes is 0.9495.

4a. $n = 200, p = 0.38$

$np = 76 \geq 5$ and $nq = 124 \geq 5$

The normal distribution can be used.

b. $\mu = np = 76$

$\sigma = \sqrt{npq} \approx 6.86$

c. $P(x \leq 85.5)$

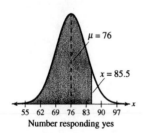

d. $z = \dfrac{x - \mu}{\sigma} = \dfrac{85.5 - 76}{6.86} \approx 1.38$

e. $P(x < 65.5) = P(z < 1.38) = 0.9162$

The probability that at most 65 people will say yes is 0.9162.

5a. $n = 200, p = 0.86$

$np = 172 \geq 5$ and $nq = 28 \geq 5$

The normal distribution can be used.

b. $\mu = np = 172$

$\sigma = \sqrt{npq} \approx 4.91$

c. $P(169.5 < x < 170.5)$

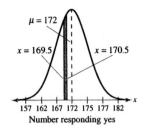

Number responding yes

d. $z = \dfrac{x - \mu}{\sigma} = \dfrac{169.5 - 172}{4.91} \approx -0.51$

$z = \dfrac{x - \mu}{\sigma} = \dfrac{170.5 - 172}{4.91} \approx -0.31$

e. $P(z < -0.51) = 0.3050$

$P(z < -0.31) = 0.3783$

$P(-0.51 < z < -0.31) = 0.3783 - 0.3050 = 0.0733$

The probability that exactly 170 people will respond yes is 0.0733.

5.5 EXERCISE SOLUTIONS

1. $np = (24)(0.85) = 20.4 \geq 5$

$nq = (24)(0.15) = 3.6 < 5$

Cannot use normal distribution.

2. $np = (15)(0.70) = 10.5 \geq 5$

$nq = (15)(0.30) = 4.5 < 5$

Cannot use the normal distribution.

3. $np = (18)(0.90) = 16.2 \geq 5$

$nq = (18)(0.10) = 1.8 < 5$

Cannot use normal distribution.

4. $np = (20)(0.65) = 13 \geq 5$

$nq = (20)(0.35) = 7 \geq 5$

Use the normal distribution.

5. $n = 10, p = 0.85, q = 0.15$

$np = 8.5 > 5, nq = 1.5 < 5$

Cannot use normal distribution because $nq < 5$.

6. $n = 20, p = 0.63, q = 0.37$

$np = 12.6 \geq 5, nq = 7.4 \geq 5$

Use the normal distribution.

$\mu = np = (20)(0.66) = 13.2$

$\sigma = \sqrt{npq} = \sqrt{(20)(0.66)(0.34)} \approx 2.12$

7. $n = 10, p = 0.99, q = 0.03$

$np = 9.9 \geq 5, nq = 0.1 < 5$

Cannot use normal distribution because $nq < 5$.

8. $n = 30, p = 0.086, q = 0.914$

$np = 2.58 < 5, nq = 27.42 \geq 5$

Cannot use the normal distribution because $np < 5$.

9. d   **10.** b   **11.** a   **12.** c

13. a   **14.** d   **15.** c   **16.** b

17. Binomial: $P(5 \leq x \leq 7) = 0.162 + 0.198 + 0.189 = 0.549$

Normal: $P(4.5 \leq x \leq 7.5) = P(-0.97 < z < 0.56) = 0.7123 - 0.1660 = 0.5463$

18. Binomial: $P(2 \leq x \leq 4) = P(2) + P(3) + P(4) \approx 0.016 + 0.054 + 0.121 = 0.191$

Normal: $P(1.5 \leq x \leq 4.5) = P(-2.60 \leq z \leq -0.87) = 0.1922 - 0.0047 = 0.1875$

19. $n = 30, p = 0.07 \rightarrow np = 2.1$ and $nq = 27.9$

Cannot use normal distribution because $np < 5$.

(a) $P(x = 10) = {}_{30}C_{10}(0.07)^{10}(0.93)^{20} \approx 0.0000199$

(b) $P(x \geq 10) = 1 - P(x < 10)$

$$= 1 - [{}_{30}C_0(0.07)^0(0.93)^{30} + {}_{30}C_1(0.07)^1(0.93)^{29} + \cdots + {}_{30}C_9(0.07)^9(0.93)^{21}]$$

$$= 1 - .999977 \approx 0.000023$$

(c) $P(x < 10) \approx 0.999977$ (see part b)

(d) $n = 100, p = 0.07 \rightarrow np = 7$ and $nq = 93$

Use normal distribution.

$$z = \frac{x - \mu}{\sigma} = \frac{4.5 - 7}{2.55} \approx -0.98$$

$$P(x < 5) = P(x < 4.5) = P(z < -0.98) = 0.1635$$

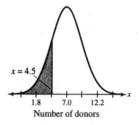

20. $n = 32, p = 0.34 \rightarrow np = 10.88$ and $nq = 21.12$

Use normal distribution.

(a) $z = \dfrac{x - \mu}{\sigma} = \dfrac{11.5 - 10.88}{2.68} \approx 0.23$

$z = \dfrac{x - \mu}{\sigma} = \dfrac{12.5 - 10.88}{2.68} \approx 0.60$

$P(x = 12) = P(11.5 < x < 12.5)$

$\qquad = P(0.23 < z < 0.60)$

$\qquad = 0.7257 - 0.5910 = 0.1347$

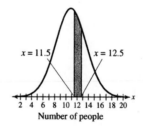

(b) $P(x \geq 12) = P(x > 11.5)$

$\qquad = P(z > 0.23)$

$\qquad = 1 - P(z < 0.23)$

$\qquad = 1 - 0.5910 = 0.4090$

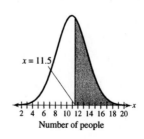

(c) $P(x < 12) = P(x < 11.5)$

$$= P(z < 0.23) = 0.5910$$

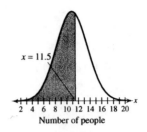
Number of people

(d) $n = 150, p = 0.34 \rightarrow np = 51$ and $nq = 99$

$$z = \frac{x - \mu}{\sigma} = \frac{59.5 - 51}{5.80} \approx 1.47$$

$$P(x < 60) = P(x < 59.5)$$

$$= P(z < 1.47)$$

$$= 0.9292$$

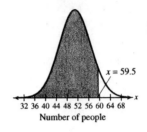
Number of people

21. $n = 250, p = 0.05, q = 0.95$

$np = 12.5 \geq 5, nq = 237.5 \geq 5$

Use the normal distribution.

(a) $z = \dfrac{x - \mu}{\sigma} = \dfrac{15.5 - 12.5}{3.45} = 0.87$

$z = \dfrac{x - \mu}{\sigma} = \dfrac{16.5 - 12.5}{3.45} = 1.16$

$P(x = 16) \approx P(15.5 \leq x \leq 16.5) = P(0.87 \leq z \leq 1.16)$

$$= 0.8770 - 0.8078 = 0.0692$$

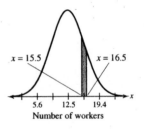
Number of workers

(b) $P(x \geq 9) \approx P(x \geq 8.5) = P(z \geq -1.16) = 1 - P(z \leq -1.16) = 1 - 0.1230 = 0.8770$

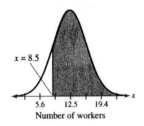
Number of workers

(c) $P(x < 16) \approx P(x \leq 15.5) = P(z \leq 0.87) = 0.8078$

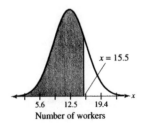
Number of workers

(d) $n = 500, p = 0.05, q = 0.95$

$np = 25 \geq 5, nq = 475 \geq 5$

Use normal distribution.

$z = \dfrac{x - \mu}{\sigma} = \dfrac{29.5 - 25}{4.87} = 0.92$

$P(x < 30) \approx P(x < 29.5) = P(z < 0.92) = 0.8212$

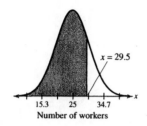

22. $n = 50, p = 0.32, q = 0.68$

$np = 16 \geq 5, nq = 34 \geq 5$

Use normal distribution.

(a) $z = \dfrac{x - \mu}{\sigma} = \dfrac{11.5 - 16}{3.30} = -1.36$

$z = \dfrac{x - \mu}{\sigma} = \dfrac{12.5 - 16}{3.30} = -1.06$

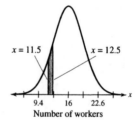

$P(x = 12) \approx P(11.5 \leq x \leq 12.5)$

$\qquad = P(-1.36 < z < -1.06)$

$\qquad = 0.1446 - 0.0869 = 0.0577$

(b) $P(x \geq 14) \approx P(x \geq 13.5)$

$\qquad = P(z \geq -0.76)$

$\qquad = 1 - P(z \leq -0.76)$

$\qquad = 1 - 0.2236 = 0.7764$

$\qquad = 0.7764$

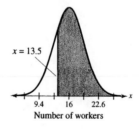

(c) $P(x < 18) \approx P(x \leq 17.5) = P(z \leq 0.45) = 0.6736$

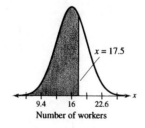

(d) $n = 150, p = 0.32, q = 0.68$

$np = 48, nq = 102$

Use normal distribution.

$$z = \frac{x - \mu}{\sigma} = \frac{29.5 - 48}{5.71} = -3.24$$

$$P(x < 30) \approx P(x \le 29.5) = P(z \le -3.24) = 0.0006$$

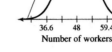

23. $n = 40, p = 0.52 \rightarrow np = 20.8$ and $nq = 19.2$

Use the normal distribution.

(a) $z = \dfrac{x - \mu}{\sigma} = \dfrac{23.5 - 20.8}{3.160} \approx 0.85$

$$P(x \le 23) = P(x < 23.5) = P(z < 0.85) = 0.8023$$

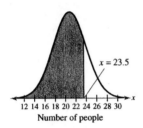

(b) $z = \dfrac{x - \mu}{\sigma} = \dfrac{17.5 - 20.8}{3.160} \approx -1.04$

$$P(x \ge 18) = P(x > 17.5) = P(z > -1.04) = 1 - P(z < -1.04) = 1 - 0.1492 = 0.8508$$

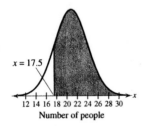

(c) $P(x > 20) = P(x > 20.5) = P(z > -0.09) = 1 - P(z < -0.09) = 1 - 0.4641 = 0.5359$

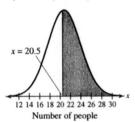

(d) $n = 650, p = 0.52 \rightarrow np = 338$ and $nq = 312$

Use normal distribution.

$$z = \frac{x - \mu}{\sigma} = \frac{350.5 - 338}{12.74} \approx 0.98$$

$P(x > 350) = P1(x > 350.5) = P(z > 0.98) = 1 - P(z < 0.98) = 1 - 0.8365 = 0.1635$

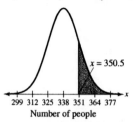

299 312 325 338 351 364 377
Number of people

24. $n = 10, p = 0.029 \rightarrow np = 0.29$ and $nq = 9.71$

Cannot use normal distribution because $np < 5$.

(a) $P(x \le 3) = P(x = 0) + P(x = 1) + P(x = 2) + P(x = 3)$

$\quad = {}_{10}C_0(0.029)^0(0.971)^{10} + {}_{10}C_1(0.029)^1(0.971)^9 + {}_{10}C_2(0.029)^2(0.971)^8$

$\quad\quad + {}_{10}C_3(0.029)^3(0.971)^7$

$\quad \approx 0.9999$

(b) $P(x \ge 1) = 1 - P(x < 1) = 1 - P(x = 0) \approx 0.2549$

(c) $P(x > 2) = 1 - P(x \le 2) \approx 1 - 0.9975 = 0.0025$

(d) $n = 50, p = 0.029 \rightarrow np = 1.45$ and $nq = 48.55$

Cannot use normal distribution.

$P(x = 0) = {}_{50}C_0(0.029)^0(0.971)^{50} \approx 0.2296$

25. (a) $n = 25, p = 0.24, q = 0.76$

$np = 6 \ge 5, nq = 19 \ge 5$

Use normal distribution.

(b) $z = \frac{x - \mu}{\sigma} = \frac{8.5 - 6}{2.135} = 1.17$

$P(x > 8) \approx P(x \ge 8.5) = P(z \ge 1.17) = 1 - P(z \le 1.17) = 1 - 0.8790 = 0.121$

(c) $z = \frac{x - \mu}{\sigma} = \frac{7.5 - 6}{2.135} = 0.70$

$z = \frac{x - \mu}{\sigma} = \frac{8.5 - 6}{2.135} = 1.17$

$P(x = 8) \approx P(7.5 \le x \le 8.5) = P(0.07 \le z \le 1.17) = 0.8790 - 0.7580 = 0.121$

It is not unusual for 8 out of 25 homeowners to say their home is too small because the z-score is within 1 standard deviation of the mean.

26. (a) $n = 40, p = 0.80, q = 0.20$

$np = 32 \geq 5, nq = 8 \geq 5$

Use normal distribution.

(b) $z = \dfrac{x - \mu}{\sigma} = \dfrac{26.5 - 32}{2.53} = -2.17$

$P(x \leq 26) \approx P(x \leq 26.5) = P(z \leq -2.17) = 0.0150$

(c) $z = \dfrac{x - \mu}{\sigma} = \dfrac{25.5 - 32}{2.53} = -2.57$

$z = \dfrac{x - \mu}{\sigma} = \dfrac{26.5 - 32}{2.53} = -2.17$

$P(x = 26) \approx P(25.5 \leq x \leq 26.5) = P(-2.57 \leq z \leq -2.17) = 0.0150 - 0.0051 = 0.0099$

Yes, because the z-score is more than two standard deviations from the mean.

27. $n = 250, p = 0.70$

60% say no $\to$ 250(0.6) = 150 say no while 100 say yes.

$z = \dfrac{x - \mu}{\sigma} = \dfrac{99.5 - 175}{7.25} = -10.41$

$P(\text{less than 100 yes}) = P(x < 100) = P(x < 99.5) = P(z < -10.41) \approx 0$

It is highly unlikely that 60% responded no. Answers will vary.

28. $n = 200, p = 0.11$

9% of 200 = 18 people

$z = \dfrac{x - \mu}{\sigma} \approx \dfrac{17.5 - 22}{4.42} \approx -1.02$

$P(x < 18) = P(x < 17.5) = P(z < -1.02) = 0.1539$

It is probable that 18 of the 200 people responded that they participate in hiking. Answers will vary.

29. $n = 100, p = 0.75$

$z = \dfrac{x - \mu}{\sigma} = \dfrac{69.5 - 75}{4.33} \approx 1.27$

$P(\text{reject claim}) = P(x < 70) = P(x < 69.5) = P(z < -1.27) = 0.1020$

30. $n = 100, p = 0.65$

$z = \dfrac{x - \mu}{\sigma} = \dfrac{69.5 - 65}{4.77} \approx 0.94$

$P(\text{accept claim}) = P(x \geq 70)$

$= P(x > 69.5) = P(z > 0.94) = 1 - P(z < 0.94) = 1 - 0.8264 = 0.1736$

CHAPTER 5 REVIEW EXERCISE SOLUTIONS

1. $\mu = 15, \sigma = 3$ **2.** $\mu = -3, \sigma = 5$

3. $x = 1.32$: $z = \dfrac{x-\mu}{\sigma} = \dfrac{1.32-1.5}{0.08} = -2.25$

$x = 1.54$: $z = \dfrac{x-\mu}{\sigma} = \dfrac{1.54-1.5}{0.08} = 0.5$

$x = 1.66$: $z = \dfrac{x-\mu}{\sigma} = \dfrac{1.66-1.5}{0.08} = 2$

$x = 1.78$: $z = \dfrac{x-\mu}{\sigma} = \dfrac{1.78-1.5}{0.08} = 3.5$

4. 1.32 and 1.78 are unusual.

5. 0.6293   **6.** 0.9946

7. 0.3936   **8.** 0.9573

9. $1 - 0.9535 = 0.0465$ **10.** $1 - 0.5478 = 0.4522$

11. $0.5 - 0.0505 = 0.4495$ **12.** $0.8508 - 0.0606 = 0.7902$

13. $0.9564 - 0.5199 = 0.4365$ **14.** $0.9750 - 0.0250 = 0.95$

15. $0.0668 + 0.0668 = 0.1336$ **16.** $0.7389 + 0.0154 = 0.7543$

17. $P(z < 1.28) = 0.8997$ **18.** $P(z > -0.74) = 0.7704$

19. $P(-2.15 < x < 1.55) = 0.9394 - 0.0158 = 0.9236$

20. $P(0.42 < z < 3.15) = 0.9992 - 0.6628 = 0.3364$

21. $P(z < -2.50 \text{ or } z > 2.50) = 2(0.0062) = 0.0124$

22. $P(z < 0 \text{ or } z > 1.68) = 0.5 + 0.0465 = 0.5465$

23. (a) $z = \dfrac{x-\mu}{\sigma} = \dfrac{1900-2200}{625} \approx -0.48$

$P(x < 1900) = P(z < -0.48) = 0.3156$

(b) $z = \dfrac{x-\mu}{\sigma} = \dfrac{2000-2200}{625} \approx -0.32$

$z = \dfrac{x-\mu}{\sigma} = \dfrac{25{,}900-2200}{625} \approx 0.48$

$P(2000 < x < 2500) = P1(-0.32 < z < 0.48) = 0.6844 - 0.3745 = 0.3099$

(c) $z = \dfrac{x-\mu}{\sigma} = \dfrac{2450-2200}{625} = -0.4$

$P(x > 2450) = P1(z > 0.4) = 0.3446$

24. (a) $z = \dfrac{x-\mu}{\sigma} = \dfrac{1-1.5}{0.25} = -2$

$z = \dfrac{x-\mu}{\sigma} = \dfrac{2-1.5}{0.25} = 2$

$P(1 < x < 2) = P(-2 < z < 2) = 0.9772 - 0.0228 = 0.9544$

(b) $z = \dfrac{x - \mu}{\sigma} = \dfrac{1.6 - 1.5}{0.25} = 0.4$

$z = \dfrac{x - \mu}{\sigma} = \dfrac{2.2 - 1.5}{0.25} = 2.8$

$P(1.6 < x < 2.2) = P(0.4 < z < 2.8) = 0.9974 - 0.6554 = 0.3420$

(c) $z = \dfrac{x - \mu}{\sigma} = \dfrac{2.2 - 1.5}{0.25} = 2.8$

$P(x > 2.2) = P(z > 2.8) = 0.0026$

25. $z = -0.07$ **26.** $z = -1.28$ **27.** $z = 1.13$

28. $z = -2.05$ **29.** $z = 1.04$ **30.** $z = -0.84$

31. $x = \mu + z\sigma = 52 + (-2.4)(2.5) = 46$ meters

32. $x = \mu + z\sigma = 52 + (1.2)(2.5) = 55$ meters

33. 95th percentile $\Rightarrow$ Area $= 0.95 \Rightarrow z = 1.645$

$x = \mu + z\sigma = 52 + (1.645)(2.5) = 56.1$ meters

34. 3rd Quartile $\Rightarrow$ Area $= 0.75 \Rightarrow z = 0.67$

$x = \mu + z\sigma = 52 + (0.67)(2.5) = 53.7$

35. Top 10% $\Rightarrow$ Area $= 0.90 \Rightarrow z = 1.28$

$x = \mu + z\sigma = 52 + (1.28)(2.5) = 55.2$ meters

36. Bottom 5% $\Rightarrow$ Area $= 0.05 \Rightarrow z = -1.645$

$x = \mu + z\sigma = 52 + (-1.645)(2.5) \approx 47.9$

37. {90 90 90, 90 90 120, 90 90 160, 90 90 210, 90 90 300, 90 120 90, 90 120 120, 90 120 160, 90 120 210, 90 120 300, 90 160 90, 90 160 120, 90 160 160, 90 160 210, 90 160 300, 90 210 90, 90 210 120, 90 210 160, 90 210 210, 90 210 300, 90 300 90, 90 300 120, 90 300 160, 90 300 210, 90 300 300, 120 90 90, 120 90 120, 120 90 160, 120 90 210, 120 90 300, 120 120 90, 120 120 120, 120 120 160, 120 120 210, 120 120 300, 120 160 90, 120 160 120, 120 160 160, 120 160 210, 120 160 300, 120 210 90, 120 210 120, 120 210 160, 120 210 210, 120 210 300, 120 300 90, 120 300 120, 120 300 160, 120 300 210, 120 300 300, 160 90 90, 160 90 120, 160 90 160, 160 90 210, 160 90 300, 160 120 90, 160 120 120, 160 120 160, 160 120 210, 160 120 300, 160 160 90, 160 160 120, 160 160 160, 160 160 210, 160 160 300, 160 210 90, 160 210 120, 160 210 160, 160 210 210, 160 210 300, 160 300 90, 160 300 120, 160 300 160, 160 300 210, 160 300 300, 210 90 90, 210 90 120, 210 90 160, 210 90 210, 210 90 300, 210 120 90, 210 120 120, 210 120 160, 210 120 210, 210 120 300, 210 160 90, 210 160 120, 210 160 160, 210 160 210, 210 160 300, 210 210 90, 210 210 120, 210 210 160, 210 210 210, 210 210 300, 210 300 90, 210 300 120, 210 300 160, 210 300 210, 210 300 300, 300 90 90, 300 90 120, 300 90 160, 300 90 210, 300 90 300, 300 120 90, 300 120 120, 300 120 160, 300 120 210, 300 120 300, 300 160 90, 300 160 120, 300 160 160, 300 160 210, 300 160 300, 300 210 90, 300 210 120, 300 210 160, 300 210 210, 300 210 300, 300 300 90, 300 300 120, 300 300 160, 300 300 210, 300 300 300}

$\mu = 176, \sigma \approx 73.919$

$\mu_{\bar{x}} = 176, \sigma_{\bar{x}} \approx 42.677$

The means are the same, but $\sigma_{\bar{x}}$ is less than σ.

38. {00, 01, 02, 03, 10, 11, 12, 13, 20, 21, 22, 23, 30, 31, 32, 33}

$\mu = 1.5, \sigma \approx 1.118$

$\mu_{\bar{x}} = 1.5, \sigma_{\bar{x}} \approx 0.791$

The means are the same, but $\sigma_{\bar{x}}$ is less than σ.

39. $\mu_{\bar{x}} = 144.3, \sigma_{\bar{x}} \dfrac{\sigma}{\sqrt{n}} = \dfrac{51.6}{\sqrt{35}} \approx 8.722$

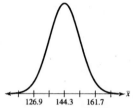

126.9   144.3   161.7
Mean consumption (in pounds)

40. $\mu_{\bar{x}} = 218.2, \sigma_{\bar{x}} = \dfrac{\sigma}{\sqrt{n}} = \dfrac{68.1}{\sqrt{40}} \approx 10.767$

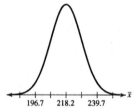

196.7   218.2   239.7
Mean consumption (in pounds)

41. (a) $z = \dfrac{\bar{x} - \mu}{\dfrac{\sigma}{\sqrt{n}}} = \dfrac{1900 - 2200}{\dfrac{625}{\sqrt{12}}} = \dfrac{-300}{180.42} \approx -1.66$

$P(\bar{x} < 1900) = P(z < -1.66) = 0.0485$

(b) $z = \dfrac{\bar{x} - \mu}{\dfrac{\sigma}{\sqrt{n}}} = \dfrac{2000 - 2200}{\dfrac{625}{\sqrt{12}}} = \dfrac{-200}{180.42} \approx -1.11$

$z = \dfrac{\bar{x} - \mu}{\dfrac{\sigma}{\sqrt{n}}} = \dfrac{2500 - 2200}{\dfrac{625}{\sqrt{12}}} = \dfrac{300}{180.42} \approx 1.66$

$P(2000 < \bar{x} < 2500) = P(-1.11 < z < 1.66) = 0.9515 - 0.1335 = 0.8180$

(c) $z = \dfrac{\bar{x} - \mu}{\dfrac{\sigma}{\sqrt{n}}} = \dfrac{2450 - 2200}{\dfrac{625}{\sqrt{12}}} = \dfrac{250}{180.42} \approx 1.39$

$P(\bar{x} > 2450) = P(z > 1.39) = 0.0823$

(a) and (c) are smaller, (b) is larger. This is to be expected because the standard error of the sample mean is smaller.

42. (a) $z = \dfrac{\bar{x} - \mu}{\frac{\sigma}{\sqrt{n}}} = \dfrac{1.0 - 1.5}{\frac{0.25}{\sqrt{7}}} = \dfrac{-0.5}{0.0945} \approx -5.29$

$z = \dfrac{\bar{x} - \mu}{\frac{\sigma}{\sqrt{n}}} = \dfrac{2.0 - 1.5}{\frac{0.25}{\sqrt{7}}} = \dfrac{0.5}{0.0945} \approx 5.29$

$P(1.0 < \bar{x} < 2.0) = P(-5.29 < z < 5.29) \approx 1$

(b) $z = \dfrac{\bar{x} - \mu}{\frac{\sigma}{\sqrt{n}}} = \dfrac{1.6 - 1.5}{\frac{0.25}{\sqrt{7}}} = \dfrac{0.1}{0.0945} \approx 1.06$

$z = \dfrac{\bar{x} - \mu}{\frac{\sigma}{\sqrt{n}}} = \dfrac{2.2 - 1.5}{\frac{0.25}{\sqrt{7}}} = \dfrac{1.7}{0.0945} \approx 7.41$

$P(1.6 < \bar{x} < 2.2) = P(1.06 < z < 7.41) \approx 1 - 0.8554 = 0.1446$

(c) $z = \dfrac{\bar{x} - \mu}{\frac{\sigma}{\sqrt{n}}} = \dfrac{2.2 - 1.5}{\frac{0.25}{\sqrt{7}}} = \dfrac{1.7}{0.0945} \approx 7.41$

$P(\bar{x} > 2.2) = P(z > 7.41) \approx 0$

(a) is larger and (b) and (c) are smaller.

43. (a) $z = \dfrac{\bar{x} - \mu}{\frac{\sigma}{\sqrt{n}}} = \dfrac{29,000 - 29,200}{\frac{1500}{\sqrt{45}}} = \dfrac{-200}{223.61} \approx -0.89$

$P(\bar{x} < 29,000) = P(z < -0.89) \approx 0.1867$

(b) $z = \dfrac{\bar{x} - \mu}{\frac{\sigma}{\sqrt{n}}} = \dfrac{31,000 - 29,200}{\frac{1500}{\sqrt{45}}} = \dfrac{1500}{223.61} \approx 8.05$

$P(\bar{x} > 31,000) = P(z > 8.05) \approx 0$

44. (a) $z = \dfrac{\bar{x} - \mu}{\frac{\sigma}{\sqrt{n}}} = \dfrac{1400 - 1300}{\frac{250}{\sqrt{36}}} = \dfrac{100}{41.67} \approx 2.4$

$P(x < 1400) = P(z < 2.4) = 0.9918$

(b) $z = \dfrac{\bar{x} - \mu}{\frac{\sigma}{\sqrt{n}}} = \dfrac{1150 - 1300}{\frac{250}{\sqrt{36}}} = \dfrac{-150}{41.67} = -3.6$

$P(x > 1150) = P(z > -3.6) \approx 1 - 0 = 0$

45. Assuming the distribution is normally distributed:

$z = \dfrac{\bar{x} - \mu}{\frac{\sigma}{\sqrt{n}}} = \dfrac{1.125 - 1.5}{\frac{.5}{\sqrt{15}}} = \dfrac{-0.375}{0.129} \approx -2.90$

$P(\bar{x} < 1.125) = P(z < -2.90) = 0.0019$

46. Assuming the distribution is normally distributed:

$$z = \frac{\bar{x} - \mu}{\frac{\sigma}{\sqrt{n}}} = \frac{5250 - 5000}{\frac{300}{\sqrt{15}}} = \frac{250}{77.46} \approx 3.23$$

$$P(\bar{x} > 5250) = P(z > 3.23) = 1 - P(z < 3.23) = 1 - 0.9994 = .0006$$

47. $n = 12, p = 0.95, q = 0.05$

$np = 11.4 > 5$, but $nq = 0.6 < 5$

Cannot use the normal distribution because $nq < 5$.

48. $n = 15, p = 0.59, q = 0.41$

$np = 8.85 > 5$ and $nq = 6.15 > 5$

Use the normal distribution.

$$\mu = np = 8.85, \sigma = \sqrt{npq} = \sqrt{15(0.59)(0.41)} \approx 1.905$$

49. $P(x \geq 25) = P(x > 24.5)$ **50.** $P(x \leq 36) = P(x < 36.5)$

51. $P(x = 45) = P(44.5 < x < 45.5)$ **52.** $P(x = 50) = P(49.5 < x < 50.5)$

53. $n = 45, p = 0.70 \rightarrow np = 31.5, nq = 13.5$

Use normal distribution.

$$\mu = np = 31.5, \sigma = \sqrt{npq} = \sqrt{45(0.70)(0.30)} \approx 3.07$$

$$z = \frac{x - \mu}{\sigma} = \frac{20.5 - 31.5}{3.07} \approx -3.58$$

$x = 20.5$

25.3   31.5   37.7

Children saying yes

$$P(x \leq 20) = P(x < 20.5) = P(z < -3.58) \approx 0$$

54. $n = 12, p = 0.33 \rightarrow np = 3.96 < 5$

Cannot use normal distribution.

$$P(x > 5) = 1 - P(x \leq 5)$$

$$= 1 - [P(x = 0) + P(x = 1) + \cdots + P(x = 5)]$$

$$= 1 - [_{12}C_0(0.33)^0(0.67)^{12} + {}_{12}C_1(0.33)^1(0.67)^{11} + {}_{12}C_2(0.33)^2(0.67)^{10}$$

$$+ {}_{12}C_3(0.33)^3(0.67)^9 + {}_{12}C_4(0.33)^4(0.67)^8 + {}_{12}C_5(0.33)^5(0.67)^7]$$

$$= 1 - 0.8289 \approx 0.1711$$

CHAPTER 5 QUIZ SOLUTIONS

1. (a) $P(z > -2.10) = 0.9821$

  (b) $P(z < 3.22) = 0.9994$

  (c) $P(-2.33 < z < 2.33) = 0.9901 - 0.0099 = 0.9802$

  (d) $P(z < -1.75 \text{ or } z > -0.75) = 0.0401 + 0.7734 = 0.8135$

2. (a) $z = \dfrac{x - \mu}{\sigma} = \dfrac{5.36 - 5.5}{0.08} \approx -1.75$

$z = \dfrac{x - \mu}{\sigma} = \dfrac{5.64 - 5.5}{0.08} \approx 1.75$

$P(5.36 < x < 5.64) = P(-1.75 < z < 1.75) = 0.9599 - 0.0401 = 0.9198$

(b) $z = \dfrac{x - \mu}{\sigma} = \dfrac{-5.00 - (-8.2)}{7.84} \approx 0.41$

$z = \dfrac{x - \mu}{\sigma} = \dfrac{0 - (-8.2)}{7.84} \approx 1.05$

$P(-5.00 < x < 0) = P(0.41 < z < 1.05) = 0.8531 - 0.6591 = 0.1940$

(c) $z = \dfrac{x - \mu}{\sigma} = \dfrac{0 - 18.5}{9.25} = -2$

$z = \dfrac{x - \mu}{\sigma} = \dfrac{37 - 18.5}{9.25} = 2$

$P(x < 0 \text{ or } x > 37) = P(z < -2 \text{ or } z > 2) = 2(0.0228) = 0.0456$

3. $z = \dfrac{x - \mu}{\sigma} = \dfrac{320 - 290}{37} \approx 0.81$

$P(x > 320) = P(z > 0.81) = 0.2090$

4. $z = \dfrac{x - \mu}{\sigma} = \dfrac{250 - 290}{37} \approx -1.08$

$z = \dfrac{x - \mu}{\sigma} = \dfrac{300 - 290}{37} \approx 0.27$

$P(250 < x < 300) = P(-1.08 < z < 0.27) = 0.6064 - 0.1401 = 0.4663$

5. $P(x > 250) = P(z > -1.08) = 0.8599 \rightarrow 85.99\%$

6. $z = \dfrac{x - \mu}{\sigma} = \dfrac{280 - 290}{37} \approx -0.27$

$P(x < 280) = P(z < -0.27) = 0.3936$

$(2000)(0.3936) = 787.2 \approx 787$ students

7. top $5\% \rightarrow z \approx 1.645$

$\mu + z\sigma = 290 + (1.645)(37) = 350.9 \approx 351$

8. bottom $25\% \rightarrow z \approx -0.67$

$\mu + z\sigma = 290 + (-0.67)(37) = 265.2 \approx 265$

9. $z = \dfrac{\bar{x} - \mu}{\dfrac{\sigma}{\sqrt{n}}} = \dfrac{300 - 290}{\dfrac{37}{\sqrt{60}}} = \dfrac{10}{4.78} \approx 2.09$

$P(\bar{x} > 300) = P(z > 2.09) \approx 0.0183$

10. $z = \dfrac{x - \mu}{\sigma} = \dfrac{300 - 290}{37} \approx 0.27$

$P(x > 300) = P(z > 0.27) = 0.3936$

$z = \dfrac{\bar{x} - \mu}{\dfrac{\sigma}{\sqrt{n}}} = \dfrac{300 - 290}{\dfrac{37}{\sqrt{15}}} = \dfrac{10}{4.78} \approx 2.09$

$P(\bar{x} > 300) = P(z > 2.09) = 0.0183$

You are more likely to select one student with a test score greater than 300 because the standard error of the mean is less than the standard deviation.

11. $n = 24, p = 0.75 \rightarrow np = 18, nq = 6$

Use normal distribution.

$\mu = np = 18 \qquad \sigma = \sqrt{npq} \approx 2.121$

12. $z = \dfrac{x - \mu}{\sigma} = \dfrac{15.5 - 18}{2.12} \approx -1.18$

$P(x \leq 15) = P(x < 15.5) = P(z < -1.18) = 0.1190$

CUMULATIVE REVIEW, CHAPTERS 3–5

1. (a) $np = 30(0.56) = 16.8 > 5$

$np = 30(0.44) = 13.2 > 5$

Use normal distribution.

(b) $\mu = np = 30(0.56) = 16.8$

$\sigma = \sqrt{npq} = \sqrt{30(0.56)(0.44)} = 2.72$

$P(x \leq 14) \approx P(x \leq 14.5)$

$= P\left(z \leq \dfrac{14.5 - 16.8}{2.72}\right)$

$= P(z \leq -0.85)$

$= 0.1977$

(c) It is not unusual for 14 out of 30 employees to say they do not use all of their vacation time because the probability is greater than 0.05.

$P(x = 14) \approx P(13.5 \leq x \leq 14.5)$

$= P\left(\dfrac{13.5 - 16.8}{2.72} \leq z \leq \dfrac{14.5 - 16.8}{2.72}\right)$

$= P(-1.21 \leq z \leq -0.85)$

$= 0.1977 - 0.1131$

$= 0.0846$

2.

x	P(x)	xP(x)	x − μ	(x − μ)²	(x − μ)²P(x)
2	0.421	0.842	−1.131	1.279	0.539
3	0.233	0.699	−0.131	0.017	0.004
4	0.202	0.808	0.869	0.755	0.153
5	0.093	0.465	1.869	3.493	0.325
6	0.033	0.198	2.869	8.231	0.272
7	0.017	0.119	3.869	14.969	0.254
		$\sum xP(x) = 3.131$			$\sum(x-\mu)^2P(x) = 1.546$

(a) $\mu = \sum xP(x) = 3.1$

(b) $\sigma^2 = \sum(x-\mu)^2P(x) = 1.5$

(c) $\sigma = \sqrt{\sigma^2} = 1.2$

(d) $\sum(x) = \mu = 3.1$

(e) The expected family household size is 3.1 people with a standard deviation of 1.2 persons.

3.

x	P(x)	xP(x)	x − μ	(x − μ)²	(x − μ)²P(x)
0	0.012	0.000	−3.596	12.931	0.155
1	0.049	0.049	−2.596	6.739	0.330
2	0.159	0.318	−1.596	2.547	0.405
3	0.256	0.768	−0.596	0.355	0.091
4	0.244	0.976	0.404	0.163	0.040
5	0.195	0.975	1.404	1.971	0.384
6	0.085	0.510	2.404	5.779	0.491
		$\sum xP(x) = 3.596$			$\sum(x-\mu)^2P(x) = 1.897$

(a) $\mu = \sum xP(x) \approx 3.6$

(b) $\sigma^2 = \sum(x-\mu)^2P(x) \approx 1.9$

(c) $\sigma = \sqrt{\sigma^2} \approx 1.4$

(d) $\sum(x) = \mu \approx 3.6$

(e) The expected number of fouls per game is 3.6 with a standard deviation of 1.4 fouls.

4. (a) $P(x < 4) = 0.012 + 0.049 + 0.159 + 0.256 = 0.476$

(b) $P(x \geq 3) = 1 - P(x \leq 2)$

$= 1 - (0.012 + 0.049 + 0.159)$

$= 0.78$

(c) $P(2 \leq x \leq 4) = 0.159 + 0.256 + 0.244 = 0.659$

5. (a) $(16)(15)(14)(13) = 43{,}680$

    (b) $\dfrac{(7)(6)(5)(4)}{(16)(15)(14)(13)} = \dfrac{840}{43{,}680} = 0.0192$

6. 0.9382 **7.** 0.0010

8. $1 - 0.2005 = 0.7995$ **9.** $0.9990 - 0.500 = 0.4990$

10. $0.3974 - 0.1112 = 0.2862$ **11.** $0.5478 + (1 - 0.9573) = 0.5905$

12. $n = 10, p = 0.78$

    (a) $P(6) = 0.1108$

    (b) $P(x \geq 6) = 0.9521$

    (c) $P(x < 6)\,1 - P(x \geq 6) = 1 - 0.952 = 0.0479$

13. $p = \dfrac{1}{200} = 0.005$

    (a) $P(x = 10) = (0.005)(0.995)^9 = 0.0048$

    (b) $P(x \leq 3) = 0.0149$

    (c) $P(x > 10) = 1 - P(x \leq 10) = 1 - 0.0489 = 0.9511$

14. (a) 0.224

    (b) 0.879

    (c) Dependent because a person can be a public school teacher and have more than 20 years experience.

    (d) $0.879 + 0.137 - 0.107 = 0.909$

    (e) $0.329 + 0.121 - 0.040 = 0.410$

15. (a) $\mu_x = 70$

$$\sigma_{\bar{x}} = \frac{\sigma}{\sqrt{n}} = \frac{1.2}{\sqrt{40}} = 0.1897$$

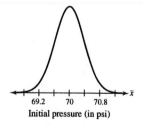

69.2   70   70.8
Initial pressure (in psi)

    (b) $P(\bar{x} \leq 69) = P\left(z \leq \dfrac{69 - 70}{\frac{1.2}{\sqrt{15}}}\right) = P(z < -3.23) = 0.0006$

16. (a) $P(x < 36) = P\left(z < \dfrac{36 - 44}{5}\right) = P(z < -1.6) = 0.0548$

(b) $P(42 < x < 60) = P\left(\dfrac{42 - 44}{5} < z < \dfrac{60 - 44}{5}\right)$

$$= P(-0.40 < z < 3.2)$$

$$= 0.9993 - 0.3446 = 0.6547$$

(c) Top 5% $\Rightarrow z = 1.645$

$$x = \mu + z\sigma = 44 + (1.645)(5) = 52.2 \text{ months}$$

17. (a) $_{12}C_4 = 495$

(b) $\dfrac{(1)(1)(1)(1)}{_{12}C_4} = 0.0020$

18. $n = 20,\ p = 0.41$

(a) $P(8) = 0.1790$

(b) $P(x \geq 6) = 1 - P(x \leq 5) = 1 - 0.108 = 0.8921$

(c) $P(x \leq 13) = 1 - P(x \geq 14) = 1 - 0.0084 = 0.9916$

Confidence Intervals

6.1 Try It Yourself Solutions

1a. $\bar{x} \approx 14.8$

b. The mean number of sentences per magazine advertisement is 14.8.

2a. $Z_c = 1.96$, $n = 30$, $s \approx 16.5$

b. $E = Z_c \dfrac{s}{\sqrt{n}} = 1.96 \dfrac{16.5}{\sqrt{30}} \approx 5.9$

c. You are 95% confident that the maximum error of the estimate is about 5.9 sentences per magazine advertisement.

3a. $\bar{x} \approx 14.8$, $E \approx 5.9$

b. $\bar{x} - E = 14.8 - 5.9 = 8.9$
$\bar{x} + E = 14.8 + 5.9 = 20.7$

c. You are 95% confident that the mean number of sentences per magazine advertisements is between 8.9 and 20.7.

4b. 75% CI: (11.6, 13.2)
85% CI: (11.4, 13.4)
99% CI: (10.6, 14.2)

c. The width of the interval increases as the level of confidence increases.

5a. $n = 30$, $\bar{x} = 22.9$, $\sigma = 1.5$, $Z_c = 1.645$

b. $E = Z_c \dfrac{\sigma}{\sqrt{n}} = 1.645 \dfrac{1.5}{\sqrt{30}} \approx 0.451 \approx 0.5$

$\bar{x} - E = 22.9 - 0.451 \approx 22.4$
$\bar{x} + E = 22.9 + 0.451 \approx 23.4$

c. You are 90% confident that the mean age of the students is between 22.4 and 23.4 years.

6a. $Z_c = 1.96$, $E = 2$, $s \approx 5.0$

b. $n = \left(\dfrac{Z_c s}{E}\right)^2 = \left(\dfrac{1.96 \cdot 5.0}{2}\right)^2 = 24.01 \rightarrow 25$

c. You should have at least 25 magazine advertisements in your sample.

6.1 EXERCISE SOLUTIONS

1. You are more likely to be correct using an interval estimate because it is unlikely that a point estimate will equal the population mean exactly.

2. b

3. d; As the level of confidence increases, z_c increases therefore creating wider intervals.

4. b; As n increases, E decreases because $\sqrt{n}$ is in the denominator of the formula for E. Therefore, the intervals become narrower.

5. 1.28   **6.** 1.44   **7.** 1.15   **8.** 2.17

9. $\bar{x} - \mu = 3.8 - 4.27 = -0.47$

10. $\bar{x} - \mu = 9.5 - 8.76 = 0.74$

11. $\bar{x} - \mu = 26.43 - 24.67 = 1.76$

12. $\bar{x} - \mu = 46.56 - 48.12 = -1.56$

13. $\bar{x} - \mu = 0.7 - 1.3 = -0.60$

14. $\bar{x} - \mu = 86.4 - 80.9 = 5.5$

15. $E = z_c \dfrac{s}{\sqrt{n}} = 1.645 \dfrac{2.5}{\sqrt{36}} \approx 0.685$

16. $E = z_c \dfrac{s}{\sqrt{n}} = 1.96 \dfrac{3.0}{\sqrt{60}} \approx 0.759$

17. $E = z_c \dfrac{s}{\sqrt{n}} = 0.93 \dfrac{1.5}{\sqrt{50}} = 0.197$

18. $E = z_c \dfrac{s}{\sqrt{n}} = 2.24 \dfrac{4.6}{\sqrt{100}} = 1.030$

19. $c = 0.88 \Rightarrow z_c = 1.55$

$\bar{x} = 57.2,\ s = 7.1,\ n = 50$

$\bar{x} \pm z_c \dfrac{s}{\sqrt{n}} = 57.2 \pm 1.55 \dfrac{7.1}{\sqrt{50}} = 57.2 \pm 1.556 \approx (55.6, 58.8)$

Answer: (c)

20. $c = 0.90 \Rightarrow z_c = 1.645$

$\bar{x} = 57.2,\ s = 7.1,\ n = 50$

$\bar{x} \pm z_c \dfrac{s}{\sqrt{n}} = 57.2 \pm 1.645 \dfrac{7.1}{\sqrt{50}} = 57.2 \pm 1.652 = (55.5, 58.9)$

Answer: (d)

21. $c = 0.95 \Rightarrow z_c = 1.96$

$\bar{x} = 57.2,\ s = 7.1,\ n = 50$

$\bar{x} \pm z_c \dfrac{s}{\sqrt{n}} = 57.2 \pm 1.96 \dfrac{7.1}{\sqrt{50}} = 57.2 \pm 1.968 \approx (55.2, 59.2)$

Answer: (b)

22. $c = 0.98 \Rightarrow z_c = 2.33$

$\bar{x} = 57.2,\ s = 7.1,\ n = 50$

$\bar{x} \pm z_c \dfrac{s}{\sqrt{n}} = 57.2 \pm 2.33 \dfrac{7.1}{\sqrt{50}} = 57.2 \pm 2.340 = (54.9, 59.5)$

Answer: (a)

23. $\bar{x} \pm z_c \dfrac{s}{\sqrt{n}} = 15.2 \pm 1.645 \dfrac{2.0}{\sqrt{60}} = 15.2 \pm 0.425 \approx (14.8, 15.6)$

24. $\bar{x} \pm z_c \dfrac{s}{\sqrt{n}} = 31.39 \pm 1.96 \dfrac{0.8}{\sqrt{82}} = 31.3 \pm 0.173 \approx (31.22, 31.56)$

25. $\bar{x} \pm z_c \dfrac{s}{\sqrt{n}} = 4.27 \pm 1.96 \dfrac{0.3}{\sqrt{42}} = 4.27 \pm 0.091 \approx (4.18, 4.36)$

26. $\bar{x} \pm z_c \dfrac{s}{\sqrt{n}} = 13.5 \pm 2.575 \dfrac{1.5}{\sqrt{100}} = 13.5 \pm 0.386 \approx (13.1, 13.9)$

27. $(0.264, 0.494) \Rightarrow 0.379 \pm 0.115 \Rightarrow \bar{x} = 0.379, E = 0.115$

28. $(3.144, 3.176) \Rightarrow 3.16 \pm 0.016 \Rightarrow \bar{x} = 3.16, E = 0.016$

29. $(1.71, 2.05) \Rightarrow 1.88 \pm 0.17 \Rightarrow \bar{x} = 1.88, E = 0.17$

30. $(21.61, 30.15) \Rightarrow 25.88 \pm 4.27 \Rightarrow \bar{x} = 25.88, E = 4.27$

31. $c = 0.90 \Rightarrow z_c = 1.645$

$n = \left(\dfrac{z_c \sigma}{E}\right)^2 = \left(\dfrac{(1.645)(6.8)}{1}\right)^2 = 125.13 \Rightarrow 126$

32. $c = 0.95 \Rightarrow z_c = 1.96$

$n = \left(\dfrac{z_c \sigma}{E}\right)^2 = \left(\dfrac{(1.96)(2.5)}{1}\right)^2 = 24.01 \Rightarrow 25$

33. $c = 0.80 \Rightarrow z_c = 1.28$

$n = \left(\dfrac{z_c \sigma}{E}\right)^2 = \left(\dfrac{(1.28)(4.1)}{2}\right)^2 = 6.89 \Rightarrow 7$

34. $c = 0.98 \Rightarrow z_c = 2.33$

$n = \left(\dfrac{z_c \sigma}{E}\right)^2 = \left(\dfrac{(2.33)(10.1)}{2}\right)^2 = 138.45 \Rightarrow 139$

35. $(2.1, 3.5) \Rightarrow 2E = 3.5 - 2.1 = 1.4 \Rightarrow E = 0.7$ and $\bar{x} = 2.1 + E$
$= 2.1 + 0.7 = 2.8$

36. $(44.07)(80.97) \Rightarrow 2E = 80.97 - 44.07 = 36.9 \Rightarrow E = 18.45$ and $\bar{x} = 44.07 + E$
$= 44.07 + 18.45 = 62.52$

37. 90% CI: $\bar{x} \pm z_c \dfrac{s}{\sqrt{n}} = 630.90 \pm 1.645 \dfrac{56.70}{\sqrt{32}} = 630.9 \pm 16.49 \approx (614.41, 647.39)$

95% CI: $\bar{x} \pm z_c \dfrac{s}{\sqrt{n}} = 630.90 \pm 1.96 \dfrac{56.70}{\sqrt{32}} = 630.9 \pm 19.65 \approx (611.25, 650.55)$

The 95% confidence interval is wider.

38. 90% CI: $\bar{x} \pm z_c \dfrac{s}{\sqrt{n}} = 23.20 \pm 1.645 \dfrac{4.34}{\sqrt{35}} = 23.20 \pm 1.207 \approx (21.99, 24.41)$

95% CI: $\bar{x} \pm z_c \dfrac{s}{\sqrt{n}} = 23.20 \pm 1.96 \dfrac{4.34}{\sqrt{35}} = 23.20 \pm 1.438 \approx (21.76, 24.64)$

The 95% CI is wider.

39. 90% CI: $\bar{x} \pm z_c \dfrac{s}{\sqrt{n}} = 99.3 \pm 1.645 \dfrac{41.5}{\sqrt{31}} = 99.3 \pm 12.26 \approx (87.0, 111.6)$

   95% CI: $\bar{x} \pm z_c \dfrac{s}{\sqrt{n}} = 99.3 \pm 1.96 \dfrac{41.5}{\sqrt{31}} = 99.3 \pm 14.61 \approx (84.7, 113.9)$

   The 95% confidence interval is wider.

40. 90% CI: $\bar{x} \pm z_c \dfrac{s}{\sqrt{n}} = 23 \pm 1.645 \dfrac{6.7}{\sqrt{36}} = 23 \pm 1.837 \approx (21, 25)$

   95% CI: $\bar{x} \pm z_c \dfrac{s}{\sqrt{n}} = 23 \pm 1.96 \dfrac{6.7}{\sqrt{36}} = 23 \pm 2.189 \approx (21, 25)$

   The 90% CI and the 95% CI have the same width.

41. $\bar{x} \pm z_c \dfrac{s}{\sqrt{n}} = 120 \pm 1.96 \dfrac{17.50}{\sqrt{40}} = 120 \pm 5.423 \approx (114.58, 125.42)$

42. $\bar{x} \pm z_c \dfrac{s}{\sqrt{n}} = 150 \pm 2.575 \dfrac{15.5}{\sqrt{60}} = 150 \pm 5.15 \approx (144.85, 155.15)$

43. $\bar{x} \pm z_c \dfrac{s}{\sqrt{n}} = 120 \pm 1.96 \dfrac{17.50}{\sqrt{80}} = 120 \pm 3.8348 \approx (116.17, 123.83)$

   $n = 40$ CI is wider because a smaller sample was taken giving less information about the population.

44. $\bar{x} \pm z_c \dfrac{s}{\sqrt{n}} = 150 \pm 2.575 \dfrac{15.5}{\sqrt{40}} = 150 \pm 6.31 \approx (143.69, 156.31)$

   $n = 40$ CI is wider because a smaller sample was taken giving less information about the population.

45. $\bar{x} \pm z_c \dfrac{s}{\sqrt{n}} = 3.12 \pm 2.575 \dfrac{0.09}{\sqrt{48}} = 3.12 \pm 0.033 \approx (3.09, 3.15)$

46. $\bar{x} \pm z_c \dfrac{s}{\sqrt{n}} = 107.05 \pm 2.575 \dfrac{28.10}{\sqrt{61}} = 107.05 \pm 9.26 \approx (97.79, 116.31)$

47. $\bar{x} \pm z_c \dfrac{s}{\sqrt{n}} = 3.12 \pm 2.575 \dfrac{0.06}{\sqrt{48}} = 3.12 \pm 0.022 \approx (3.10, 3.14)$

   $s = 0.09$ CI is wider because of the increased variability within the sample.

48. $\bar{x} \pm z_c \dfrac{s}{\sqrt{n}} = 107.05 \pm 2.575 \dfrac{32.50}{\sqrt{61}} = 107.05 \pm 10.72 \approx (96.33, 117.77)$

   $s = 32.10$ CI is wider because of the increased variability within the population.

49. (a) An increase in the level of confidence will widen the confidence interval.

   (b) An increase in the sample size will narrow the confidence interval.

   (c) An increase in the standard deviation will widen the confidence interval.

50. Answers will vary.

51. $\bar{x} = \dfrac{\Sigma x}{n} = \dfrac{136}{15} \approx 9.1$

90% CI: $\bar{x} \pm z_c \dfrac{\sigma}{\sqrt{n}} = 9.1 \pm 1.645 \dfrac{1.5}{\sqrt{15}} = 9.1 \pm 0.637 \approx (8.4, 9.7)$

99% CI: $\bar{x} \pm z_c \dfrac{\sigma}{\sqrt{n}} = 9.1 \pm 2.575 \dfrac{1.5}{\sqrt{15}} = 9.1 \pm 0.997 \approx (8.1, 10.1)$

99% CI is wider.

52. 90% CI: $\bar{x} \pm z_c = \dfrac{\sigma}{\sqrt{n}} = 1.7 \pm 1.645 \dfrac{0.6}{\sqrt{21}} = 1.7 \pm 0.215 \approx (1.5, 1.9)$

99% CI: $\bar{x} \pm z_c = \dfrac{\sigma}{\sqrt{n}} = 1.7 \pm 2.575 \dfrac{0.6}{\sqrt{21}} = 1.7 \pm 0.337 \approx (1.4, 2.0)$

99% CI is wider.

53. $n = \left(\dfrac{z_c \sigma}{E}\right)^2 = \left(\dfrac{1.96 \cdot 4.8}{1}\right)^2 \approx 88.510 \rightarrow 89$

54. $n = \left(\dfrac{z_c \sigma}{E}\right)^2 = \left(\dfrac{2.575 \cdot 1.4}{2}\right)^2 \approx 3.249 \rightarrow 4$

55. (a) $n = \left(\dfrac{z_c \sigma}{E}\right)^2 = \left(\dfrac{1.96 \cdot 2.8}{0.5}\right)^2 \approx 120.473 \rightarrow 121$ servings

(b) $n = \left(\dfrac{z_c \sigma}{E}\right)^2 = \left(\dfrac{2.575 \cdot 2.8}{0.5}\right)^2 \approx 207.936 \rightarrow 208$ servings

99% CI requires larger sample because more information is needed from the population to be 99% confident.

56. (a) $n = \left(\dfrac{z_c \sigma}{E}\right)^2 = \left(\dfrac{1.645 \cdot 1.2}{1}\right)^2 \approx 3.897 \rightarrow 4$ students

(b) $n = \left(\dfrac{z_c \sigma}{E}\right)^2 = \left(\dfrac{2.575 \cdot 1.2}{1}\right)^2 \approx 9.548 \rightarrow 10$ students

99% CI requires larger sample because more information is needed from the population to be 99% confident.

57. (a) $n = \left(\dfrac{z_c \sigma}{E}\right)^2 = \left(\dfrac{1.645 \cdot 0.85}{0.25}\right)^2 \approx 31.282 \rightarrow 32$ cans

(b) $n = \left(\dfrac{z_c \sigma}{E}\right)^2 = \left(\dfrac{1.645 \cdot 0.85}{0.15}\right)^2 \approx 86.893 \rightarrow 87$ cans

$E = 0.15$ requires a larger sample size. As the error size decreases, a larger sample must be taken to obtain enough information from the population to ensure desired accuracy.

58. (a) $n = \left(\dfrac{z_c \sigma}{E}\right)^2 = \left(\dfrac{1.96 \cdot 3}{1}\right)^2 \approx 34.574 \rightarrow 35$ bottles

(b) $n = \left(\dfrac{z_c \sigma}{E}\right)^2 = \left(\dfrac{1.96 \cdot 3}{2}\right)^2 \approx 8.644 \rightarrow 9$ bottles

$E = 1$ requires a larger sample size. As the error size decreases, a larger sample must be taken to obtain enough information from the population to ensure desired accuracy.

59. $n = \left(\dfrac{z_c \sigma}{E}\right)^2 = \left(\dfrac{1.960 \cdot 0.25}{0.125}\right)^2 = 15.3664 \to 16$ sheets

$n = \left(\dfrac{z_c \sigma}{E}\right)^2 = \left(\dfrac{1.960 \cdot 0.25}{0.0625}\right)^2 = 61.4656 \to 62$ sheets

$E = 0.0625$ requires a larger sample size. As the error size decreases, a larger sample must be taken to obtain enough information from the population to ensure desired accuracy.

60. (a) $n = \left(\dfrac{z_c \sigma}{E}\right)^2 = \left(\dfrac{1.645 \cdot 0.15}{0.0425}\right)^2 \approx 33.708 \to 34$ units

(b) $n = \left(\dfrac{z_c \sigma}{E}\right)^2 = \left(\dfrac{1.645 \cdot 0.15}{0.02125}\right)^2 \approx 134.833 \to 135$ units

$E = 0.02125$ requires a larger sample size. As the error size decreases, a larger sample must be taken to obtain enough information from the population to ensure desired accuracy.

61. (a) $n = \left(\dfrac{z_c \sigma}{E}\right)^2 = \left(\dfrac{2.575 \cdot 0.25}{0.1}\right)^2 \approx 41.441 \to 42$ soccer balls

(b) $n = \left(\dfrac{z_c \sigma}{E}\right)^2 = \left(\dfrac{2.575 \cdot 0.30}{0.1}\right)^2 \approx 59.676 \to 60$ soccer balls

$\sigma = 0.30$ requires a larger sample size. Due to the increased variability in the population, a larger sample size is needed to ensure the desired accuracy.

62. (a) $n = \left(\dfrac{z_c \sigma}{E}\right)^2 = \left(\dfrac{2.575 \cdot 0.20}{0.15}\right)^2 \approx 11.788 \to 12$ soccer balls

(b) $n = \left(\dfrac{z_c \sigma}{E}\right)^2 = \left(\dfrac{2.575 \cdot 0.10}{0.15}\right)^2 \approx 2.947 \to 3$ soccer balls

$\sigma = 0.20$ requires a larger sample size. Due to the decreased variability in the population, a smaller sample is needed to ensure desired accuracy.

63. (a) An increase in the level of confidence will increase the minimum sample size required.

(b) An increase (larger E) in the error tolerance will decrease the minimum sample size required.

(c) An increase in the population standard deviation will increase the minimum sample size required.

64. A 99% CI may not be practical to use in all situations. It may produce a CI so wide that it has no practical application.

65. $\bar{x} = 238.77, s \approx 13.20, n = 31$

$\bar{x} \pm z_c \dfrac{s}{\sqrt{n}} = 238.77 \pm 1.96 \dfrac{13.20}{\sqrt{31}} = 238.77 \pm 4.65 \approx (234.1, 243.4)$

66. $\bar{x} = 99.08, s \approx 12.22, n = 37$

$\bar{x} \pm z_c \dfrac{s}{\sqrt{n}} = 99.08 \pm 1.96 \dfrac{12.22}{\sqrt{37}} = 99.08 \pm 3.94 \approx (95.1, 103.0)$

67. $\bar{x} = 15.783, s \approx 2.464, n = 30$

$\bar{x} \pm z_c \dfrac{s}{\sqrt{n}} = 15.783 \pm 1.96 \dfrac{2.464}{\sqrt{30}} = 15.783 \pm 0.88173 \approx (14.902, 16.665)$

68. $\bar{x} = 16.656, s \approx 4.235, n = 33$

$$\bar{x} \pm z_c \frac{s}{\sqrt{n}} = 16.656 \pm 1.96 \frac{4.235}{\sqrt{33}} = 16.656 \pm 1.445 \approx (15.211, 18.101)$$

69. (a) $\sqrt{\dfrac{N-n}{N-1}} = \sqrt{\dfrac{1000-500}{1000-1}} \approx 0.707$ (b) $\sqrt{\dfrac{N-n}{N-1}} = \sqrt{\dfrac{1000-100}{1000-1}} \approx 0.949$

  (c) $\sqrt{\dfrac{N-n}{N-1}} = \sqrt{\dfrac{1000-75}{1000-1}} \approx 0.962$ (d) $\sqrt{\dfrac{N-n}{N-1}} = \sqrt{\dfrac{1000-50}{1000-1}} \approx 0.975$

  (e) The finite population correction factor approaches 1 as the sample size decreases while the population size remains the same.

70. (a) $\sqrt{\dfrac{N-n}{N-1}} = \sqrt{\dfrac{100-50}{100-1}} \approx 0.711$ (b) $\sqrt{\dfrac{N-n}{N-1}} = \sqrt{\dfrac{400-50}{400-1}} \approx 0.937$

  (c) $\sqrt{\dfrac{N-n}{N-1}} = \sqrt{\dfrac{700-50}{700-1}} \approx 0.964$ (d) $\sqrt{\dfrac{N-n}{N-1}} = \sqrt{\dfrac{2000-50}{2000-1}} \approx 0.988$

  (e) The finite population correction factor approaches 1 as the population size increases while the sample size remains the same.

71. $E = \dfrac{z_c \sigma}{\sqrt{n}} \to \sqrt{n} = \dfrac{z_c \sigma}{E} \to n = \left(\dfrac{z_c \sigma}{E}\right)^2$

6.2 CONFIDENCE INTERVALS FOR THE MEAN (SMALL SAMPLES)

6.2 Try It Yourself Solutions

1a. d.f. $= n - 1 = 22 - 1 = 21$

b. $c = 0.90$

c. 1.721

2a. 90% CI: $t_c = 1.753$

$$E = t_c \frac{s}{\sqrt{n}} = 1.753 \frac{10}{\sqrt{16}} \approx 4.383 \approx 4.4$$

99% CI: $t_c = 2.947$

$$E = t_c \frac{s}{\sqrt{n}} = 2.947 \frac{10}{\sqrt{16}} \approx 7.368 \approx 7.4$$

b. 90% CI: $\bar{x} \pm E = 162 \pm 4.383 \approx (157.6, 166.4)$
   99% CI: $\bar{x} \pm E = 162 \pm 7.368 \approx (154.6, 169.4)$

c. You are 90% confident that the mean temperature of coffee sold is between 157.6° and 166.4°.
   You are 99% confident that the mean temperature of coffee sold is between 154.6° and 169.4°.

3a. 90% CI: $t_c = 1.729$

$$E = t_c \frac{s}{\sqrt{n}} = 1.729 \frac{0.42}{\sqrt{20}} \approx 0.162 \approx 0.16$$

95% CI: $t_c = 2.093$

$$E = t_c \frac{s}{\sqrt{n}} = 2.093 \frac{0.42}{\sqrt{20}} \approx 0.197 \approx 0.20$$

b. 90% CI: $\bar{x} \pm E = 6.22 \pm 0.162 \approx (6.06, 6.38)$

95% CI: $\bar{x} \pm E = 6.22 \pm 0.197 \approx (6.02, 6.42)$

c. You are 90% confident that the mean mortgage interest rate is contained between 6.06% and 6.38%.

You are 95% confident that the mean mortgage interest rate is contained between 6.02% and 6.42%.

4a. Is $n \geq 30$? No

Is the population normally distributed? Yes

Is σ known? No

Use the t-distribution to construct the 90% CI.

6.2 EXERCISE SOLUTIONS

1. 1.833   **2.** 2.201   **3.** 2.947   **4.** 2.539

5. $E = t_c \frac{s}{\sqrt{n}} = 2.131 \frac{5}{\sqrt{16}} \approx 2.664 \approx 2.7$ **6.** $E = t_c \frac{s}{\sqrt{n}} = 4.032 \frac{3}{\sqrt{6}} \approx 4.938 \approx 4.9$

7. $E = t_c \frac{s}{\sqrt{n}} = 1.796 \frac{2.4}{\sqrt{12}} = 1.244 \approx 1.2$ **8.** $E = t_c \frac{s}{\sqrt{n}} = 2.896 \frac{4.7}{\sqrt{9}} = 4.537 \approx 4.5$

9. (a) $\bar{x} \pm t_c \frac{s}{\sqrt{n}} = 12.5 \pm 2.015 \frac{2.0}{\sqrt{6}} = 12.5 \pm 1.645 \approx (10.9, 14.1)$

(b) $\bar{x} \pm t_c \frac{s}{\sqrt{n}} = 12.5 \pm 1.645 \frac{2.0}{\sqrt{6}} = 12.5 \pm 1.343 \approx (11.2, 13.8)$

t-CI is wider.

10. (a) $\bar{x} \pm t_c \frac{s}{\sqrt{n}} = 13.4 \pm 2.365 \frac{0.85}{\sqrt{8}} = 13.4 \pm 0.711 \approx (12.7, 14.1)$

(b) $\bar{x} \pm z_c \frac{s}{\sqrt{n}} = 13.4 \pm 1.96 \frac{0.85}{\sqrt{8}} = 13.4 \pm 0.589 \approx (12.8, 14.0)$

t-CI is wider.

11. (a) $\bar{x} \pm t_c \frac{s}{\sqrt{n}} = 4.3 \pm 2.650 \frac{0.34}{\sqrt{14}} = 4.3 \pm 0.241 \approx (4.1, 4.5)$

(b) $\bar{x} \pm z_c \frac{s}{\sqrt{n}} = 4.3 \pm 2.326 \frac{0.34}{\sqrt{14}} = 4.3 \pm 0.211 \approx (4.1, 4.5)$

Both CIs have the same width.

12. (a) $\bar{x} \pm t_c \dfrac{s}{\sqrt{n}} = 24.7 \pm 3.250 \dfrac{4.6}{\sqrt{10}} = 24.7 \pm 4.728 \approx (20.0, 29.4)$

  (b) $\bar{x} \pm z_c \dfrac{s}{\sqrt{n}} = 24.7 \pm 2.575 \dfrac{4.6}{\sqrt{10}} = 24.7 \pm 3.746 \approx (21.0, 28.4)$

  t-CI is wider.

13. $\bar{x} \pm t_c \dfrac{s}{\sqrt{n}} = 75 \pm 2.776 \dfrac{12.50}{\sqrt{5}} = 75 \pm 15.518 \approx (59.48, 90.52)$

  $E = t_c \dfrac{s}{\sqrt{n}} = 2.776 \dfrac{12.50}{\sqrt{5}} \approx 15.518$

14. $\bar{x} \pm t_c \dfrac{s}{\sqrt{n}} = 100 \pm 2.447 \dfrac{42.50}{\sqrt{7}} = 100 \pm 39.307 \approx (60.69, 139.31)$

  $E = t_c \dfrac{s}{\sqrt{n}} = 2.447 \dfrac{42.50}{\sqrt{7}} \approx 39.307 \approx 39.31$

15. $\bar{x} \pm z_c \dfrac{\sigma}{\sqrt{n}} = 75 \pm 1.96 \dfrac{15}{\sqrt{5}} = 75 \pm 13.148 \approx (61.85, 88.15)$

  $E = z_c \dfrac{\sigma}{\sqrt{n}} = 1.96 \dfrac{15}{\sqrt{5}} \approx 13.148 \approx 13.15$

  t-CI is wider.

16. $\bar{x} \pm t_c \dfrac{\sigma}{\sqrt{n}} = 100 \pm 1.96 \dfrac{50}{\sqrt{7}} = 100 \pm 37.041 \approx (62.96, 137.04)$

  $E = z_c \dfrac{\sigma}{\sqrt{n}} = 1.96 \dfrac{50}{\sqrt{7}} \approx 37.041 \approx 37.04$

  t-CI is wider.

17. (a) $\bar{x} \pm t_c \dfrac{s}{\sqrt{n}} = 4.54 \pm 1.833 \dfrac{1.21}{\sqrt{10}} = 4.54 \pm 0.701 \approx (3.84, 5.24)$

  (b) $\bar{x} \pm z_c \dfrac{s}{\sqrt{n}} = 4.54 \pm 1.645 \dfrac{1.21}{\sqrt{500}} = 4.54 \pm 0.089 \approx (4.45, 4.63)$

  t-CI is wider.

18. (a) $\bar{x} \pm t_c \dfrac{s}{\sqrt{n}} = 1.46 \pm 1.796 \dfrac{0.28}{\sqrt{12}} = 1.46 \pm 0.145 \approx (1.31, 1.61)$

  (b) $\bar{x} \pm z_c \dfrac{s}{\sqrt{n}} = 1.46 \pm 1.645 \dfrac{0.28}{\sqrt{600}} = 1.46 \pm 0.019 \approx (1.44, 1.48)$

  t-CI is wider.

19. (a) $\bar{x} = 4460.16$ (b) $s \approx 146.143$

  (c) $\bar{x} \pm t_c \dfrac{s}{\sqrt{n}} = 4460.16 \pm 3.250 \dfrac{146.143}{\sqrt{10}} = 4460.16 \pm 150.197 \approx (4309.96, 4610.36)$

20. (a) $\bar{x} \approx 5627.652$ (b) $s \approx 336.186$

  (c) $\bar{x} \pm t_c \dfrac{s}{\sqrt{n}} = 5627.652 \pm 3.012 \dfrac{336.186}{\sqrt{14}} = 5627.652 \pm 270.627 \approx (5357.025, 5898.279)$

21. (a) $\bar{x} \approx 1767.7$

    (b) $s \approx 252.23 \approx 252.2$

    (c) $\bar{x} \pm t_c \dfrac{s}{\sqrt{n}} = 1767.7 \pm 3.106 \dfrac{252.23}{\sqrt{12}} = 1767.7 \pm 226.16 \approx (1541.5, 1993.8)$

22. (a) $\bar{x} \approx 2.35$

    (b) $s \approx 1.03$

    (c) $\bar{x} \pm t_c \dfrac{s}{\sqrt{n}} = 2.35 \pm 2.977 \dfrac{1.03}{\sqrt{15}} = 2.35 \pm 0.793 \approx (1.56, 3.14)$

23. $n \geq 30 \rightarrow$ use normal distribution

$$\bar{x} \pm z_c \dfrac{s}{\sqrt{n}} = 1.25 \pm 1.96 \dfrac{0.05}{\sqrt{70}} = 1.25 \pm 0.012 \approx (1.24, 1.26)$$

24. $\bar{x} = 57.79, s = 19.05, n < 30, \sigma$ unknown, and pop normally distributed $\rightarrow$ use t-distribution

$$\bar{x} \pm t_c \dfrac{s}{\sqrt{n}} = 57.79 \pm 2.201 \dfrac{19.05}{\sqrt{12}} = 57.79 \pm 12.104 \approx (45.69, 69.89)$$

25. $\bar{x} = 21.9, s = 3.46, n < 30, \sigma$ known, and pop normally distributed $\rightarrow$ use t-distribution

$$\bar{x} \pm t_c \dfrac{s}{\sqrt{n}} = 21.9 \pm 2.064 \dfrac{3.46}{\sqrt{25}} = 21.9 \pm 1.43 \approx (20.5, 23.3)$$

26. $\bar{x} = 20.8, s \approx 4.5, n < 30, \sigma$ known, and pop normally distributed $\rightarrow$ use normal distribution

$$\bar{x} \pm z_c \dfrac{\sigma}{\sqrt{n}} = 20.8 \pm 1.96 \dfrac{4.7}{\sqrt{20}} = 20.8 \pm 2.06 \approx (18.7, 22.9)$$

27. $n < 30, \sigma$ unknown, and pop *not* normally distributed $\rightarrow$ cannot use either the normal or t-distributions.

28. $n < 30, \sigma$ unknown, and pop normally distributed $\rightarrow$ use t-distribution

$$\bar{x} \pm t_c \dfrac{s}{\sqrt{n}} = 144.19 \pm 2.110 \dfrac{61.32}{\sqrt{18}} = 144.19 \pm 30.496 \approx (113.69, 174.69)$$

29. $n = 25, \bar{x} = 56.0, s = 0.25$

    $\pm t_{0.99} \rightarrow 99\%$ t-CI

$$\bar{x} \pm t_c \dfrac{s}{\sqrt{n}} = 56.0 \pm 2.797 \dfrac{0.25}{\sqrt{25}} = 56.0 \pm 0.140 \approx (55.9, 56.1)$$

They are not making good tennis balls because desired bounce height of 55.5 inches is not contained between 55.9 and 56.1 inches.

30. $n = 16, \bar{x} = 1015, s = 25$

    $\pm t_{0.99} \rightarrow 99\%$ t-CI

$$\bar{x} \pm t_c \dfrac{s}{\sqrt{n}} = 1015 \pm 2.947 \dfrac{25}{\sqrt{16}} = 1015 \pm 18.419 \approx (997, 1033)$$

They are making good light bulbs because the desired bulb life of 1000 hours is contained between 997 and 1033 hours.

6.3 CONFIDENCE INTERVALS FOR POPULATION PROPORTIONS

6.3 Try It Yourself Solutions

1a. $x = 181, n = 1006$

b. $\hat{p} = \dfrac{181}{1006} \approx 0.180$

2a. $\hat{p} \approx 0.180, \hat{q} \approx 0.820$

b. $n\hat{p} = (1006)(0.180) = 181.08 > 5$

$n\hat{q} = (1006)(0.820) = 824.92 > 5$

c. $z_c = 1.645$

$$E = z_c \sqrt{\dfrac{\hat{p}\hat{q}}{n}} = 1.645 \sqrt{\dfrac{0.180 \cdot 0.820}{1006}} \approx 0.020$$

d. $\hat{p} \pm E = 0.180 \pm 0.020 \approx (0.160, 0.200)$

e. You are 90% confident that the proportion of adults say that Abraham Lincoln was the greatest president is contained between 16.0% and 20.0%.

3a. $n = 900, \hat{p} \approx 0.33$

b. $\hat{q} = 1 - \hat{p} = 1 - 0.33 \approx 0.67$

c. $n\hat{p} = 900 \cdot 0.33 \approx 297 > 5$

$n\hat{q} = 900 \cdot 0.67 \approx 603 > 5$

Distribution of $\hat{p}$ is approximately normal.

d. $z_c = 2.575$

e. $\hat{p} \pm z_c \sqrt{\dfrac{\hat{p}\hat{q}}{n}} = 0.33 \pm 2.575 \sqrt{\dfrac{0.33 \cdot 0.67}{900}} = 0.33 \pm 0.04 \approx (0.290, 0.370)$

f. You are 99% confident that the proportion of adults who think that people over 75 are more dangerous drivers is contained between 29.0% and 37.0%.

4a. (1) $\hat{p} = 0.50, \hat{q} = 0.50$

$z_c = 1.645, E = 0.02$

(2) $\hat{p} = 0.064, \hat{q} = 0.936$

$z_c = 1.645, E = 0.02$

b. (1) $n = \hat{p}\hat{q}\left(\dfrac{z_c}{E}\right)^2 = (0.50)(0.50)\left(\dfrac{1.645}{0.02}\right)^2 = 1691.266 \rightarrow 1692$

(2) $n = \hat{p}\hat{q}\left(\dfrac{z_c}{E}\right)^2 = 0.064 \cdot 0.936 \left(\dfrac{1.645}{0.02}\right)^2 \approx 405.25 \rightarrow 406$

c. (1) At least 1692 males should be included in the sample.

(2) At least 406 males should be included in the sample.

6.3 EXERCISE SOLUTIONS

1. False. To estimate the value of p, the population proportion of successes, use the point estimate $\hat{p} = \dfrac{x}{n}$.

2. True

3. $\hat{p} = \dfrac{x}{n} = \dfrac{752}{1002} \approx 0.750$

   $\hat{q} = 1 - \hat{p} \approx 0.250$

4. $\hat{p} = \dfrac{x}{n} = \dfrac{2439}{2939} \approx 0.830$

   $\hat{q} = 1 - \hat{p} \approx 0.170$

5. $\hat{p} = \dfrac{x}{n} = \dfrac{2938}{4431} \approx 0.663$

   $\hat{q} = 1 - \hat{p} \approx 0.337$

6. $\hat{p} = \dfrac{x}{n} = \dfrac{224}{458} \approx 0.489$

   $\hat{q} = 1 - \hat{p} \approx 0.511$

7. $\hat{p} = \dfrac{x}{n} = \dfrac{144}{848} \approx 0.170$

   $\hat{q} = 1 - \hat{p} \approx 0.830$

8. $\hat{p} = \dfrac{x}{n} = \dfrac{110}{1003} \approx 0.110$

   $\hat{q} = 1 - \hat{p} \approx 0.890$

9. $\hat{p} = \dfrac{x}{n} = \dfrac{204}{284} \approx 0.718$

   $\hat{q} = 1 - \hat{p} \approx 0.282$

10. $\hat{p} = \dfrac{x}{n} = \dfrac{662}{1003} \approx 0.660$

    $\hat{q} = 1 - \hat{p} = 0.340$

11. $\hat{p} = \dfrac{x}{n} = \dfrac{230}{1000} = 0.230$

    $q = 1 - \hat{p} = 1 - 0.230 = 0.770$

12. $\hat{p} = \dfrac{x}{n} = \dfrac{1515}{3224} = 0.470$

    $\hat{q} = 1 - \hat{p} = 1 - 0.470 = 0.530$

13. $\hat{p} = 0.48$, $E = 0.03$

    $\hat{p} \pm E = 0.48 \pm 0.03 = (0.45, 0.51)$

14. $\hat{p} = 0.15$, $E = 0.052$

    $\hat{p} \pm E = 0.51 \pm 0.052 = (0.458, 0.562)$

Note: Exercises 15–20 may have slightly different answers from the text due to rounding.

15. 95% CI: $\hat{p} \pm z_c \sqrt{\dfrac{\hat{p}\hat{q}}{n}} = 0.750 \pm 1.96 \sqrt{\dfrac{0.750 \cdot 0.25}{1002}} = 0.750 \pm 0.027 \approx (0.723, 0.777)$

    99% CI: $\hat{p} \pm z_c \sqrt{\dfrac{\hat{p}\hat{q}}{n}} = 0.750 \pm 2.575 \sqrt{\dfrac{0.750 \cdot 0.250}{1002}} = 0.75 \pm 0.035 \approx (0.715, 0.785)$

    99% CI is wider.

16. 95% CI: $\hat{p} \pm z_c \sqrt{\dfrac{\hat{p}\hat{q}}{n}} = 0.830 \pm 1.96 \sqrt{\dfrac{0.830 \cdot 0.170}{2939}} = 0.830 \pm 0.014 \approx (0.816, 0.845)$

    99% CI: $\hat{p} \pm z_c \sqrt{\dfrac{\hat{p}\hat{q}}{n}} = 0.830 \pm 2.575 \sqrt{\dfrac{0.830 \cdot 0.170}{2939}} = 0.830 \pm 0.018 \approx (0.812, 0.848)$

    99% CI is wider.

17. 95% CI: $\hat{p} \pm z_c \sqrt{\dfrac{\hat{p}\hat{q}}{n}} = 0.663 \pm 1.96 \sqrt{\dfrac{0.663 \cdot 0.337}{4431}} = 0.663 \pm 0.014 \approx (0.649, 0.677)$

    99% CI: $\hat{p} \pm z_c \sqrt{\dfrac{\hat{p}\hat{q}}{n}} = 0.663 \pm 2.575 \sqrt{\dfrac{0.663 \cdot 0.337}{4431}} = 0.663 \pm 0.018 \approx (0.645, 0.681)$

    99% CI is wider.

18. 95% CI: $\hat{p} \pm z_c \sqrt{\dfrac{\hat{p}\hat{q}}{n}} = 0.489 \pm 1.96 \sqrt{\dfrac{0.489 \cdot 0.511}{458}} = 0.489 \pm 0.046 \approx (0.433, 0.535)$

99% CI: $\hat{p} \pm z_c \sqrt{\dfrac{\hat{p}\hat{q}}{n}} = 0.489 \pm 2.575 \sqrt{\dfrac{0.489 \cdot 0.511}{458}} = 0.489 \pm 0.060 \approx (0.429, 0.549)$

99% CI is wider.

19. 95% CI: $\hat{p} \pm z_c \sqrt{\dfrac{\hat{p}\hat{q}}{n}} = 0.170 \pm 1.96 \sqrt{\dfrac{0.170 \cdot 0.830}{848}} = 0.170 \pm 0.025 \approx (0.145, 0.195)$

99% CI: $\hat{p} \pm z_c \sqrt{\dfrac{\hat{p}\hat{q}}{n}} = 0.170 \pm 2.575 \sqrt{\dfrac{0.170 \cdot 0.830}{848}} = 0.170 \pm 0.033 \approx (0.137, 0.203)$

99% CI is wider.

20. 95% CI: $\hat{p} \pm z_c \sqrt{\dfrac{\hat{p}\hat{q}}{n}} = 0.110 \pm 1.96 \sqrt{\dfrac{0.110 \cdot 0.890}{1003}} = 0.110 \pm 0.019 \approx (0.091, 0.129)$

99% CI: $\hat{p} \pm z_c \sqrt{\dfrac{\hat{p}\hat{q}}{n}} = 0.110 \pm 2.575 \sqrt{\dfrac{0.110 \cdot 0.890}{1003}} = 0.110 \pm 0.025 \approx (0.085, 0.135)$

99% CI is wider.

21. (a) $n = \hat{p}\hat{q}\left(\dfrac{z_c}{E}\right)^2 = 0.5 \cdot 0.5\left(\dfrac{1.96}{0.03}\right)^2 \approx 1067.111 \rightarrow 1068$ vacationers

(b) $n = \hat{p}\hat{q}\left(\dfrac{z_c}{E}\right)^2 = 0.26 \cdot 0.74\left(\dfrac{1.96}{0.03}\right)^2 \approx 821.249 \rightarrow 822$ vacationers

(c) Having an estimate of the proportion reduces the minimum sample size needed.

22. (a) $n = \hat{p}\hat{q}\left(\dfrac{z_c}{E}\right)^2 = 0.5 \cdot 0.5\left(\dfrac{2.33}{0.04}\right)^2 \approx 848.27 \rightarrow 849$ vacationers

(b) $n = \hat{p}\hat{q}\left(\dfrac{z_c}{E}\right)^2 = 0.3 \cdot 0.7\left(\dfrac{2.33}{0.04}\right)^2 \approx 712.54 \rightarrow 713$ vacationers

(c) Having an estimate of the proportion reduces the minimum sample size needed.

23. (a) $n = \hat{p}\hat{q}\left(\dfrac{z_c}{E}\right)^2 = 0.5 \cdot 0.5\left(\dfrac{2.05}{0.025}\right)^2 = 1681$ camcorders

(b) $n = \hat{p}\hat{q}\left(\dfrac{z_c}{E}\right)^2 = 0.25 \cdot 0.75\left(\dfrac{2.05}{0.025}\right)^2 \approx 1260.75 \rightarrow 1261$ camcorders

(c) Having an estimate of the proportion reduces the minimum sample size needed.

24. (a) $n = \hat{p}\hat{q}\left(\dfrac{z_c}{E}\right)^2 = 0.5 \cdot 0.5\left(\dfrac{2.17}{0.035}\right)^2 = 961$ computers

(b) $n = \hat{p}\hat{q}\left(\dfrac{z_c}{E}\right)^2 = 0.19 \cdot 0.81\left(\dfrac{2.17}{0.035}\right)^2 \approx 591.592 \rightarrow 592$ computers

(c) Having an estimate of the proportion reduces the minimum sample size needed.

25. (a) $\hat{p} = 0.36, n = 400$

$$\hat{p} \pm z_c \sqrt{\frac{\hat{p}\hat{q}}{n}} = 0.36 \pm 2.575 \sqrt{\frac{0.36 \cdot 0.64}{400}} = 0.36 \pm 0.062 \approx (0.298, 0.422)$$

(b) $\hat{p} = 0.32, n = 400$

$$\hat{p} \pm z_c \sqrt{\frac{\hat{p}\hat{q}}{n}} = 0.32 \pm 2.575 \sqrt{\frac{0.32 \cdot 0.68}{400}} = 0.32 \pm 0.060 \approx (0.260, 0.380)$$

It is possible that the two proportions are equal because the confidence intervals estimating the proportions overlap.

26. (a) $\hat{p} = 0.26, n = 400$

$$\hat{p} \pm z_c \sqrt{\frac{\hat{p}\hat{q}}{n}} = 0.26 \pm 2.575 \sqrt{\frac{0.26 \cdot 0.74}{400}} = 0.26 \pm 0.056 \approx (0.204, 0.316)$$

(b) $\hat{p} = 0.56, n = 400$

$$\hat{p} \pm z_c \sqrt{\frac{\hat{p}\hat{q}}{n}} = 0.56 \pm 2.575 \sqrt{\frac{0.56 \cdot 0.44}{400}} = 0.56 \pm 0.064 \approx (0.496, 0.624)$$

It is unlikely that the two proportions are equal because the confidence intervals estimating the proportions do not overlap.

27. (a) $\hat{p} = 0.65, \hat{q} = 0.35, n = 2563$

$$\hat{p} \pm z_c \sqrt{\frac{\hat{p}\hat{q}}{n}} = 0.65 \pm 2.575 \sqrt{\frac{(0.65)(0.35)}{2563}} = 0.65 \pm 0.024 = (0.626, 0.674)$$

(b) $\hat{p} = 0.88, \hat{q} = 0.12, n = 1125$

$$\hat{p} \pm z_c \sqrt{\frac{\hat{p}\hat{q}}{n}} = 0.88 \pm 2.575 \sqrt{\frac{(0.88)(0.12)}{1125}} = 0.88 \pm 0.025 = (0.855, 0.905)$$

(c) $\hat{p} = 0.92, \hat{q} = 0.08, n = 1086$

$$\hat{p} \pm z_c \sqrt{\frac{\hat{p}\hat{q}}{n}} = 0.92 \pm 2.575 \sqrt{\frac{(0.92)(0.08)}{1086}} = 0.92 \pm 0.021 = (0.899, 0.941)$$

28. (a) The two proportions are possibly unequal because the 2 CI's (0.626, 0.674) and (0.855, 0.905) do not overlap.

(b) The two proportions are possibly equal because the 2 CI's (0.855, 0.905) and (0.899, 0.941) overlap.

(c) The two proportions are possibly unequal because the 2 CI's (0.626, 0.674) and (0.899, 0.941) do not overlap.

29. $31.4\% \pm 1\% \rightarrow (30.4\%, 32.4\%) \rightarrow (0.304, 0.324)$

$$E = z_c \sqrt{\frac{\hat{p}\hat{q}}{n}} \rightarrow z_c = E \sqrt{\frac{n}{\hat{p}\hat{q}}} = 0.01 \sqrt{\frac{8451}{0.314 \cdot 0.686}} \approx 1.981 \rightarrow z_c = 1.98 \rightarrow c = 0.952$$

(30.4%, 32.4%) is approximately a 95.2% CI.

30. 27% ± 3% → (24%, 30%) → (0.24, 0.30)

$$E = z_c \sqrt{\frac{\hat{p}\hat{q}}{n}} \rightarrow z_c = E \sqrt{\frac{n}{\hat{p}\hat{q}}} = 0.03 \sqrt{\frac{1001}{0.27 \cdot 0.73}} \approx 2.138 \rightarrow z_c = 2.14 \rightarrow c = 0.968$$

(24%, 30%) is approximately a 96.8% CI.

31. If $n\hat{p} < 5$ or $n\hat{q} < 5$, the sampling distribution of $\hat{p}$ may not be normally distributed, therefore preventing the use of z_c when calculating the confidence interval.

32. $E = z_c \sqrt{\dfrac{\hat{p}\hat{q}}{n}} \rightarrow \dfrac{E}{z_c} = \sqrt{\dfrac{\hat{p}\hat{q}}{n}} \rightarrow \left(\dfrac{E}{z_c}\right)^2 = \dfrac{\hat{p}\hat{q}}{n} \rightarrow n = \hat{p}\hat{q}\left(\dfrac{z_c}{E}\right)^2$

33.

$\hat{p}$	$\hat{q} = 1 - \hat{p}$	$\hat{p}\hat{q}$	$\hat{p}$	$\hat{q} = 1 - \hat{p}$	$\hat{p}\hat{q}$
0.0	1.0	0.00	0.45	0.55	0.2475
0.1	0.9	0.09	0.46	0.54	0.2484
0.2	0.8	0.16	0.47	0.53	0.2491
0.3	0.7	0.21	0.48	0.52	0.2496
0.4	0.6	0.24	0.49	0.51	0.2499
0.5	0.5	0.25	0.50	0.50	0.2500
0.6	0.4	0.24	0.51	0.49	0.2499
0.7	0.3	0.21	0.52	0.48	0.2496
0.8	0.2	0.16	0.53	0.47	0.2491
0.9	0.1	0.09	0.54	0.46	0.2484
1.0	0.0	0.00	0.55	0.45	0.2475

$\hat{p} = 0.5$ give the maximum value of $\hat{p}\hat{q}$.

6.4 CONFIDENCE INTERVALS FOR VARIANCE AND STANDARD DEVIATION

6.4 Try It Yourself Solutions

1a. d.f. = $n - 1 = 24$

level of confidence = 0.95

b. Area to the right of χ_R^2 is 0.025.

Area to the left of χ_L^2 is 0.975.

c. $\chi_R^2 = 39.364$, $\chi_L^2 = 12.401$

2a. 90% CI: $\chi_R^2 = 42.557$, $\chi_L^2 = 17.708$

95% CI: $\chi_R^2 = 45.722$, $\chi_L^2 = 16.047$

b. 90% CI for σ^2: $\left(\dfrac{(n-1)s^2}{\chi_R^2}, \dfrac{(n-1)s^2}{\chi_L^2}\right) = \left(\dfrac{29 \cdot (1.2)^2}{42.557}, \dfrac{29 \cdot (1.2)^2}{17.708}\right) \approx (0.98, 2.36)$

95% CI for σ^2: $\left(\dfrac{(n-1)s^2}{\chi_R^2}, \dfrac{(n-1)s^2}{\chi_L^2}\right) = \left(\dfrac{29 \cdot (1.2)^2}{45.722}, \dfrac{29 \cdot (1.2)^2}{16.047}\right) \approx (0.91, 2.60)$

c. 90% CI for σ: $\left(\sqrt{0.981},\ \sqrt{2.358}\right) = (0.99, 1.54)$

95% CI for σ: $\left(\sqrt{0.913},\ \sqrt{2.602}\right) = (0.96, 1.61)$

d. You are 90% confident that the population variance is between 0.98 and 2.36, and that the population standard deviation is between 0.99 and 1.54. You are 95% confident that the population variance is between 0.91 and 2.60, and that the population standard deviation is between 0.96 and 1.61.

6.4 EXERCISE SOLUTIONS

1. $\chi_R^2 = 16.919,\ \chi_L^2 = 3.325$ **2.** $\chi_R^2 = 28.299,\ \chi_L^2 = 3.074$ **3.** $\chi_R^2 = 35.479,\ \chi_L^2 = 10.283$

4. $\chi_R^2 = 44.314,\ \chi_L^2 = 11.524$ **5.** $\chi_R^2 = 52.336,\ \chi_L^2 = 13.121$ **6.** $\chi_R^2 = 37.916,\ \chi_L^2 = 18.939$

7. (a) $s = 0.00843$

$$\left(\frac{(n-1)s^2}{\chi_R^2},\ \frac{(n-1)s^2}{\chi_L^2}\right) = \left(\frac{13 \cdot (0.00843)^2}{22.362},\ \frac{13 \cdot (0.00843)^2}{5.892}\right) \approx (0.0000413, 0.000157)$$

(b) $\left(\sqrt{0.0000413},\ \sqrt{0.000157}\right) \approx (0.00643, 0.0125)$

8. (a) $s = 0.0321$

$$\left(\frac{(n-1)s^2}{\chi_R^2},\ \frac{(n-1)s^2}{\chi_L^2}\right) \approx \left(\frac{14 \cdot (0.0321)^2}{23.685},\ \frac{14 \cdot (0.0321)^2}{6.571}\right) \approx (0.000610, 0.00220)$$

(b) $\left(\sqrt{0.000609},\ \sqrt{0.00220}\right) \approx (0.0247, 0.0469)$

9. (a) $s = 0.253$

$$\left(\frac{(n-1)s^2}{\chi_R^2},\ \frac{(n-1)s^2}{\chi_L^2}\right) = \left(\frac{17 \cdot (0.253)^2}{35.718},\ \frac{17 \cdot (0.253)^2}{5.697}\right) \approx (0.0305, 0.191)$$

(b) $\left(\sqrt{0.0305},\ \sqrt{0.191}\right) \approx (0.175, 0.437)$

10. (a) $s = 0.0918$

$$\left(\frac{(n-1)s^2}{\chi_R^2},\ \frac{(n-1)s^2}{\chi_L^2}\right) = \left(\frac{16 \cdot (0.0918)^2}{28.845},\ \frac{16 \cdot (0.0918)^2}{6.908}\right) \approx (0.00467, 0.0195)$$

(b) $\left(\sqrt{0.00467},\ \sqrt{0.0195}\right) \approx (0.0683, 0.140)$

11. (a) $\left(\frac{(n-1)s^2}{\chi_R^2},\ \frac{(n-1)s^2}{\chi_L^2}\right) = \left(\frac{11 \cdot (3.25)^2}{26.757},\ \frac{11 \cdot (3.25)^2}{2.603}\right) \approx (4.34, 44.64)$

(b) $\left(\sqrt{4.342},\ \sqrt{44.636}\right) \approx (2.08, 6.68)$

12. (a) $\left(\frac{(n-1)s^2}{\chi_R^2},\ \frac{(n-1)s^2}{\chi_L^2}\right) = \left(\frac{13 \cdot (123)^2}{24.736},\ \frac{13 \cdot (123)^2}{5.009}\right) \approx (7951, 39{,}265)$

(b) $\left(\sqrt{7951},\ \sqrt{39{,}265}\right) \approx (89, 198)$

13. (a) $\left(\frac{(n-1)s^2}{\chi_R^2},\ \frac{(n-1)s^2}{\chi_L^2}\right) = \left(\frac{9 \cdot (26)^2}{16.919},\ \frac{9 \cdot (26)^2}{3.325}\right) \approx (359.6, 1829.8)$

(b) $\left(\sqrt{359.596},\ \sqrt{1829.774}\right) \approx (19.0, 42.8)$

14. (a) $\left(\dfrac{(n-1)s^2}{\chi_R^2}, \dfrac{(n-1)s^2}{\chi_L^2}\right) = \left(\dfrac{15(6.42)^2}{27.488}, \dfrac{15(6.42)^2}{6.262}\right) = (22.5, 98.7)$

   (b) $\left(\sqrt{22.5}, \sqrt{98.7}\right) = (4.7, 9.9)$

15. (a) $\left(\dfrac{(n-1)s^2}{\chi_R^2}, \dfrac{(n-1)s^2}{\chi_L^2}\right) = \left(\dfrac{18 \cdot (15)^2}{31.526}, \dfrac{18 \cdot (15)^2}{8.231}\right) \approx (128, 492)$

   (b) $\left(\sqrt{128.465}, \sqrt{492.042}\right) \approx (11, 22)$

16. (a) $\left(\dfrac{(n-1)s^2}{\chi_R^2}, \dfrac{(n-1)s^2}{\chi_L^2}\right) = \left(\dfrac{29(3600)^2}{42.557}, \dfrac{29(3600)^2}{17.708}\right) = (8{,}831{,}450, 21{,}224{,}305)$

   (b) $\left(\sqrt{8{,}831{,}450}, \sqrt{21{,}224{,}305}\right) = (2972, 4607)$

17. (a) $\left(\dfrac{(n-1)s^2}{\chi_R^2}, \dfrac{(n-1)s^2}{\chi_L^2}\right) = \left(\dfrac{13 \cdot (342)^2}{24.736}, \dfrac{13 \cdot (342)^2}{5.009}\right) \approx (61{,}470, 303{,}559)$

   (b) $\left(\sqrt{61{,}470.41}, \sqrt{303{,}559.99}\right) \approx (248, 551)$

18. (a) $\left(\dfrac{(n-1)s^2}{\chi_R^2}, \dfrac{(n-1)s^2}{\chi_L^2}\right) = \left(\dfrac{29 \cdot (2.46)^2}{52.336}, \dfrac{29 \cdot (2.46)^2}{13.121}\right) \approx (3.35, 13.38)$

   (b) $\left(\sqrt{3.35}, \sqrt{13.38}\right) \approx (1.83, 3.66)$

19. (a) $\left(\dfrac{(n-1)s^2}{\chi_R^2}, \dfrac{(n-1)s^2}{\chi_L^2}\right) = \left(\dfrac{(21)(3.6)^2}{38.932}, \dfrac{(21)(3.6)^2}{8.897}\right) \approx (7.0, 30.6)$

   (b) $\left(\sqrt{6.99}, \sqrt{30.59}\right) = (2.6, 5.5)$

20. (a) $\left(\dfrac{(n-1)s^2}{\chi_R^2}, \dfrac{(n-1)s^2}{\chi_L^2}\right) = \left(\dfrac{19(3900)^2}{30.144}, \dfrac{19(3900)^2}{10.117}\right) \approx (9{,}586{,}982, 28{,}564{,}792)$

   (b) $\left(\sqrt{9{,}586{,}982}, \sqrt{28{,}564{,}792}\right) \approx (3096, 5345)$

21. 90% CI for σ: (0.00643, 0.0125)

   Yes, because the confidence interval is below 0.015.

22. 90% CI for σ: (0.0247, 0.0469) No, because the majority of the confidence interval is above 0.025.

CHAPTER 6 REVIEW EXERCISE SOLUTIONS

1. (a) $\bar{x} \approx 103.5$

   (b) $s \approx 34.663$

   $E = z_c \dfrac{s}{\sqrt{n}} = 1.645 \dfrac{34.663}{\sqrt{40}} \approx 9.0$

2. (a) $\bar{x} \approx 9.5$

   (b) $s \approx 7.1$

   $E = z_c \dfrac{s}{\sqrt{n}} = 1.645 \dfrac{7.1}{\sqrt{32}} \approx 2.1$

3. $\bar{x} \pm z_c \dfrac{s}{\sqrt{n}} = 10.3 \pm 1.96 \dfrac{0.277}{\sqrt{100}} = 10.3 \pm 0.054 \approx (10.2, 10.4)$

4. $\bar{x} \pm z_c \dfrac{s}{\sqrt{n}} = 0.0925 \pm 1.645 \dfrac{0.0013}{\sqrt{45}} = 0.0925 \pm 0.0003 \approx (0.0922, 0.0928)$

5. $s = 34.663$

$$n = \left(\frac{z_c \sigma}{E}\right)^2 = \left(\frac{1.96 \cdot 34.663}{10}\right)^2 \approx 46.158 \Rightarrow 47 \text{ people}$$

6. $n = \left(\dfrac{z_c \sigma}{E}\right)^2 = \left(\dfrac{2.575 \cdot 34.663}{2}\right)^2 \approx 1991.713 \Rightarrow 1992 \text{ people}$

7. $n = \left(\dfrac{z_c \sigma}{E}\right)^2 = \left(\dfrac{(1.96)(7.098)}{2}\right)^2 \approx 48.39 \Rightarrow 49 \text{ people}$

8. $n = \left(\dfrac{z_c \sigma}{E}\right)^2 = \left(\dfrac{(2.33)(7.098)}{0.5}\right)^2 \approx 1094.06 \Rightarrow 1095 \text{ people}$

9. $t_c = 2.365$ **10.** $t_c = 1.721$ **11.** $t_c = 2.624$ **12.** $t_c = 2.756$

13. $E = t_c \dfrac{s}{\sqrt{n}} = 1.753 \dfrac{25.6}{\sqrt{16}} \approx 11.2$ **14.** $E = t_c \dfrac{s}{\sqrt{n}} = 2.064 \dfrac{1.1}{\sqrt{25}} \approx 0.5$

15. $E = t_c \dfrac{s}{\sqrt{n}} = 2.718\left(\dfrac{0.9}{\sqrt{12}}\right) \approx 0.7$ **16.** $E = t_c \dfrac{s}{\sqrt{n}} = 2.861\left(\dfrac{16.5}{\sqrt{20}}\right) \approx 10.6$

17. $\bar{x} \pm z_c \dfrac{s}{\sqrt{n}} = 72.1 \pm 1.753 \dfrac{25.6}{\sqrt{16}} = 72.1 \pm 11.219 \approx (60.9, 83.3)$

18. $\bar{x} \pm t_c \dfrac{s}{\sqrt{n}} = 3.5 \pm 2.064 \dfrac{1.1}{\sqrt{25}} = 3.5 \pm 0.454 \approx (3, 4)$

19. $\bar{x} \pm t_c \dfrac{s}{\sqrt{n}} = 6.8 \pm 2.718\left(\dfrac{0.9}{\sqrt{12}}\right) = 6.8 \pm 0.706 = (6.1, 7.5)$

20. $\bar{x} \pm t_c \dfrac{s}{\sqrt{n}} = 25.2 \pm 2.861\left(\dfrac{6.5}{\sqrt{20}}\right) = 25.2 \pm 10.556 = (14.6, 35.8)$

21. $\bar{x} \pm t_c \dfrac{s}{\sqrt{n}} = 80 \pm 1.761 \dfrac{14}{\sqrt{15}} = 80 \pm 6.366 \approx (74, 86)$

22. $\bar{x} \pm t_c \dfrac{s}{\sqrt{n}} = 80 \pm 2.977 \dfrac{14}{\sqrt{15}} = 80 \pm 10.761 \approx (69, 91)$

23. $\hat{p} = \dfrac{x}{n} = \dfrac{560}{2000} = 0.28, \hat{q} = 0.72$ **24.** $\hat{p} = \dfrac{x}{n} = \dfrac{425}{500} = 0.85, \hat{q} = 0.15$

25. $\hat{p} = \dfrac{x}{n} = \dfrac{442}{2010} = 0.220, \hat{q} = 0.780$ **26.** $\hat{p} = \dfrac{x}{n} = \dfrac{90}{800} = 0.113, \hat{q} = 0.887$

27. $\hat{p} = \dfrac{x}{n} = \dfrac{116}{644} = 0.180, \hat{q} = 0.820$ **28.** $\hat{p} = \dfrac{x}{n} = \dfrac{594}{1007} = 0.590, \hat{q} = 0.410$

29. $\hat{p} = \dfrac{x}{n} = \dfrac{2021}{4813} = 0.420, \hat{q} = 0.580$ **30.** $\hat{p} = \dfrac{x}{n} = \dfrac{1230}{2365} = 0.520, \hat{q} = 0.480$

31. $\hat{p} \pm z_c \sqrt{\dfrac{\hat{p}\hat{q}}{n}} = 0.28 \pm 1.96 \sqrt{\dfrac{0.28 \cdot 0.72}{2000}} = 0.28 \pm 0.020 \approx (0.260, 0.300)$

32. $\hat{p} \pm z_c \sqrt{\dfrac{\hat{p}\hat{q}}{n}} = 0.85 \pm 2.575 \sqrt{\dfrac{0.85 \cdot 0.15}{500}} = 0.85 \pm 0.041 \approx (0.809, 0.891)$

33. $\hat{p} \pm z_c \sqrt{\dfrac{\hat{p}\hat{q}}{n}} = 0.220 \pm 1.645 \sqrt{\dfrac{0.220 \cdot 0.780}{2010}} = 0.220 \pm 0.015 \approx (0.205, 0.235)$

34. $\hat{p} \pm z_c \sqrt{\dfrac{\hat{p}\hat{q}}{n}} = 0.113 \pm 2.326 \sqrt{\dfrac{0.113 \cdot 0.887}{800}} = 0.113 \pm 0.026 \approx (0.087, 0.139)$

35. $\hat{p} \pm z_c \sqrt{\dfrac{\hat{p}\hat{q}}{n}} = 0.180 \pm 2.575 \sqrt{\dfrac{0.180 \cdot 0.820}{644}} = 0.180 \pm 0.039 = (0.141, 0.219)$

36. $\hat{p} \pm z_c \sqrt{\dfrac{\hat{p}\hat{q}}{n}} = 0.590 \pm 1.645 \sqrt{\dfrac{0.590 \cdot 0.410}{1007}} = 0.590 \pm 0.025 \approx (0.565, 0.615)$

37. $\hat{p} \pm z_c \sqrt{\dfrac{\hat{p}\hat{q}}{n}} = 0.420 \pm 1.96 \sqrt{\dfrac{(0.420)(0.580)}{4813}} = 0.420 \pm 0.014 = (0.406, 0.434)$

38. $\hat{p} \pm z_c \sqrt{\dfrac{\hat{p}\hat{q}}{n}} = 0.52 \pm 2.326 \sqrt{\dfrac{(0.52)(0.48)}{2365}} = 0.52 \pm 0.024 = (0.496, 0.544)$

39. (a) $n = \hat{p}\hat{q}\left(\dfrac{z_c}{E}\right)^2 = 0.50 \cdot 0.50 \left(\dfrac{1.96}{0.05}\right)^2 \approx 384.16 \rightarrow 385$ adults

(b) $n = \hat{p}\hat{q}\left(\dfrac{z_c}{E}\right)^2 = 0.63 \cdot 0.37 \left(\dfrac{1.96}{0.05}\right)^2 \approx 358.19 \rightarrow 359$ adults

(c) The minimum sample size needed is smaller when a preliminary estimate is available.

40. $n = \hat{p}\hat{q}\left(\dfrac{z_c}{E}\right)^2 = 0.63 \cdot 0.37 \left(\dfrac{2.575}{0.025}\right)^2 \approx 2472.96 \rightarrow 2473$ adults

The sample size is larger.

41. $\chi_R^2 = 23.337, \chi_L^2 = 4.404$ **42.** $\chi_R^2 = 42.980, \chi_L^2 = 10.856$

43. $\chi_R^2 = 14.067, \chi_L^2 = 2.167$ **44.** $\chi_R^2 = 23.589, \chi_L^2 = 1.735$

45. $s = 0.0727$

95% CI for σ^2: $\left(\dfrac{(n-1)s^2}{\chi_R^2}, \dfrac{(n-1)s^2}{\chi_L^2}\right) = \left(\dfrac{15 \cdot (0.0727)^2}{27.488}, \dfrac{15 \cdot (0.0727)^2}{6.262}\right) \approx (0.0029, 0.0127)$

95% CI for σ: $\left(\sqrt{0.00288}, \sqrt{0.01266}\right) \approx (0.0537, 0.1125)$

46. 99% CI for σ^2: $\left(\dfrac{(n-1)s^2}{\chi_R^2}, \dfrac{(n-1)s^2}{\chi_L^2}\right) = \left(\dfrac{15 \cdot (0.0727)^2}{32.801}, \dfrac{15 \cdot (0.0727)^2}{4.601}\right) \approx (0.0024, 0.0172)$

99% CI for σ: $\left(\sqrt{0.00242}, \sqrt{0.01723}\right) \approx (0.0492, 0.1313)$

47. $s = 1.125$

90% CI for σ^2: $\left(\dfrac{(n-1)s^2}{\chi_R^2}, \dfrac{(n-1)s^2}{\chi_L^2}\right) = \left(\dfrac{(23)(1.125)^2}{35.172}, \dfrac{(23)(1.125)^2}{13.091}\right) = (0.83, 2.22)$

90% CI for σ: $\left(\sqrt{0.83}, \sqrt{2.22}\right) \approx (0.91, 1.49)$

48. 95% CI for σ^2: $\left(\dfrac{(n-1)s^2}{\chi_R^2}, \dfrac{(n-1)s^2}{\chi_L^2}\right) = \left(\dfrac{(23)(1.125)^2}{38.076}, \dfrac{(23)(1.125)^2}{11.689}\right) = (0.76, 2.49)$

95% CI for σ: $\left(\sqrt{0.76}, \sqrt{2.49}\right) \approx (0.87, 0.158)$

CHAPTER 6 QUIZ SOLUTIONS

1. (a) $\bar{x} \approx 98.110$

(b) $s \approx 24.722$

$$E = t_c \frac{s}{\sqrt{n}} = 1.960 \frac{24.722}{\sqrt{30}} \approx 8.847$$

(c) $\bar{x} \pm t_c \frac{s}{\sqrt{n}} = 98.110 \pm 1.960 \frac{24.722}{\sqrt{30}} = 98.110 \pm 8.847 \approx (89.263, 106.957)$

You are 95% confident that the population mean repair costs is contained between $89.26 and $106.96.

2. $n = \left(\frac{z_c \sigma}{E}\right)^2 = \left(\frac{2.575 \cdot 22.50}{10}\right) \approx 33.568 \rightarrow 34$ dishwashers

3. (a) $\bar{x} = 12.96$

(b) $s \approx 1.35$

(c) $\bar{x} \pm t_c \frac{s}{\sqrt{n}} = 12.96 \pm 1.833 \frac{1.35}{\sqrt{10}} = 12.96 \pm 0.783 \approx (12.18, 13.74)$

(d) $\bar{x} \pm z_c \frac{\sigma}{\sqrt{n}} = 12.96 \pm 1.645 \frac{2.63}{\sqrt{10}} = 12.96 \pm 1.368 \approx (11.59, 14.33)$

The z-CI is wider because $\sigma = 2.63$ while $s = 1.35$.

4. $\bar{x} \pm t_c \frac{s}{\sqrt{n}} = 6824 \pm 2.447 \frac{340}{\sqrt{7}} = 6824 \pm 314.46 \approx (6510, 7138)$

5. (a) $\hat{p} = \frac{x}{n} = \frac{643}{1037} = 0.620$

(b) $\hat{p} \pm z_c \sqrt{\frac{\hat{p}\hat{q}}{n}} = 0.620 \pm 1.645 \sqrt{\frac{0.620 \cdot 0.38}{1037}} = 0.620 \pm 0.025 \approx (0.595, 0.645)$

(c) $n = \hat{p}\hat{q}\left(\frac{z_c}{E}\right)^2 = 0.620 \cdot 0.38\left(\frac{2.575}{0.04}\right)^2 \approx 976.36 \rightarrow 977$ adults

Note: The answer for Exercise 6 may differ slightly from the text answer due to rounding.

6. (a) $\left(\frac{(n-1)s^2}{\chi_R^2}, \frac{(n-1)s^2}{\chi_L^2}\right) = \left(\frac{29 \cdot (24.722)^2}{45.722}, \frac{29 \cdot (24.722)^2}{16.047}\right) \approx (387.650, 1104.514)$

(b) $\left(\sqrt{387.650}, \sqrt{1104.514}\right) \approx (19.689, 33.234)$

7.1 Try It Yourself Solutions

1a. (1) The mean . . . is not 74 months.
$$\mu \neq 74$$

(2) The variance . . . is less than or equal to 3.5.
$$\sigma^2 \leq 3.5$$

(3) The proportion . . . is greater than 39%.
$$p > 0.39$$

b. (1) $\mu = 74$ (2) $\sigma^2 > 3.5$ (3) $p \leq 0.39$

c. (1) $H_0: \mu = 74$; $H_a: \mu \neq 74$; (claim)

(2) $H_0: \sigma^2 \leq 3.5$ (claim); $H_a: \sigma^2 > 3.5$

(3) $H_0: p \leq 0.39$; $H_a: p > 0.39$ (claim)

2a. $H_0: p \leq 0.01$; $H_a: p > 0.01$

b. Type I error will occur if the actual proportion is less than or equal to 0.01, but you reject H_0.

Type II error will occur if the actual proportion is greater than 0.01, but you fail to reject H_0.

c. Type II error is more serious because you would be misleading the consumer, possibly causing serious injury or death.

3a. (1) $H_0: \mu = 74$; $H_a: \mu \neq 74$
(2) $H_0: p \leq 0.39$; $H_a: p > 0.39$

b. (1) Two-tailed   (2) Right-tailed

c. (1) 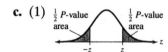   (2)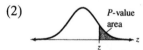

4a. There is enough evidence to support the radio station's claim.

b. There is not enough evidence to support the radio station's claim.

5a. (1) Support claim.   (2) Reject claim.

b. (1) $H_0: \mu \geq 650$; $H_a: \mu < 650$ (claim)

(2) $H_0: \mu = 98.6$ (claim); $H_a: \mu \neq 98.6$

7.1 EXERCISE SOLUTIONS

1. Null hypothesis (H_0) and alternative hypothesis (H_a). One represents the claim, the other, its complement.

2. Type I Error: The null hypothesis is rejected when it is true.
   Type II Error: The null hypothesis is not rejected when it is false.

3. False. In a hypothesis test, you assume the null hypothesis is true.

4. False. A statistical hypothesis is a statement about a population.

5. True

6. True

7. False. A small P-value in a test will favor a rejection of the null hypothesis.

8. False. If you want to support a claim, write it as your alternative hypothesis.

9. $H_0: \mu \le 645$ (claim); $H_a: \mu > 645$

10. $H_0: \mu \ge 128$; $H_a: \mu < 128$ (claim)

11. $H_0: \sigma = 5$; $H_a: \sigma \ne 5$ (claim)

12. $H_0: \sigma^2 \ge 1.2$ (claim); $H_a: \sigma^2 < 1.2$

13. $H_0: p \ge 0.45$; $H_a: p < 0.45$ (claim)

14. $H_0: p = 0.21$; $H_a: p \ne 0.21$

15. c, $H_0: \mu \le 3$

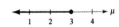

16. d, $H_a: \mu \ge 3$

17. b, $H_0: \mu = 3$

18. a, $H_a: \mu \le 2$

19. Right-tailed

20. Left-tailed

21. Two-tailed

22. Two-tailed

23. $\mu > 750$
    $H_0: \mu \le 750$; $H_a: \mu > 750$ (claim)

24. $\sigma < 3$
    $H_0: \sigma \ge 3$; $H_a: \sigma < 3$ (claim)

25. $\sigma \le 320$
    $H_0: \sigma \le 320$ (claim); $H_a: \sigma > 320$

26. $p = 0.28$
    $H_0: p = 0.28$ (claim); $H_a: p \ne 0.28$

27. $p = 0.81$
    $H_0: p = 0.81$ (claim); $H_a: p \ne 0.81$

28. $\mu < 45$
    $H_0: \mu \ge 45$; $H_a: \mu < 45$ (claim)

29. Type I: Rejecting $H_0: p \ge 0.60$ when actually $p \ge 0.60$.
    Type II: Not rejecting $H_0: p \ge 0.60$ when actually $p < 0.60$.

30. Type I: Rejecting $H_0: p = 0.05$ when actually $p = 0.05$.
    Type II: Not rejecting $H_0: p = 0.05$ when actually $p \ne 0.05$.

31. Type I: Rejecting $H_0: \sigma \le 12$ when actually $\sigma \le 12$.
    Type II: Not rejecting $H_0: \sigma \le 12$ when actually $\sigma > 12$.

32. Type I: Rejecting $H_0: p = 0.50$ when actually $p = 0.50$.
    Type II: Not rejecting $H_0: p = 0.50$ when actually $p \ne 0.50$.

33. Type I: Rejecting $H_0: p = 0.88$ when actually $p = 0.88$.
    Type II: Not rejecting $H_0: p = 0.88$ when actually $p \ne 0.88$.

34. Type I: Rejecting $p = 0.30$ when actually $p = 0.30$.

Type II: Not rejecting H_0: $p = 0.30$ when actually $p \neq 0.30$.

35. The null hypothesis is H_0: $p \geq 0.14$, the alternative hypothesis is H_a: $p < 0.14$.
Therefore, because the alternative hypothesis contains <, the test is a left-tailed test.

36. The null hypothesis is H_0: $\mu \leq 0.02$, the alternative hypothesis is H_a: $\mu > 0.02$.
Therefore, because the alternative hypothesis contains >, the test is a right-tailed test.

37. The null hypothesis is H_0: $p = 0.87$, the alternative hypothesis is H_a: $p \neq 0.87$.
Therefore, because the alternative hypothesis contains ≠, the test is a two-tailed test.

38. The null hypothesis is H_0: $\mu \geq 80,000$, the alternative hypothesis is H_a: $\mu < 80,000$.
Therefore, because the alternative hypothesis contains <, the test is a left-tailed test.

39. The null hypothesis is H_0: $p = 0.053$, the alternative hypothesis is H_a: $p \neq 0.053$.
Therefore, because the alternative hypothesis contains ≠, the test is a two-tailed test.

40. The null hypothesis is H_0: $\mu \leq 10$, the alternative hypothesis is H_a: $\mu > 10$.
Therefore, because the alternative hypothesis contains >, the test is a right-tailed test.

41. (a) There is enough evidence to support the company's claim.

(b) There is not enough evidence to support the company's claim.

42. (a) There is enough evidence to reject the government worker's claim.

(b) There is not enough evidence to reject the government worker's claim.

43. (a) There is enough evidence to support the Department of Labor's claim.

(b) There is not enough evidence to support the Department of Labor's claim.

44. (a) There is enough evidence to reject the manufacturer's claim.

(b) There is not enough evidence to reject the manufacturer's claim.

45. (a) There is enough evidence to support the manufacturer's claim.

(b) There is not enough evidence to support the manufacturer's claim.

46. (a) There is enough evidence to reject the soft-drink maker's claim.

(b) There is not enough evidence to reject the soft-drink maker's claim.

47. H_0: $\mu \geq 60$; H_a: $\mu < 60$

48. H_0: $\mu = 21$; H_a: $\mu \neq 21$

49. (a) H_0: $\mu \geq 15$; H_a: $\mu < 15$

(b) H_0: $\mu \leq 15$ H_a: $\mu > 15$

50. (a) H_0: $\mu \leq 28$; H_a: $\mu > 28$

(b) H_0: $\mu \geq 28$; H_a: $\mu < 28$

51. If you decrease α, you are decreasing the probability that you reject H_0. Therefore, you are increasing the probability of failing to reject H_0. This could increase β, the probability of failing to reject H_0 when H_0 is false.

52. If $\alpha = 0$, the null hypothesis cannot be rejected and the hypothesis test is useless.

53. (a) Fail to reject H_0 because the CI includes values greater than 70.

(b) Reject H_0 because the CI is located below 70.

(c) Fail to reject H_0 because the CI includes values greater than 70.

54. (a) Fail to reject H_0 because the CI includes values less than 54.

(b) Fail to reject H_0 because the CI includes values less than 54.

(c) Reject H_0 because the CI is located to the right of 54.

55. (a) Reject H_0 because the CI is located to the right of 0.20.

(b) Fail to reject H_0 because the CI includes values less than 0.20.

(c) Fail to reject H_0 because the CI includes values less than 0.20.

56. (a) Fail to reject H_0 because the CI includes values greater than 0.73.

(b) Reject H_0 because the CI is located to the left of 0.73.

(c) Fail to reject H_0 because the CI includes values greater than 0.73.

7.2 HYPOTHESIS TESTING FOR THE MEAN (LARGE SAMPLES)

7.2 Try It Yourself Solutions

1a. (1) $P = 0.0347 > 0.01 = \alpha$
 (2) $P = 0.0347 < 0.05 = \alpha$

b. (1) Fail to reject H_0 because $0.0347 > 0.01$.
 (2) Reject H_0 because $0.0347 < 0.05$.

2a.

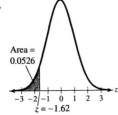

Area = 0.0526

$z = -1.62$

b. $P = 0.0526$

c. Fail to reject H_0 because $P = 0.0526 > 0.05 = \alpha$.

3a. Area that corresponds to $z = 2.31$ is 0.9896.

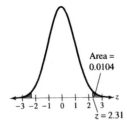

Area = 0.0104

$z = 2.31$

b. $P = 2 \text{ (area)} = 2(0.0104) = 0.0208$

c. Fail to reject H_0 because $P = 0.0208 > 0.01 = \alpha$.

4a. The claim is "the mean speed is greater than 35 miles per hour."

$H_0: \mu \le 35;\ H_a: \mu > 35$ (claim)

b. $\alpha = 0.05$

c. $z = \dfrac{\bar{x} - \mu}{\frac{s}{\sqrt{n}}} = \dfrac{36 - 35}{\frac{4}{\sqrt{100}}} = \dfrac{1}{0.4} = 2.500$

d. P-value = Area right of $z = 2.50 = 0.0062$

e. Reject H_0 because P-value $= 0.0062 < 0.05 = \alpha$.

f. Because you reject H_0, there is enough evidence to claim the average speed limit is greater than 35 miles per hour.

5a. The claim is "one of your distributors reports an average of 150 sales per day."

$H_0: \mu = 150$ (claim); $H_a: \mu \ne 150$

b. $\alpha = 0.01$

c. $z = \dfrac{\bar{x} - \mu}{\frac{\sigma}{\sqrt{n}}} = \dfrac{143 - 150}{\frac{15}{\sqrt{35}}} = \dfrac{-7}{2.535} \approx -2.76$

d. P-value $= 0.0058$

e. Reject H_0 because P-value $= 0.0058 < 0.01 = \alpha$.

f. There is enough evidence to reject the claim.

6a. $P = 0.0440 > 0.01 = \alpha$ **b.** Fail to reject H_0.

7a.

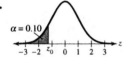

b. Area $= 0.1003$

c. $z_0 = -1.28$

d. $z < -1.28$

8a.

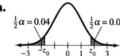

b. 0.0401 and 0.9599

c. $z_0 = -1.75$ and 1.75

d. $z < -1.75,\ z > 1.75$

9a. The claim is "the mean work day of the firm's accountants is less than 8.5 hours."

$H_0: \mu > 8.5;\ H_a: \mu < 8.5$ (claim)

b. $\alpha = 0.01$

c. $z_0 = -2.33$; Rejection region: $z < -2.33$

d. $z = \dfrac{x - \mu}{\frac{s}{\sqrt{n}}} = \dfrac{8.2 - 8.5}{\frac{0.5}{\sqrt{35}}} = \dfrac{-0.300}{0.0845} \approx -3.550$

e.

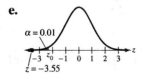

Reject H_0.

f. There is enough evidence to support the claim.

10a. $\alpha = 0.01$

   b. $\pm z_0 = \pm 2.575$; Rejection regions: $z < -2.575$, $z > 2.575$

   c. Fail to reject H_0.

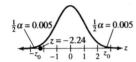

   d. There is not enough evidence to support the claim that the mean cost is significantly different from \$10,460 at the 1% level of significance.

7.2 EXERCISE SOLUTIONS

1.

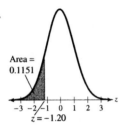

$P = 0.1151$; Fail to reject H_0 because $P = 0.1151 > 0.10 = \alpha$.

2.

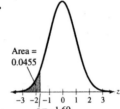

$P = 0.0455$; Reject H_0 because $P = 0.0455 < 0.05 = \alpha$.

3.

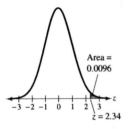

$P = 0.0096$; Reject H_0 because $P = 0.0096 < 0.01 = \alpha$.

4.

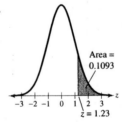

$P = 0.1093$; Fail to reject H_0 because $P = 0.1093 > 0.10 = \alpha$.

5.

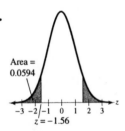

$P = 2(\text{Area}) = 2(0.0594) = 0.1188$; Fail to reject H_0 because $P = 0.1188 > 0.05 = \alpha$.

6.

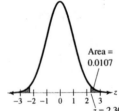

$P = 2(\text{Area}) = 2(0.0107) = 0.0214$; Fail to reject H_0 because $P = 0.0214 > 0.01 = \alpha$.

7. c   **8.** d   **9.** e   **10.** f   **11.** b   **12.** a

13. (a) Fail to reject H_0.

    (b) Reject H_0 ($P = 0.0461 < 0.05 = \alpha$).

14. (a) Fail to reject H_0 ($P = 0.0691 > 0.01 = \alpha$).

(b) Fail to reject H_0 ($P = 0.0691 > 0.05 = \alpha$).

15. 1.645   **16.** 1.41   **17.** -1.88

18. -1.34 **19.** ± 2.33 **20.** ± 1.645

21. Right-tailed ($\alpha = 0.01$)   **22.** Two-tailed ($\alpha = 0.05$)

23. Two-tailed ($\alpha = 0.10$)   **24.** Left-tailed ($\alpha = 0.05$)

25. (a) Fail to reject H_0 because $-1.645 < z < 1.645$.

(b) Reject H_0 because $z > 1.645$.

(c) Fail to reject H_0 because $-1.645 < z < 1.645$.

(d) Reject H_0 because $z < -1.645$.

26. (a) Reject H_0 because $z > 1.96$.

(b) Fail to reject H_0 because $-1.96 < z < 1.96$.

(c) Fail to reject H_0 because $-1.96 < z < 1.96$.

(d) Reject H_0 because $z < -1.96$.

27. (a) Fail to reject H_0 because $z < 1.285$.

(b) Fail to reject H_0 because $z < 1.285$.

(c) Fail to reject H_0 because $z < 1.285$.

(d) Reject H_0 because $z > 1.285$.

28. (a) Fail to reject H_0 because $-2.575 < z < 2.575$.

(b) Reject H_0 because $z < -2.575$.

(c) Reject H_0 because $z > 2.575$.

(d) Fail to reject H_0 because $-2.575 < z < 2.575$.

29. $H_0: \mu = 40; H_a: \mu \neq 40$

$\mu = 0.05 \rightarrow z_0 = \pm 1.96$

$$z = \frac{\bar{x} - \mu}{\frac{s}{\sqrt{n}}} = \frac{39.2 - 40}{\frac{3.23}{\sqrt{75}}} = \frac{-0.8}{0.373} \approx -2.145$$

Reject H_0. There is enough evidence to reject the claim.

30. $H_0: \mu \leq 1030$ and $H_a: \mu > 1030$

$\alpha = 0.05 \rightarrow z_0 = 1.645$

$$z = \frac{\bar{x} - \mu}{\frac{s}{\sqrt{n}}} = \frac{1035 - 1030}{\frac{23}{\sqrt{50}}} = \frac{5}{3.253} \approx 1.537$$

Fail to reject H_0. There is not enough evidence to support the claim.

31. $H_0: \mu = 6000; H_a: \mu \neq 6000$

$\alpha = 0.01 \rightarrow z_0 = \pm 2.575$

$z = \dfrac{\bar{x} - \mu}{\dfrac{s}{\sqrt{n}}} = \dfrac{5800 - 6000}{\dfrac{350}{\sqrt{35}}} = \dfrac{-200}{59.161} \approx -3.381$

Reject H_0. There is enough evidence to support the claim.

32. $H_0: \mu \leq 22{,}500$ and $H_a: \mu > 22{,}500$

$\alpha = 0.01 \rightarrow z_0 = 2.33$

$z = \dfrac{\bar{x} - \mu}{\dfrac{s}{\sqrt{n}}} = \dfrac{23{,}250 - 22{,}500}{\dfrac{1200}{\sqrt{45}}} = \dfrac{750}{178.885} \approx 4.193$

Reject H_0. There is enough evidence to reject the claim.

33. (a) $H_0: \mu \leq 275; H_a: \mu > 275$ (claim)

(b) $z = \dfrac{\bar{x} - \mu}{\dfrac{s}{\sqrt{n}}} = \dfrac{282 - 275}{\dfrac{35}{\sqrt{85}}} = \dfrac{7}{3796} \approx 1.84$ Area $= 0.9671$

(c) P-value $= \{$Area to right of $z = 1.84\} = 0.0329$

(d) Reject H_0. There is sufficient evidence at the 4% level of significance to support the claim that the mean score for Illinois' eighth grades is more than 275.

34. (a) $H_0: \mu \geq 135$ (claim); $H_a: \mu < 135$

(b) $z = \dfrac{\bar{x} - \mu}{\dfrac{s}{\sqrt{n}}} = \dfrac{133 - 135}{\dfrac{3.3}{\sqrt{32}}} = \dfrac{-2}{0.583} \approx -3.43$

Area $= 0.0003$

(c) P-value $= \{$Area to left of $z = -3.43\} = 0.0003$

(d) Reject H_0.

(e) There is sufficient evidence at the 10% level of significance to reject the claim that the average activating temperature is at least 135°F.

35. (a) $H_0: \mu \leq 8; H_a: \mu > 8$ (claim)

(b) $z = \dfrac{\bar{x} - \mu}{\dfrac{s}{\sqrt{n}}} = \dfrac{7.9 - 8}{\dfrac{2.67}{\sqrt{100}}} = \dfrac{-0.1}{0.267} \approx -0.37$ Area $= 0.3557$

(c) P-value $= \{$Area to right of $z = -0.37\} = 0.6443$

(d) Fail to reject H_0.

(e) There is insufficient evidence at the 7% level of significance to support the claim that the mean consumption of tea by a person in the United States is more than 8 gallons per year.

36. (a) $H_0: \mu = 3.1$ (claim); $H_a: \mu \neq 3.1$

(b) $z = \dfrac{\bar{x} - \mu}{\frac{s}{\sqrt{n}}} = \dfrac{2.9 - 3.1}{\frac{0.94}{\sqrt{60}}} = \dfrac{-0.2}{0.121} \approx -1.65$ Area $= 0.0495$

(c) P-value $= 2\{$Area to left of $z = -1.65\} = 2\{0.0495\} = 0.099$

(d) Fail to reject H_0.

(e) There is insufficient evidence at the 8% level to reject the claim that the mean tuna consumed by a person in the United States is 3.1 pounds per year.

37. (a) $H_0: \mu = 15$ (claim); $H_a: \mu \neq 15$

(b) $\bar{x} \approx 14.834$ $s \approx 4.288$

$z = \dfrac{\bar{x} - \mu}{\frac{s}{\sqrt{n}}} = \dfrac{14.834 - 15}{\frac{4.288}{\sqrt{32}}} = \dfrac{-0.166}{0.758} \approx -0.219$ Area $= 0.4129$

(c) P-value $= 2\{$Area to left of $z = -0.22\} = 2\{0.4129\} = 0.8258$

(d) Fail to reject H_0.

(e) There is insufficient evidence at the 5% level of significance to reject the claim that the mean time it takes smokers to quit smoking permanently is 15 years.

38. (a) $H_0: \mu \leq \$100,800$; $H_a: \mu > \$100,800$ (claim)

(b) $\bar{x} \approx 94,891.47$, $s \approx 5239.516$

$z = \dfrac{\bar{x} - \mu}{\frac{s}{\sqrt{n}}} = \dfrac{94,891.47 - 100,800}{\frac{5239.516}{\sqrt{34}}} = \dfrac{-5908.53}{898.570} \approx -6.58$ Area ≈ 0

(c) P-value $= \{$Area to right of $z = -6.58\} \approx 1$

(d) Fail to reject H_0.

(e) There is insufficient evidence at the 3% level to support the claim that the mean annual salary for engineering managers in Alabama is at least \$100,800.

39. (a) $H_0: \mu = 40$ (claim); $H_a: \mu \neq 40$

(b) $z_0 = \pm 2.575$;

Rejection regions: $z < -2.575$ and $z > 2.575$

(c) $z = \dfrac{\bar{x} - \mu}{\frac{s}{\sqrt{n}}} = \dfrac{39.2 - 40}{\frac{7.5}{\sqrt{30}}} = \dfrac{-0.8}{1.369} \approx -0.584$

(d) Fail to reject H_0.

(e) There is insufficient evidence at the 1% level of significance to reject the claim that the mean caffeine content per one 12-ounce bottle of cola is 40 milligrams.

40. (a) H_0: $\mu = 140$ (claim); H_a: $\mu \neq 140$

(b) $z_0 = \pm 1.96$; Rejection regions: $z < -1.96$ and $z > 1.96$

(c) $z = \dfrac{\bar{x} - \mu}{\dfrac{s}{\sqrt{n}}} = \dfrac{146 - 140}{\dfrac{22}{\sqrt{42}}} = \dfrac{6}{3.395} \approx 1.77$

(d) Fail to reject H_0.

(e) There is insufficient evidence at the 5% level to reject the claim that the mean caffeine content is 140 milligrams per 8 ounces.

41. (a) H_0: $\mu \geq 750$ (claim); H_a: $\mu < 750$

(b) $z_0 = -2.05$; Rejection region: $z < -2.05$

(c) $z = \dfrac{\bar{x} - \mu}{\dfrac{s}{\sqrt{n}}} = \dfrac{745 - 750}{\dfrac{60}{\sqrt{36}}} = \dfrac{-5}{10} \approx -0.500$

(d) Fail to reject H_0.

(e) There is insufficient evidence at the 2% level of significance to reject the claim that the mean life of the bulb is at least 750 hours.

42. (a) H_0: $\mu \leq 230$; H_a: $\mu > 230$ (claim)

(b) $z_0 = 1.75$; Rejection region: $z > 1.75$

(c) $z = \dfrac{\bar{x} - \mu}{\dfrac{s}{\sqrt{n}}} = \dfrac{232 - 230}{\dfrac{10}{\sqrt{52}}} = \dfrac{2}{1.387} \approx 1.44$

(d) Fail to reject H_0.

(e) There is insufficient evidence at the 4% level to support the claim that the mean sodium content per serving of cereal is greater than 230 milligrams.

43. (a) H_0: $\mu \leq 32$; H_a: $\mu > 32$ (claim)

(b) $z_0 = 1.55$; Rejection region: $z > 1.55$

(c) $\bar{x} \approx 29.676$ $s \approx 9.164$

$z = \dfrac{\bar{x} - \mu}{\dfrac{s}{\sqrt{n}}} = \dfrac{29.676 - 32}{\dfrac{9.164}{\sqrt{34}}} = \dfrac{-2.324}{1.572} \approx -1.478$

(d) Fail to reject H_0.

(e) There is insufficient evidence at the 6% level of significance to support the claim that the mean nitrogen dioxide level in Calgary is greater than 32 parts per billion.

44. (a) $H_0: \mu \geq 10{,}000$ (claim); $H_a: \mu < 10{,}000$

(b) $z_0 = -1.34$; Rejection region: $z < -1.34$

(c) $\bar{x} = 9580.9$, $s = 1722.4$

$$z = \frac{\bar{x} - \mu}{\frac{s}{\sqrt{n}}} = \frac{9580.9 - 10{,}000}{\frac{1722.4}{\sqrt{32}}} = \frac{-419.1}{304.5} = -1.38$$

(d) Reject H_0.

(e) There is sufficient evidence at the 9% level to reject the claim that the mean life of fluorescent lamps is at least 10,000 hours.

45. (a) $H_0: \mu \geq 10$ (claim); $H_a: \mu < 10$

(b) $z_0 = -1.88$; Rejection region: $z < -1.88$

(c) $\bar{x} \approx 9.780$, $s \approx 2.362$

$$x = \frac{\bar{x} - \mu}{\frac{s}{\sqrt{n}}} = \frac{9.780 - 10}{\frac{2.362}{\sqrt{30}}} = \frac{-0.22}{0.431} \approx -0.51$$

(d) Fail to reject H_0.

(e) There is insufficient evidence at the 3% level to reject the claim that the mean weight loss after 1 month is at least 10 pounds.

46. (a) $H_0: \mu \geq 60$; $H_a: \mu < 60$ (claim)

(b) $z_0 = -2.33$; Rejection region: $z < -2.33$

(c) $\bar{x} = 49$, $s = 21.51$

$$z = \frac{\bar{x} - \mu}{\frac{s}{\sqrt{n}}} = \frac{49 - 60}{\frac{21.51}{\sqrt{50}}} = \frac{-11}{3.042} = -3.62$$

(d) Fail to reject H_0.

(e) There is sufficient evidence at the 1% level to support the claim that the mean time it takes an employee to evacuate a building during a fire drill is less than 60 seconds.

47. $z = \frac{\bar{x} - \mu}{\frac{s}{\sqrt{n}}} = \frac{11{,}400 - 11{,}500}{\frac{320}{\sqrt{30}}} = \frac{-100}{58.424} \approx -1.71$

P-value = {Area left of $z = -1.71$} = 0.0436
Fail to reject H_0 because the standardized test statistic $z = -1.71$ is greater than the critical value $z_0 = -2.33$.

48. $z = \frac{\bar{x} - \mu}{\frac{s}{\sqrt{n}}} = \frac{22{,}200 - 22{,}000}{\frac{775}{\sqrt{36}}} = \frac{200}{129.2} = 1.548$

P-value = {Area right of $z = 1.55$} = 0.0606
Fail to reject H_0 because P-value = 0.0606 > 0.05 = α.

49. (a) $\alpha = 0.02$; Fail to reject H_0.

(b) $\alpha = 0.05$; Reject H_0.

(c) $z = \dfrac{\bar{x} - \mu}{\frac{s}{\sqrt{n}}} = \dfrac{11{,}400 - 11{,}500}{\frac{320}{\sqrt{50}}} = \dfrac{-100}{45.254} \approx -2.21$

P-value = {Area left of $z = -2.21$} = 0.0136 → Fail to reject H_0.

(d) $z = \dfrac{\bar{x} - \mu}{\frac{s}{\sqrt{n}}} = \dfrac{11{,}400 - 11{,}500}{\frac{320}{\sqrt{100}}} = \dfrac{-100}{32} \approx -3.13$

P-value = {Area left of $z = -3.13$} < 0.0009 → Reject H_0.

50. (a) $\alpha = 0.06$; Fail to reject H_0.

(b) $\alpha = 0.07$; Reject H_0.

(c) $z = \dfrac{\bar{x} - \mu}{\frac{s}{\sqrt{n}}} = \dfrac{22{,}200 - 20{,}000}{\frac{775}{\sqrt{40}}} = \dfrac{200}{122.5} = 1.63$

P-value = {Area right of $z = 1.63$} = 0.0516; Fail to reject H_0.

(d) $z = \dfrac{\bar{x} - \mu}{\frac{s}{\sqrt{n}}} = \dfrac{22{,}200 - 20{,}000}{\frac{775}{\sqrt{80}}} = \dfrac{200}{86.6} \approx 2.31$

P-value = {Area right of $z = 2.31$} = 0.0104; Reject H_0.

51. Using the classical z-test, the test statistic is compared to critical values. The z-test using a P-value compares the P-value to the level of significance α.

7.3 HYPOTHESIS TESTING FOR THE MEAN (SMALL SAMPLES)

7.3 Try It Yourself Solutions

1a. 2.650   **b.** $t_0 = -2.650$

2a. 1.860   **b.** $t_0 = +1.860$

3a. 2.947   **b.** $t_0 = \pm 2.947$

4a. The claim is "the mean cost of insuring a 2005 Honda Pilot LX is at least $1350."

$H_0: \mu \geq \$1350$ (claim); $H_a: \mu < \$1350$

b. $\alpha = 0.01$ and d.f. = $n - 1 = 8$

c. $t_0 = -2.896$; Reject H_0 if $t \leq -2.896$.

d. $t = \dfrac{\bar{x} - \mu}{\frac{s}{\sqrt{n}}} = \dfrac{1290 - 1350}{\frac{70}{\sqrt{9}}} = \dfrac{-60}{23.333} \approx -2.571$

e. Fail to reject H_0.

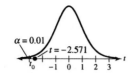

f. There is not enough evidence to reject the claim.

5a. The claim is "the mean conductivity of the river is 1890 milligrams per liter."

H_0: $\mu = 1890$ (claim); H_a: $\mu \neq 1890$

b. $\alpha = 0.01$ and d.f. $= n - 1 = 18$

c. $t_0 = \pm 2.878$; Reject H_0 if $t < -2.878$ or $t > 2.878$.

d. $t = \dfrac{\bar{x} - \mu}{\frac{s}{\sqrt{n}}} = \dfrac{2500 - 1890}{\frac{700}{\sqrt{19}}} = \dfrac{610}{160.591} \approx 3.798$

e. Reject H_0.

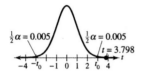

f. There is enough evidence to reject the company's claim.

6a. $t = \dfrac{\bar{x} - \mu}{\frac{s}{\sqrt{n}}} = \dfrac{172 - 185}{\frac{15}{\sqrt{6}}} = \dfrac{-13}{6.124} \approx -2.123$

P-value $= \{$Area left of $t = -2.123\} \approx 0.0436$

b. P-value $= 0.0436 < 0.05 = \alpha$

c. Reject H_0.

d. There is enough evidence to reject the claim.

7.3 EXERCISE SOLUTIONS

1. Identify the level of significance α and the degrees of freedom, d.f. $= n - 1$. Find the critical value(s) using the t-distribution table in the row with $n - 1$ d.f. If the hypothesis test is:

   (1) Left-tailed, use "One Tail, α" column with a negative sign.

   (2) Right-tailed, use "One Tail, α" column with a positive sign.

   (3) Two-tailed, use "Two Tail, α" column with a negative and a positive sign.

2. Identify the claim. State H_0 and H_a. Specify the level of significance. Identify the degrees of freedom and sketch the sampling distribution. Determine the critical value(s) and rejection region(s). Find the standardized test statistic. Make a decision and interpret it in the context of the original claim. The population must be normal or nearly normal.

3. $t_0 = 1.717$ 4. $t_0 = 2.764$ 5. $t_0 = -2.101$ 6. $t_0 = -1.771$

7. $t_0 = \pm 2.779$ 8. $t_0 = \pm 2.262$ 9. 1.328 10. 1.895

11. -2.473 12. -3.106 13. ± 3.747 14. ± 1.721

15. (a) Fail to reject H_0 because $t > -2.086$.

    (b) Fail to reject H_0 because $t > -2.086$.

    (c) Fail to reject H_0 because $t > -2.086$.

    (d) Reject H_0 because $t < -2.086$.

16. (a) Fail to reject H_0 because $-1.372 < t < 1.372$.

    (b) Reject H_0 because $t < -1.372$.

    (c) Reject H_0 because $t > 1.372$.

    (d) Fail to reject H_0 because $-1.372 < t < 1.372$.

17. (a) Fail to reject H_0 because $-2.602 < t < 2.602$.

    (b) Fail to reject H_0 because $-2.602 < t < 2.602$.

    (c) Reject H_0 because $t > 2.602$.

    (d) Reject H_0 because $t < -2.602$.

18. (a) Fail to reject H_0 because $-1.725 < t < 1.725$.

    (b) Reject H_0 because $t < -1.725$.

    (c) Fail to reject H_0 because $-1.725 < t < 1.725$.

    (d) Reject H_0 because $t > 1.725$.

19. $H_0: \mu = 15$ (claim); $H_a: \mu \neq 15$

$\alpha = 0.01$ and d.f. $= n - 1 = 5$

$t_0 = \pm 4.032$

$t = \dfrac{\bar{x} - \mu}{\frac{s}{\sqrt{n}}} = \dfrac{13.9 - 15}{\frac{3.23}{\sqrt{6}}} = \dfrac{-1.1}{1.319} \approx -0.834$

Fail to reject H_0. There is not enough evidence to reject the claim.

20. $H_0: \mu \leq 25$; $H_a: \mu > 25$ (claim)

$\alpha = 0.05$ and d.f. $= n - 1 = 16$

$t_0 = 1.746$

$t = \dfrac{\bar{x} - \mu}{\frac{s}{\sqrt{n}}} = \dfrac{26.2 - 25}{\frac{2.32}{\sqrt{17}}} = \dfrac{1.2}{0.563} \approx 2.133$

Reject H_0. There is enough evidence to support the claim.

21. H_0: $\mu \geq 8000$ (claim); H_a: $\mu < 8000$

$\alpha = 0.01$ and d.f. $= n - 1 = 24$

$t_0 = -2.492$

$t = \dfrac{\bar{x} - \mu}{\dfrac{s}{\sqrt{n}}} = \dfrac{7700 - 8000}{\dfrac{450}{\sqrt{25}}} = \dfrac{-300}{90} \approx -3.333$

Reject H_0. There is enough evidence to reject the claim.

22. H_0: $\mu = 52{,}200$; H_a: $\mu \neq 52{,}200$ (claim)

$\alpha = 0.05$ and d.f. $= n - 1 = 3$

$t_0 = \pm 3.182$

$t = \dfrac{\bar{x} - \mu}{\dfrac{s}{\sqrt{n}}} = \dfrac{53{,}220 - 52{,}200}{\dfrac{1200}{\sqrt{4}}} = \dfrac{1020}{600} = 1.7$

Fail to reject H_0. There is not enough evidence to support the claim.

23. (a) H_0: $\mu > 100$; H_a: $\mu < 100$ (claim)

(b) $t_0 = -3.747$; Reject H_0 if $t < -3.747$.

(c) $t = \dfrac{\bar{x} - \mu}{\dfrac{s}{\sqrt{n}}} = \dfrac{75 - 100}{\dfrac{12.50}{\sqrt{5}}} = \dfrac{-25}{5.590} \approx -4.472$

(d) Reject H_0.

(e) There is sufficient evidence at the 1% significance level to support the claim that the mean repair cost for damaged microwave ovens is less than $100.

24. (a) H_0: $\mu \leq \$95$; H_a: $\mu > \$95$ (claim)

(b) $t_0 = 3.143$; Reject H_0 if $t > 3.143$.

(c) $t = \dfrac{\bar{x} - \mu}{\dfrac{s}{\sqrt{n}}} = \dfrac{100 - 95}{\dfrac{42.50}{\sqrt{7}}} = \dfrac{5}{16.0635} \approx 0.311$

(d) Fail to reject H_0.

(e) There is not enough evidence at the 1% significance level to support the claim that the mean repair cost for damaged computers is more than $95.

25. (a) H_0: $\mu \leq 1$; H_a: $\mu > 1$ (claim)

(b) $t_0 = 1.796$; Reject H_0 if $t > 1.796$.

(c) $t = \dfrac{\bar{x} - \mu}{\dfrac{s}{\sqrt{n}}} = \dfrac{1.46 - 1}{\dfrac{0.28}{\sqrt{12}}} = \dfrac{0.46}{0.081} \approx 5.691$

(d) Reject H_0.

(e) There is sufficient evidence at the 5% significance level to support the claim that the mean waste recycled by adults in the United States is more than 1 pound per person per day.

26. (a) H_0: $\mu \leq 4$; H_a: $\mu > 4$ (claim)

(b) $t_0 = 1.833$; Reject H_0 if $t > 1.833$.

(c) $t = \dfrac{\bar{x} - \mu}{\dfrac{s}{\sqrt{n}}} = \dfrac{4.54 - 4}{\dfrac{1.21}{\sqrt{10}}} = \dfrac{0.54}{0.383} = 1.411$

(d) Fail to reject H_0.

(e) There is not enough evidence at the 5% significance level to support the claim that the mean waste generated by adults in the U.S. is more than 4 pounds per day.

27. (a) H_0: $\mu = \$25,000$ (claim); H_a: $\mu \neq \$25,000$

(b) $t_0 = \pm 2.262$; Reject H_0 if $t < -2.262$ or $t > 2.262$.

(c) $\bar{x} \approx 25,852.2$ $s \approx \$3197.1$

$t = \dfrac{\bar{x} - \mu}{\dfrac{s}{\sqrt{n}}} = \dfrac{25,852.2 - 25,000}{\dfrac{3197.1}{\sqrt{10}}} = \dfrac{-852.2}{1011.0} \approx 0.843$

(d) Fail to reject H_0.

(e) There is insufficient evidence at the 5% significance level to reject the claim that the mean salary for full-time male workers over age 25 without a high school diploma is $25,000.

28. (a) H_0: $\mu = \$19,100$ (claim); H_a: $\mu \neq \$19,100$

(b) $t_0 = \pm 2.201$; Reject H_0 if $t < -2.201$ or $t > 2.201$.

(c) $\bar{x} \approx \$18,886.5$, $s \approx \$1397.4$

$t = \dfrac{\bar{x} - \mu}{\dfrac{s}{\sqrt{n}}} = \dfrac{18,886.5 - 19,100}{\dfrac{1397.4}{\sqrt{12}}} = \dfrac{-213.5}{403.4} \approx -0.529$

(d) Fail to reject H_0.

(e) There is not enough evidence at the 5% significance level to reject the claim that the mean annual pay for full-time female workers over age 25 without high school diplomas is $19,100.

29. (a) H_0: $\mu \geq 3.0$; H_a: $\mu < 3.0$ (claim)

(b) $\bar{x} = 1.925$ $x = 0.654$

$t = \dfrac{\bar{x} - \mu}{\dfrac{s}{\sqrt{n}}} = \dfrac{1.925 - 3.0}{\dfrac{0.654}{\sqrt{20}}} = \dfrac{-1.075}{0.146} \approx -7.351$

P-value = {Area left of $t = -7.351$} ≈ 0

(c) Reject H_0.

(e) There is sufficient evidence at the 5% significance level to support the claim that teenage males drink fewer than three 12-ounce servings of soda per day.

30. (a) H_0: $\mu \leq \$550$; H_a: $\mu > \$550$ (claim)

(b) $\bar{x} = 605$, $s = 150.8$

$$t = \frac{\bar{x} - \mu}{\frac{s}{\sqrt{n}}} = \frac{605 - 550}{\frac{150.8}{\sqrt{24}}} = \frac{55}{30.782} \approx 1.787$$

P-value = {Area right of $t = 1.787$} = 0.0436

(c) Reject H_0.

(d) There is sufficient evidence at the 5% significance level to support the claim that teachers spend a mean of more than \$550 of their own money on school supplies in a year.

31. (a) H_0: $\mu \geq 32$; H_a: $\mu < 32$ (claim)

(b) $\bar{x} = 30.167$ $s = 4.004$

$$t = \frac{\bar{x} - \mu}{\frac{s}{\sqrt{n}}} = \frac{30.167 - 32}{\frac{4.004}{\sqrt{18}}} = \frac{-1.833}{0.944} \approx -1.942$$

P-value = {Area left of $t = -1.942$} ≈ 0.0344

(c) Fail to reject H_0.

(e) There is insufficient evidence at the 1% significance level to support the claim that the mean class size for full-time faculty is fewer than 32.

32. (a) H_0: $\mu = 11.0$ (claim); H_a: $\mu \neq 11.0$

(b) $\bar{x} = 10.050$, $s = 2.485$

$$t = \frac{\bar{x} - \mu}{\frac{s}{\sqrt{n}}} = \frac{10.050 - 11.0}{\frac{2.485}{\sqrt{8}}} = \frac{-.95}{0.879} \approx -1.081$$

P-value = 2{area left of $t = -1.081$} = 2(0.15775) = 0.3155

(c) Fail to reject H_0.

(d) There is not enough evidence at the 1% significance level to reject the claim that the mean number of classroom hours per week for full-time faculty is 11.0.

33. (a) H_0: $\mu = \$2634$ (claim); H_a: $\mu \neq \$2634$

(b) $\bar{x} = \$2785.6$ $s = \$759.3$

$$t = \frac{\bar{x} - \mu}{\frac{s}{\sqrt{n}}} = \frac{2785.6 - 2634}{\frac{759.3}{\sqrt{12}}} = \frac{151.6}{219.19} = 0.692$$

P-value = 2{Area right of $t = 0.692$} = 2{0.2518} = 0.5036

(c) Fail to reject H_0.

(e) There is insufficient evidence at the 2% significance level to reject the claim that the typical household in the U.S. spends a mean amount of \$2634 per year on food away from home.

34. (a) H_0: $\mu = \$152$ (claim); H_a: $\mu \neq \$152$

(b) $\bar{x} = \$142.8$, $s = \$37.52$

$$t = \frac{\bar{x} - \mu}{\frac{s}{\sqrt{n}}} = \frac{142.8 - 152}{\frac{37.52}{\sqrt{10}}} = \frac{-9.2}{11.865} = -0.775$$

P-value $= 2\{$Area left of $t = -0.78\}$
$\quad\quad\quad = 2\{0.229\}$
$\quad\quad\quad = 0.4580$

(c) Fail to reject H_0.

(d) There is insufficient evidence at the 2% significance level to reject the claim that the daily lodging costs for a family in the U.S. is $152.

35. H_0: $\mu \leq \$2328$; H_a: $\mu > 2328$ (claim)

$$t = \frac{\bar{x} - \mu}{\frac{s}{\sqrt{n}}} = \frac{2528 - 2328}{\frac{325}{\sqrt{6}}} = \frac{200}{132.681} \approx 1.507$$

P-value $= \{$Area right of t $= 1.507\} \approx 0.096$

Because $0.096 > 0.01 = \alpha$, fail to reject H_0.

36. (a) Because $0.096 > 0.05 = \alpha$, fail to reject H_0.

(b) Because $0.096 < 0.10 = \alpha$, reject H_0.

(c) $$t = \frac{\bar{x} - \mu}{\frac{s}{\sqrt{n}}} = \frac{2528 - 2328}{\frac{325}{\sqrt{12}}} = \frac{200}{93.819} = 2.132$$

P-value $= \{$Area right of $t = 2.132\} \approx 0.028$

Because $0.028 > 0.01 = \alpha$, fail to reject H_0.

(d) $$t = \frac{\bar{x} - \mu}{\frac{s}{\sqrt{n}}} = \frac{2528 - 2328}{\frac{325}{\sqrt{24}}} = \frac{200}{66.340} = 3.015$$

P-value $= \{$Area right of $t = 3.015\} \approx 0.003$

Since $0.003 < 0.01 = \alpha$, reject H_0.

37. Because σ is unknown, $n < 30$, and the gas mileage is normally distributed, use the t-distribution.

H_0: $\mu \geq 23$ (claim); H_a: $\mu < 23$

$$t = \frac{\bar{x} - \mu}{\frac{s}{\sqrt{n}}} = \frac{22 - 23}{\frac{4}{\sqrt{5}}} = \frac{-1}{1.789} \approx -0.559$$

P-value $= \{$Area left of $t = -0.559\} = 0.303$

Fail to reject H_0. There is insufficient evidence at the 5% significance level to reject the claim that the mean gas mileage for the luxury sedan is at least 23 miles per gallon.

38. Because σ is unknown and $n \geq 30$, use the z-distribution.

$H_0: \mu \geq 23,000; H_a: \mu < 23,000$ (claim)

$$z = \frac{\bar{x} - \mu}{\frac{s}{\sqrt{n}}} = \frac{21,856 - 23,000}{\frac{3163}{\sqrt{50}}} = \frac{-1144}{447.32} \approx -2.557$$

P-value $= 2\{\text{Area left of } z = -2.557\} = 2(0.0026) = 0.0052$

Reject H_0. There is enough evidence at the 1% significance level to reject the claim that the mean price for 1 year of graduate school for a full-time student in a master's degree program at a public institution is less than $23,000.

7.4 HYPOTHESIS TESTING FOR PROPORTIONS

7.4 Try It Yourself Solutions

1a. $np = (86)(0.30) = 25.8 > 5, nq = (86)(0.70) = 60.2 > 5$

b. The claim is "less than 30% of cellular phone users whose phone can connect to the Internet have done so while at home."

$H_0: p \geq 0.30; H_a: p < 0.30$ (claim)

c. $\alpha = 0.05$

d. $z_0 = -1.645$; Reject H_0 if $z < -1.645$.

e. $z = \dfrac{\hat{p} - p}{\sqrt{\dfrac{pq}{n}}} = \dfrac{0.20 - 0.30}{\sqrt{\dfrac{(0.30)(0.70)}{86}}} = \dfrac{-0.1}{0.0494} \approx -2.024$

f. Reject H_0.

g. There is enough evidence to support the claim.

2a. $np = (250)(0.05) = 12.5 > 5, nq = (250)(0.95) = 237.5 > 5$

b. The claim is "5% of U.S. adults have had vivid dreams about UFOs."

$H_0: p = 0.05$ (claim); $H_a: p \neq 0.05$

c. $\alpha = 0.01$

d. $z_0 = \pm 2.575$; Reject H_0 if $z < -2.575$ or $z > 2.575$.

e. $z = \dfrac{\hat{p} - p}{\sqrt{\dfrac{pq}{n}}} = \dfrac{0.08 - 0.05}{\sqrt{\dfrac{(0.05)(0.95)}{250}}} = \dfrac{0.03}{0.0138} \approx 2.176$

f. Fail to reject H_0.

g. There is not enough evidence to reject the claim.

3a. $np = (75)(0.30) = 22.5 > 5, nq = (75)(0.70) = 52.5 > 5$

b. The claim is "more than 30% of U.S. adults regularly watch the Weather Channel."

$H_0: p \leq 0.30; H_a: p > 0.30$ (claim)

c. $\alpha = 0.01$

d. $z_0 = 2.33$; Reject H_0 if $z > 2.33$.

e. $\hat{p} = \dfrac{x}{n} = \dfrac{27}{75} = 0.360$

$$z = \frac{\hat{p} - p}{\sqrt{\dfrac{pq}{n}}} = \frac{0.360 - 0.30}{\sqrt{\dfrac{(0.30)(0.70)}{75}}} = \frac{0.06}{0.053} \approx 1.13$$

f. Fail to reject H_0.

g. There is not enough evidence to support the claim.

7.4 EXERCISE SOLUTIONS

1. Verify that $np \geq 5$ and $nq \geq 5$. State H_0 and H_a. Specify the level of significance α. Determine the critical value(s) and rejection region(s). Find the standardized test statistic. Make a decision and interpret in the context of the original claim.

2. If $np \geq 5$ and $nq \geq 5$, the normal distribution can be used.

3. $np = (105)(0.25) = 26.25 > 5$

$nq = (105)(0.75) = 78.75 > 5 \rightarrow$ use normal distribution

$H_0: p = 0.25$; $H_a: p \neq 0.25$ (claim)

$z_0 \pm 1.96$

$$z = \frac{\hat{p} - p}{\sqrt{\dfrac{pq}{n}}} = \frac{0.239 - 0.25}{\sqrt{\dfrac{(0.25)(0.75)}{105}}} = \frac{-0.011}{0.0423} \approx -0.260$$

Fail to reject H_0. There is not enough evidence to support the claim.

4. $np = (500)(0.30) = 150 \geq 5$

$nq = (500)(0.70) = 350 \geq 5 \rightarrow$ use normal distribution

$H_0: p \leq 0.30$ (claim); $H_a: p > 0.30$

$z_0 = 1.645$

$$z = \frac{\hat{p} - p}{\sqrt{\dfrac{pq}{n}}} = \frac{0.35 - 0.30}{\sqrt{\dfrac{(0.30)(0.70)}{500}}} = \frac{0.05}{0.0205} \approx 2.440$$

Reject H_0. There is enough evidence to reject the claim.

5. $np = (20)(0.12) = 2.4 < 5$

$nq = (20)(0.88) = 17.6 \geq 5 \rightarrow$ cannot use normal distribution

6. $np = (45)(0.125) = 5.625 \geq 5$

$nq = (45)(0.875) = 39.375 \geq 5 \rightarrow$ use normal distribution

$H_0: p \leq 0.125$; $H_a: p > 0.125$ (claim)

$z_0 = 2.33$

$z = \dfrac{\hat{p} - p}{\sqrt{\dfrac{pq}{n}}} = \dfrac{0.2325 - 0.125}{\sqrt{\dfrac{(0.125)(0.875)}{45}}} = \dfrac{0.1075}{0.0493} \approx 2.180$

Fail to reject H_0. There is not enough evidence to support the claim.

7. $np = (70)(0.48) = 33.6 \geq 5$

$nq = (70)(0.52) = 36.4 \geq 5 \rightarrow$ use normal distribution

$H_0: p \geq 0.48$ (claim); $H_a: p < 0.48$

$z_0 = -1.29$

$z = \dfrac{\hat{p} - p}{\sqrt{\dfrac{pz}{n}}} = \dfrac{0.40 - 0.48}{\sqrt{\dfrac{(0.48)(0.52)}{70}}} = \dfrac{-0.08}{0.060} \approx -1.34$

Reject H_0. There is enough evidence to reject the claim.

8. $np = (16)(0.80) = 12.8 \geq 5$

$nq = (16)(0.20) = 3.2 < 5 \rightarrow$ cannot use normal distribution

9. (a) $H_0: p \geq 0.20$ (claim); $H_a: p < 0.20$

(b) $z_0 = -2.33$; Reject H_0 if $z < -2.33$.

(c) $z = \dfrac{\hat{p} - p}{\sqrt{\dfrac{pq}{n}}} = \dfrac{0.185 - 0.20}{\sqrt{\dfrac{(0.20)(0.80)}{200}}} = \dfrac{-0.015}{0.0283} \approx -0.53$

(d) Fail to reject H_0.

(e) There is insufficient evidence at the 1% significance level to reject the claim that at least 20% of U.S. adults are smokers.

10. (a) $H_0: p \leq 0.40$ (claim); $H_a: p > 0.40$

(b) $z_0 = 2.33$; Reject H_0 if $z > 2.33$.

(c) $z = \dfrac{\hat{p} - p}{\sqrt{\dfrac{pq}{n}}} = \dfrac{0.416 - 0.40}{\sqrt{\dfrac{(0.40)(0.60)}{250}}} = \dfrac{0.016}{0.0310} \approx 0.52$

(d) Fail to reject H_0.

(e) There is not enough evidence at the 1% significance level to reject the claim that no more than 40% of U.S. adults eat breakfast every day.

11. (a) $H_0: p \le 0.30$; $H_a: p > 0.30$ (claim)

    (b) $z_0 = 1.88$; Reject H_0 if $z > 1.88$.

    (c) $z = \dfrac{\hat{p} - p}{\sqrt{\dfrac{pq}{n}}} = \dfrac{0.32 - 0.3}{\sqrt{\dfrac{(0.30)(0.70)}{1050}}} = \dfrac{0.02}{0.0141} \approx 1.41$

    (d) Fail to reject H_0.

    (e) There is insufficient evidence at the 3% significance level to support the claim that more than 30% of U.S. consumers have stopped buying the product because the manufacturing of the product pollutes the environment.

12. (a) $H_0: p \le 0.60$; $H_a: p > 0.60$ (claim)

    (b) $z_0 = 1.28$; Reject H_0 if $z > 1.28$.

    (c) $z = \dfrac{\hat{p} - p}{\sqrt{\dfrac{pq}{n}}} = \dfrac{0.65 - 0.60}{\sqrt{\dfrac{(0.60)(0.40)}{100}}} = \dfrac{0.05}{0.0490} = 1.02$

    (d) Fail to reject H_0.

    (e) There is not enough evidence at the 10% significance level to support the claim that more than 60% of British consumers are concerned about the use of genetic modification in food production and want to avoid genetically modified foods.

13. (a) $H_0: p = 0.44$ (claim); $H_a: p \ne 0.44$

    (b) $z_0 = \pm 2.33$; Reject H_0 if $z < -2.33$ or $z > 2.33$.

    (c) $\hat{p} = \dfrac{722}{1762} \approx 0.410$

        $z = \dfrac{\hat{p} - p}{\sqrt{\dfrac{pq}{n}}} = \dfrac{0.410 - 0.44}{\sqrt{\dfrac{(0.44)(0.56)}{1762}}} = \dfrac{-0.03024}{0.01183} \approx -2.537$

    (d) Reject H_0.

    (e) There is sufficient evidence at the 2% significance level to reject the claim that 44% of home buyers find their real estate agent through a friend.

14. (a) $H_0: p = 0.24$ (claim); $H_a: p \ne 0.24$

    (b) $z_0 = \pm 1.96$; Reject H_0 if $z < -1.96$ or $z > 1.96$.

    (c) $\hat{p} = \dfrac{292}{1075} \approx 0.2716$

        $z = \dfrac{\hat{p} - p}{\sqrt{\dfrac{pq}{n}}} = \dfrac{0.2716 - 0.240}{\sqrt{\dfrac{(0.240)(0.760)}{1075}}} = \dfrac{0.316}{0.013} \approx 2.43$

    (d) Reject H_0.

    (e) There is sufficient evidence at the 5% significance level to reject the claim that 24% of adults in the United States are afraid to fly.

15. $H_0: p \geq 0.52$ (claim); $H_a: p < 0.52$

$z_0 = -1.645$; Rejection region: $z < -1.645$

$z = \dfrac{\hat{p} - p}{\sqrt{\dfrac{pq}{n}}} = \dfrac{0.48 - 0.52}{\sqrt{\dfrac{(0.52)(0.48)}{50}}} = \dfrac{-0.04}{0.0707} \approx -0.566$

Fail to reject H_0. There is insufficient evidence to reject the claim.

16. The company should continue the use of giveaways because there is not enough evidence to say that less than 52% of the adults would be more likely to buy a product when there are free samples.

17. $H_0: p = 0.44$ (claim); $H_a: p \neq 0.44$

$z = \dfrac{x - np}{\sqrt{npq}} = \dfrac{722 - (1762)(0.44)}{\sqrt{(1762)(0.44)(0.56)}} = \dfrac{-53.28}{20.836} \approx -2.56$

Reject H_0. The results are the same.

18. $z = \dfrac{\hat{p} - p}{\sqrt{\dfrac{pq}{n}}} \Rightarrow \dfrac{\left(\dfrac{x}{n}\right) - p}{\sqrt{\dfrac{pq}{n}}} \Rightarrow \dfrac{\left(\dfrac{x}{n}\right) - p}{\dfrac{\sqrt{pq}}{\sqrt{n}}} \Rightarrow \dfrac{\left[\left(\dfrac{x}{n}\right) - p\right]\sqrt{n}}{\sqrt{pq}} \cdot \dfrac{\sqrt{n}}{\sqrt{n}} \Rightarrow \dfrac{\left[\left(\dfrac{x}{n}\right) - p\right]n}{\sqrt{pqn}} \Rightarrow \dfrac{x - np}{\sqrt{pqn}}$

7.5 HYPOTHESIS TESTING FOR VARIANCE AND STANDARD DEVIATION

7.5 Try It Yourself Solutions

1a. $\chi_0^2 = 33.409$

2a. $\chi_0^2 = 17.708$

3a. $\chi_R^2 = 31.526$ **b.** $\chi_L^2 = 8.231$

4a. The claim is "the variance of the amount of sports drink in a 12-ounce bottle is no more than 0.40."

$H_0: \sigma^2 \leq 0.40$ (claim); $H_a: \sigma^2 > 0.40$

b. $\alpha = 0.01$ and d.f. $= n - 1 = 30$

c. $\chi_0^2 = 50.892$; Reject H_0 if $\chi^2 > 50.892$.

d. $\chi^2 = \dfrac{(n-1)s^2}{\sigma^2} = \dfrac{(30)(0.75)}{0.40} = 56.250$

e. Reject H_0.

f. There is enough evidence to reject the claim.

5a. The claim is "the standard deviation in the length of response times is less than 3.7 minutes."

H_0: $\sigma \geq 3.7$; H_a: $\sigma < 3.7$ (claim)

b. $\alpha = 0.05$ and d.f. $= n - 1 = 8$

c. $\chi_0^2 = 2.733$; Reject H_0 if $\chi^2 < 2.733$.

d. $\chi^2 = \dfrac{(n-1)s^2}{\sigma^2} = \dfrac{(8)(3.0)^2}{(3.7)^2} \approx 5.259$

e. Fail to reject H_0.

f. There is not enough evidence to support the claim.

6a. The claim is "the variance of the diameters in a certain tire model is 8.6."

H_0: $\sigma^2 = 8.6$ (claim); H_a: $\sigma^2 \neq 8.6$

b. $\alpha = 0.01$ and d.f. $= n - 1 = 9$

c. $\chi_L^2 = 1.735$ and $\chi_R^2 = 23.589$

Reject H_0 if $\chi^2 > 23.589$ or $\chi^2 < 1.735$.

d. $\chi^2 = \dfrac{(n-1)s^2}{\sigma^2} = \dfrac{(9)(4.3)}{(8.6)} = 4.50$

e. Fail to reject H_0.

f. There is not enough evidence to reject the claim.

7.5 EXERCISE SOLUTIONS

1. Specify the level of significance α. Determine the degrees of freedom. Determine the critical values using the χ^2 distribution. If (a) right-tailed test, use the value that corresponds to d.f. and α. (b) left-tailed test, use the value that corresponds to d.f. and $1 - \alpha$; and (c) two-tailed test, use the value that corresponds to d.f. and $\frac{1}{2}\alpha$ and $1 - \frac{1}{2}\alpha$.

2. State H_0 and H_a. Specify the level of significance. Determine the degrees of freedom. Determine the critical value(s) and rejection region(s). Find the standardized test statistic. Make a decision and interpret in the context of the original claim.

3. $\chi_0^2 = 38.885$ 4. $\chi_0^2 = 14.684$ 5. $\chi_0^2 = 0.872$ 6. $\chi_0^2 = 13.091$

7. $\chi_L^2 = 7.261, \chi_R^2 = 24.996$ 8. $\chi_L^2 = 12.461, \chi_R^2 = 50.993$

9. (a) Fail to reject H_0.   10. (a) Fail to reject H_0.

(b) Fail to reject H_0.   (b) Fail to reject H_0.

(c) Fail to reject H_0.   (c) Reject H_0.

(d) Reject H_0.   (d) Reject H_0.

11. (a) Fail to reject H_0.

    (b) Reject H_0.

    (c) Reject H_0.

    (d) Fail to reject H_0.

12. (a) Fail to reject H_0.

    (b) Fail to reject H_0.

    (c) Fail to reject H_0.

    (d) Reject H_0.

13. $H_0: \sigma^2 = 0.52$ (claim); $H_a: \sigma^2 \neq 0.52$

$\chi_L^2 = 7.564, \chi_R^2 = 30.191$

$$\chi^2 = \frac{(n-1)s^2}{\sigma^2} = \frac{(17)(0.508)^2}{(0.52)} \approx 16.608$$

Fail to reject H_0. There is insufficient evidence to reject the claim.

14. $H_0: \sigma \geq 40; H_a: \sigma < 40$ (claim)

$\chi_0^2 = 3.053$

$$\chi^2 = \frac{(n-1)s^2}{\sigma^2} = \frac{(11)(40.8)^2}{(40)^2} \approx 11.444$$

Fail to reject H_0. There is insufficient evidence to support the claim.

15. (a) $H_0: \sigma^2 = 3$ (claim); $H_a: \sigma^2 \neq 3$

    (b) $\chi_L^2 = 13.844, \chi_R^2 = 41.923$; Reject H_0 if $\chi^2 > 41.923$ or $\chi^2 < 13.844$.

    (c) $\chi^2 = \dfrac{(n-1)s^2}{\sigma^2} = \dfrac{(26)(2.8)}{3} \approx 24.267$

    (d) Fail to reject H_0.

    (e) There is insufficient evidence at the 5% level of significance to reject the claim that the variance of the life of the appliances is 3.

16. (a) $H_0: \sigma^2 = 6$ (claim); $H_a: \sigma^2 \neq 6$

    (b) $\chi_L^2 = 14.573, \chi_R^2 = 43.194$; Reject H_0 if $\chi^2 > 43.194$ or $\chi^2 < 14.573$.

    (c) $\chi^2 = \dfrac{(n-1)s^2}{\sigma^2} = \dfrac{(27)(4.25)}{6} = 19.125$

    (d) Fail to reject H_0.

    (e) There is not enough evidence at the 5% significance level to reject the claim that the variance of the gas mileage is 6.

17. (a) $H_0: \sigma \geq 36; H_a: \sigma < 36$ (claim)

    (b) $\chi_0^2 = 13.240$; Reject H_0 if $\chi^2 < 13.240$.

    (c) $\chi^2 = \dfrac{(n-1)s^2}{\sigma^2} = \dfrac{(21)(33.4)^2}{(36)^2} \approx 18.076$

    (d) Fail to reject H_0.

    (e) There is insufficient evidence at the 10% significance level to support the claim that the standard deviation for eighth graders on the examination is less than 36.

204   CHAPTER 7 | HYPOTHESIS TESTING WITH ONE SAMPLE

18. (a) H_0: $\sigma \geq 30$; H_a: $\sigma < 30$ (claim)

(b) $\chi_0^2 = 6.408$; Reject H_0 if $\chi^2 < 6.408$.

(c) $\chi^2 = \dfrac{(n-1)s^2}{\sigma^2} = \dfrac{(17)(33.6)^2}{(30)^2} \approx 21.325$

(d) Fail to reject H_0.

(e) There is not enough evidence at the 1% significance level to support the claim that the standard deviation of test scores for eighth grade students who took a U.S. history assessment test is less than 30 points.

19. (a) H_0: $\sigma \leq 0.5$ (claim); H_a: $\sigma > 0.5$

(b) $\chi_0^2 = 33.196$; Reject H_0 if $\chi^2 > 33.196$.

(c) $\chi^2 = \dfrac{(n-1)s^2}{\sigma^2} = \dfrac{(24)(0.7)^2}{(0.5)^2} = 47.04$

(d) Reject H_0.

(e) There is sufficient evidence at the 10% significance level to reject the claim that the standard deviation of waiting times is no more than 0.5 minute.

20. (a) H_0: $\sigma = 6.14$ (claim); H_a: $\sigma \neq 6.14$

(b) $\chi_L^2 = 8.907$, $\chi_R^2 = 32.852$; Reject H_0 if $\chi^2 < 8.907$ or $\chi^2 > 32.852$.

(c) $\chi^2 = \dfrac{(n-1)s^2}{\sigma^2} = \dfrac{(19)(6.5)^2}{(6.14)^2} = 21.293$

(d) Fail to reject H_0.

(e) There is insufficient evidence at the 5% significance level to reject the claim that the standard deviation of the lengths of stay is 6.14 days.

21. (a) H_0: $\sigma \geq \$3500$; H_a: $\sigma < \$3500$ (claim)

(b) $\chi_0^2 = 18.114$; Reject H_0 if $\chi^2 < 18.114$.

(c) $\chi^2 = \dfrac{(n-1)s^2}{\sigma^2} = \dfrac{(27)(4100)^2}{(3500)^2} \approx 37.051$

(d) Fail to reject H_0.

(e) There is insufficient evidence at the 10% significance level to support the claim that the standard deviation of the total charge for patients involved in a crash where the vehicle struck a construction baracade is less than $3500.

22. (a) H_0: $\sigma \leq \$30$ (claim); H_a: $\sigma > \$30$

(b) $\chi_0^2 = 37.566$; Reject H_0 if $\chi^2 < 37.566$.

(c) $\chi^2 = \dfrac{(n-1)s^2}{\sigma^2} = \dfrac{(20)(35.25)^2}{(30)^2} = 27.613$

(d) Fail to reject H_0.

(e) There is not enough evidence at the 1% significance level to reject the claim that the standard deviation of the room rates of hotels in the city is no more than $30.

© 2009 Pearson Education, Inc., Upper Saddle River, NJ. All rights reserved. This material is protected under all copyright laws as they currently exist. No portion of this material may be reproduced, in any form or by any means, without permission in writing from the publisher.

23. (a) H_0: $\sigma \le \$20,000$; H_a: $\sigma > \$20,000$ (claim)

(b) $\chi_0^2 = 24.996$; Reject H_0 if $\chi^2 > 24.996$.

(c) $s = 20,826.145$

$$\chi^2 = \frac{(n-1)s^2}{\sigma^2} = \frac{(15)(20,826.145)^2}{(20,000)^2} \approx 16.265$$

(d) Fail to reject H_0.

(e) There is insufficient evidence at the 5% significance level to support the claim that the standard deviation of the annual salaries for actuaries is more than $20,000.

24. (a) H_0: $\sigma \ge \$14,500$ (claim); H_a: $\sigma < \$14,500$

(b) $\chi_0^2 = 10.085$; Reject H_0 if $\chi^2 < 10.085$.

(c) $s = 13,950.604$

$$\chi^2 = \frac{(n-1)s^2}{\sigma^2} = \frac{(17)(13,950.604)^2}{(14,500)^2} \approx 15.736$$

(d) Fail to reject H_0.

(e) There is not enough evidence at the 10% significance level to reject the claim that the standard deviation of the annual salaries for public relations managers is at least $14,500.

25. $\chi^2 = 37.051$

P-value = {Area left of $\chi^2 = 37.051$} = 0.9059

Fail to reject H_0 because P-value = $0.9059 > 0.10 = \alpha$.

26. $\chi^2 = 27.613$

P-value = {Area right of $\chi^2 = 27.613$} = 0.1189

Fail to reject H_0 because P-value = $0.1189 > 0.01 = \alpha$.

27. $\chi^2 = 16.265$

P-value = {Area right of $\chi^2 = 16.265$} = 0.3647

Fail to reject H_0 because P-value = $0.3647 > 0.05 = \alpha$.

28. $\chi^2 = 15.736$

P-value = {Area left of $\chi^2 = 15.736$} = 0.4574

Fail to reject H_0 because P-value = $0.4574 > 0.10 = \alpha$.

CHAPTER 7 REVIEW EXERCISE SOLUTIONS

1. H_0: $\mu \le 1479$ (claim); H_a: $\mu > 1479$

2. H_0: $\mu = 95$ (claim); H_a: $\mu \ne 95$

3. H_0: $p \ge 0.205$; H_a: $p < 0.205$ (claim)

4. H_0: $\mu = 150,020$; H_a: $\mu \ne 150,020$ (claim)

5. H_0: $\sigma \le 6.2$; H_a: $\sigma > 6.2$ (claim)

6. H_0: $p \ge 0.78$ (claim); H_a: $p < 0.78$

7. (a) H_0: $p = 0.73$ (claim); H_a: $p \neq 0.73$

   (b) Type I error will occur if H_0 is rejected when the actual proportion of college students that occasionally or frequently come late to class is 0.63.

   Type II error if H_0 is not rejected when the actual proportion of college students that occasionally or frequently come late to class is not 0.63.

   (c) Two-tailed, because hypothesis compares "= vs ≠".

   (d) There is enough evidence to reject the claim.

   (e) There is not enough evidence to reject the claim.

8. (a) H_0: $\mu \geq 30{,}000$ (claim); H_a: $\mu < 30{,}000$

   (b) Type I error will occur if H_0 is rejected when the actual mean tire life is at least 30,000 miles.

   Type II error if H_0 is not rejected when the actual mean tire life is less than 30,000 miles.

   (c) Left-tailed, because hypothesis compares "≥ vs <".

   (d) There is enough evidence to reject the claim.

   (e) There is not enough evidence to reject the claim.

9. (a) H_0: $\mu \leq 50$ (claim); H_a: $\mu > 50$

   (b) Type I error will occur if H_0 is rejected when the actual standard deviation sodium content is no more than 50 milligrams.

   Type II error if H_0 is not rejected when the actual standard deviation sodium content is more than 50 milligrams.

   (c) Right-tailed, because hypothesis compares "≤ vs >".

   (d) There is enough evidence to reject the claim.

   (e) There is not enough evidence to reject the claim.

10. (a) H_0: $\mu \geq 25$; H_a: $\mu < 25$ (claim)

   (b) Type I error will occur if H_0 is rejected when the actual mean number of grams of carbohydrates in one bar is greater than or equal to 25.

   Type II error if H_0 is not rejected when the actual mean number of grams of carbohydrates in one bar is less than 25.

   (c) Left-tailed, because hypothesis compares "≥ vs <".

   (d) There is enough evidence to support the claim.

   (e) There is not enough evidence to support the claim.

11. $z_0 \approx -2.05$ **12.** $z_0 = \pm 2.81$ **13.** $z_0 = 1.96$ **14.** $z_0 = \pm 1.75$

15. H_0: $\mu \leq 45$ (claim); H_a: $\mu > 45$

$z_0 = 1.645$

$$z = \frac{\bar{x} - \mu}{\frac{s}{\sqrt{n}}} = \frac{47.2 - 45}{\frac{6.7}{\sqrt{42}}} = \frac{2.2}{1.0338} \approx 2.128$$

Reject H_0. There is enough evidence to reject the claim.

16. H_0: $\mu = 0$; H_a: $\mu \neq 0$ (claim)

$z_0 = \pm 1.96$

$$z = \frac{\bar{x} - \mu}{\frac{s}{\sqrt{n}}} = \frac{-0.69 - 0}{\frac{2.62}{\sqrt{60}}} = \frac{-0.69}{0.338} \approx -2.040$$

Reject H_0. There is enough evidence to support the claim.

17. H_0: $\mu \geq 5.500$; H_a: $\mu < 5.500$ (claim)

$z_0 = -2.33$

$$z = \frac{\bar{x} - \mu}{\frac{s}{\sqrt{n}}} = \frac{5.497 - 5.500}{\frac{0.011}{\sqrt{36}}} = \frac{-0.003}{0.00183} \approx -1.636$$

Fail to reject H_0. There is not enough evidence to support the claim.

18. H_0: $\mu = 7450$ (claim); H_a: $\mu \neq 7450$

$z_0 = \pm 1.96$

$$z = \frac{\bar{x} - \mu}{\frac{s}{\sqrt{n}}} = \frac{7512 - 7450}{\frac{243}{\sqrt{57}}} = \frac{62}{32.186} \approx 1.926$$

Fail to reject H_0. There is not enough evidence to reject the claim.

19. H_0: $\mu \leq 0.05$ (claim); H_a: $\mu > 0.05$

$$z = \frac{\bar{x} - \mu}{\frac{s}{\sqrt{n}}} = \frac{0.057 - 0.05}{\frac{0.018}{\sqrt{32}}} = \frac{0.007}{0.00318} \approx 2.20$$

P-value = {Area right of $z = 2.20$} = 0.0139

$\alpha = 0.10$; Reject H_0.

$\alpha = 0.05$; Reject H_0.

$\alpha = 0.01$; Fail to reject H_0.

20. H_0: $\mu = 230$; H_a: $\mu \neq 230$ (claim)

$$z = \frac{\bar{x} - \mu}{\frac{s}{\sqrt{n}}} = \frac{216.5 - 230}{\frac{17.3}{\sqrt{48}}} = \frac{-13.5}{2.497} \approx -5.41$$

P-value = 2{Area left of $z = -5.41$} ≈ 0

$\alpha = 0.10$; Reject H_0.

$\alpha = 0.05$; Reject H_0.

$\alpha = 0.01$; Reject H_0.

21. $H_0: \mu = 326$ (claim); $H_a: \mu \neq 326$

$$z = \frac{\bar{x} - \mu}{\frac{s}{\sqrt{n}}} = \frac{318 - 326}{\frac{25}{\sqrt{50}}} = \frac{-8}{3.536} \approx -2.263$$

P-value $= 2\{$Area left of $z = -2.263\} = 2\{0.012\} = 0.024$

Reject H_0. There is sufficient evidence to reject the claim.

22. $H_0: \mu \leq \$650$ (claim); $H_a: \mu > \$650$

$$z = \frac{\bar{x} - \mu}{\frac{s}{\sqrt{n}}} = \frac{657 - 650}{\frac{40}{\sqrt{45}}} = \frac{7}{5.963} \approx 1.174$$

P-value $= \{$Area right of $z = 1.17\} = 0.1210$

Fail to reject H_0. There is not enough evidence to reject the claim.

23. $t_0 = \pm 2.093$ **24.** $t_0 = 2.998$ **25.** $t_0 = -1.345$ **26.** $t_0 = \pm 2.201$

27. $H_0: \mu = 95$; $H_a: \mu \neq 95$ (claim)

$t_0 = \pm 2.201$

$$t = \frac{\bar{x} - \mu}{\frac{s}{\sqrt{n}}} = \frac{94.1 - 95}{\frac{1.53}{\sqrt{12}}} = \frac{-0.9}{0.442} \approx -2.038$$

Fail to reject H_0. There is not enough evidence to support the claim.

28. $H_0: \mu \leq 12,700$; $H_a: \mu > 12,700$ (claim)

$t_0 = 1.725$

$$t = \frac{\bar{x} - \mu}{\frac{s}{\sqrt{n}}} = \frac{12,804 - 12,700}{\frac{248}{\sqrt{21}}} = \frac{104}{54.118} \approx 1.922$$

Reject H_0. There is enough evidence to support the claim.

29. $H_0: \mu \geq 0$ (claim); $H_a: \mu < 0$

$t_0 = -1.341$

$$t = \frac{\bar{x} - \mu}{\frac{s}{\sqrt{n}}} = \frac{-0.45 - 0}{\frac{1.38}{\sqrt{16}}} = \frac{-0.45}{0.345} \approx -1.304$$

Fail to reject H_0. There is not enough evidence to reject the claim.

30. $H_0: \mu = 4.20$ (claim); $H_a: \mu \neq 4.20$

$t_0 = \pm 2.896$

$$t = \frac{\bar{x} - \mu}{\frac{s}{\sqrt{n}}} = \frac{4.41 - 4.20}{\frac{0.26}{\sqrt{9}}} = \frac{0.21}{0.0867} \approx 2.423$$

Fail to reject H_0. There is not enough evidence to reject the claim.

31. H_0: $\mu \le 48$ (claim); H_a: $\mu > 48$

$t_0 = 3.148$

$$t = \frac{\bar{x} - \mu}{\frac{s}{\sqrt{n}}} = \frac{52 - 48}{\frac{2.5}{\sqrt{7}}} = \frac{4}{0.945} \approx 4.233$$

Reject H_0. There is enough evidence to reject the claim.

32. H_0: $\mu \ge 850$; H_a: $\mu < 850$ (claim)

$t_0 = -2.160$

$$t = \frac{\bar{x} - \mu}{\frac{s}{\sqrt{n}}} = \frac{875 - 850}{\frac{25}{\sqrt{14}}} = \frac{25}{6.682} \approx 3.742$$

Fail to reject H_0. There is not enough evidence to support the claim.

33. H_0: $\mu = \$25$ (claim); H_a: $\mu \ne \$25$

$t_0 = \pm 1.740$

$$t = \frac{\bar{x} - \mu}{\frac{s}{\sqrt{n}}} = \frac{26.25 - 25}{\frac{3.23}{\sqrt{18}}} = \frac{1.25}{0.761} \approx 1.642$$

Fail to reject H_0. There is not enough evidence to reject the claim.

34. H_0: $\mu \le 10$ (claim); H_a: $\mu > 10$

$t_0 = 1.397$

$$t = \frac{\bar{x} - \mu}{\frac{s}{\sqrt{n}}} = \frac{13.5 - 10}{\frac{5.8}{\sqrt{9}}} = \frac{3.5}{1.933} \approx 1.810$$

Reject H_0. There is enough evidence to reject the claim.

35. H_0: $\mu \ge \$10,200$ (claim); H_a: $\mu < \$10,200$

$t_0 = -2.602$

$\bar{x} = 9895.8$ $s = 490.88$

$$t = \frac{\bar{x} - \mu}{\frac{s}{\sqrt{n}}} = \frac{9895.8 - 10,200}{\frac{490.88}{\sqrt{16}}} = \frac{-304.2}{122.72} \approx -2.479$$

P-value ≈ 0.0128

Fail to reject H_0. There is not enough evidence to reject the claim.

36. H_0: $\mu \le 9$; H_a: $\mu > 9$ (claim)

$\bar{x} = 9.982$, $s = 2.125$

$$t = \frac{\bar{x} - \mu}{\frac{s}{\sqrt{n}}} = \frac{9.982 - 9}{\frac{2.125}{\sqrt{11}}} = \frac{0.982}{0.641} \approx 1.532$$

P-value $= 0.078$

Fail to reject H_0. There is not enough evidence to support the claim.

37. H_0: $p = 0.15$ (claim); H_a: $p \neq 0.15$

$z_0 = \pm 1.96$

$$z = \frac{\hat{p} - p}{\sqrt{\dfrac{pq}{n}}} = \frac{0.09 - 0.15}{\sqrt{\dfrac{(0.15)(0.85)}{40}}} = \frac{-0.06}{0.0565} \approx -1.063$$

Fail to reject H_0. There is not enough evidence to reject the claim.

38. H_0: $p \geq 0.70$; H_a: $p < 0.70$ (claim)

$z_0 = -2.33$

$$z = \frac{\hat{p} - p}{\sqrt{\dfrac{pq}{n}}} = \frac{0.50 - 0.70}{\sqrt{\dfrac{(0.70)(0.30)}{68}}} = \frac{-0.2}{0.0556} \approx -3.599$$

Reject H_0. There is enough evidence to support the claim.

39. Because $np = 3.6$ is less than 5, the normal distribution cannot be used to approximate the binomial distribution.

40. H_0: $p = 0.50$ (claim); H_a: $p \neq 0.50$

$z_0 = \pm 1.645$

$$z = \frac{\hat{p} - p}{\sqrt{\dfrac{pq}{n}}} = \frac{0.71 - 0.50}{\sqrt{\dfrac{(0.50)(0.50)}{129}}} = \frac{0.21}{0.0440} \approx 4.770$$

Reject H_0. There is enough evidence to reject the claim.

41. Because $np = 1.2 < 5$, the normal distribution cannot be used to approximate the binomial distribution.

42. H_0: $p = 0.34$; H_a: $p \neq 0.34$ (claim)

$z_0 = \pm 2.575$

$$z = \frac{\hat{p} - p}{\sqrt{\dfrac{pq}{n}}} = \frac{0.29 - 0.34}{\sqrt{\dfrac{(0.34)(0.66)}{60}}} = \frac{-0.05}{0.061} \approx 0.820$$

Fail to reject H_0. There is not enough evidence to support the claim.

43. H_0: $p = 0.20$; H_a: $p \neq 0.20$ (claim)

$z_0 = \pm 2.575$

$$z = \frac{\hat{p} - p}{\sqrt{\dfrac{pq}{n}}} = \frac{0.23 - 0.20}{\sqrt{\dfrac{(0.20)(0.80)}{56}}} = \frac{0.03}{0.0534} \approx 0.561$$

Fail to reject H_0. There is not enough evidence to support the claim.

44. H_0: $p \leq 0.80$ (claim); H_a: $p > 0.80$

$z_0 = 1.28$

$$z = \frac{\hat{p} - p}{\sqrt{\dfrac{pq}{n}}} = \frac{0.85 - 0.80}{\sqrt{\dfrac{(0.80)(0.20)}{43}}} = \frac{0.05}{0.061} \approx 0.820$$

Fail to reject H_0. There is not enough evidence to reject the claim.

45. H_0: $p \leq 0.40$; H_a: $p > 0.40$ (claim)

$z_0 = 1.28$

$$\hat{p} = \frac{x}{n} = \frac{1130}{2730} \approx 0.414$$

$$z = \frac{\hat{p} - p}{\sqrt{\dfrac{pq}{n}}} = \frac{0.414 - 0.40}{\sqrt{\dfrac{(0.40)(0.60)}{2730}}} = \frac{0.14}{0.0094} \approx 1.493$$

Reject H_0. There is enough evidence to support the claim.

46. H_0: $p = 0.02$ (claim); H_a: $p \neq 0.02$

$z_0 = \pm 1.96$

$$\hat{p} = \frac{x}{n} = \frac{3}{300} \approx 0.01$$

$$z = \frac{\hat{p} - p}{\sqrt{\dfrac{pq}{n}}} = \frac{0.01 - 0.02}{\sqrt{\dfrac{(0.02)(0.98)}{300}}} = \frac{-0.01}{0.0081} \approx -1.24$$

Fail to reject H_0. There is not enough evidence to reject the claim.

47. $\chi_R^2 = 30.144$

48. $\chi_L^2 = 3.565$, $\chi_R^2 = 29.819$

49. $\chi_R^2 = 33.196$

50. $\chi_0^2 = 1.145$

51. H_0: $\sigma^2 \leq 2$; H_a: $\sigma^2 > 2$ (claim)

$\chi_0^2 = 24.769$

$$\chi^2 = \frac{(n-1)s^2}{\sigma^2} = \frac{(17)(2.95)}{(2)} = 25.075$$

Reject H_0. There is enough evidence to support the claim.

52. H_0: $\sigma^2 \leq 60$ (claim); H_a: $\sigma^2 > 60$

$\chi_0^2 = 26.119$

$$\chi^2 = \frac{(n-1)s^2}{\sigma^2} = \frac{(14)(72.7)}{(60)} \approx 16.963$$

Fail to reject H_0. There is not enough evidence to reject the claim.

53. $H_0: \sigma^2 = 1.25$ (claim); $H_a: \sigma^2 \neq 1.25$

$\chi_L^2 = 0.831$, $\chi_R^2 = 12.833$

$$\chi^2 = \frac{(n-1)s^2}{\sigma^2} = \frac{(5)(1.03)^2}{(1.25)^2} \approx 3.395$$

Fail to reject H_0. There is not enough evidence to reject the claim.

54. $H_0: \sigma = 0.035$; $H_a: \sigma \neq 0.035$ (claim)

$\chi_L^2 = 4.601$, $\chi_R^2 = 32.801$

$$\chi^2 = \frac{(n-1)s^2}{\sigma^2} = \frac{(15)(0.026)^2}{(0.035)^2} \approx 8.278$$

Fail to reject H_0. There is not enough evidence to support the claim.

55. $H_0: \sigma^2 \leq 0.01$ (claim); $H_a: \sigma^2 > 0.01$

$\chi_0^2 = 49.645$

$$\chi^2 = \frac{(n-1)s^2}{\sigma^2} = \frac{(27)(0.064)}{(0.01)} = 172.800$$

Reject H_0. There is enough evidence to reject the claim.

56. $H_0: \sigma \leq 0.0025$ (claim); $H_a: \sigma > 0.0025$

$\chi_0^2 = 27.688$

$$\chi^2 = \frac{(n-1)s^2}{\sigma^2} = \frac{(13)(0.0031)^2}{(0.0025)^2} \approx 19.989$$

Fail to reject H_0. There is not enough evidence to reject the claim.

CHAPTER 7 QUIZ SOLUTIONS

1. (a) $H_0: \mu \geq 22$ (claim); $H_a: \mu < 22$

(b) "$\geq$ vs $<$" $\rightarrow$ Left-tailed

σ is unknown and $n \geq 30 \rightarrow z$-test.

(c) $z_0 = -2.05$; Reject H_0 if $z < -2.05$.

(d) $z = \dfrac{\bar{x} - \mu}{\dfrac{s}{\sqrt{n}}} = \dfrac{21.6 - 22}{\dfrac{7}{\sqrt{103}}} = \dfrac{-0.4}{0.690} \approx -0.580$

(e) Fail to reject H_0. There is insufficient evidence at the 2% significance level to reject the claim that the mean utilization of fresh citrus fruits by people in the U.S. is at least 22 pounds per year.

2. (a) $H_0: \mu \geq 20$ (claim); $H_a: \mu < 20$

   (b) "$\geq$ vs $<$" $\rightarrow$ Left-tailed

   σ is unknown, the population is normal, and $n < 30 \rightarrow t$-test.

   (c) $t_0 = -1.895$; Reject H_0 if $t < -1.895$.

   (d) $z = \dfrac{\bar{x} - \mu}{\frac{s}{\sqrt{n}}} = \dfrac{18 - 20}{\frac{5}{\sqrt{8}}} = \dfrac{-2}{1.768} \approx -1.131$

   (e) Fail to reject H_0. There is insufficient evidence at the 5% significance level to reject the claim that the mean gas mileage is at least 20 miles per gallon.

3. (a) $H_0: p \leq 0.10$ (claim); $H_a: p > 0.10$

   (b) "$\leq$ vs $>$" $\rightarrow$ Right-tailed

   $np \geq 5$ and $nq \geq 5 \rightarrow z$-test

   (c) $z_0 = 1.75$; Reject H_0 if $z > 1.75$.

   (d) $z = \dfrac{\hat{p} - p}{\sqrt{\frac{pq}{n}}} = \dfrac{0.13 - 0.10}{\sqrt{\frac{(0.10)(0.90)}{57}}} = \dfrac{0.03}{0.0397} \approx 0.75$

   (e) Fail to reject H_0. There is insufficient evidence at the 4% significance level to reject the claim that no more than 10% of microwaves need repair during the first five years of use.

4. (a) $H_0: \sigma = 113$ (claim); $H_a: \sigma \neq 113$

   (b) "$=$ vs $\neq$" $\rightarrow$ Two-tailed

   Assuming the scores are normally distributed and you are testing the hypothesized standard deviation $\rightarrow \chi^2$ test.

   (c) $\chi_L^2 = 3.565$, $\chi_R^2 = 29.819$; Reject H_0 if $\chi^2 < 3.565$ or if $\chi^2 > 29.819$.

   (d) $\chi^2 = \dfrac{(n-1)s^2}{\sigma^2} = \dfrac{(13)(108)^2}{(113)^2} \approx 11.875$

   (e) Fail to reject H_0. There is insufficient evidence at the 1% significance level to reject the claim that the standard deviation of the SAT critical reading scores for the state is 105.

5. (a) $H_0: \mu = \$48{,}718$ (claim); $H_a: \mu \neq \$48{,}718$

   (b) "$=$ vs $\neq$" $\rightarrow$ Two-tailed

   σ is unknown, $n < 30$, and assuming the salaries are normally distributed $\rightarrow t$-test.

   (c) not applicable

   (d) $t = \dfrac{\bar{x} - \mu}{\frac{s}{\sqrt{n}}} = \dfrac{47{,}164 - 48{,}718}{\frac{6500}{\sqrt{12}}} = \dfrac{-1554}{1876.388} \approx -0.828$

   P-value $= 2\{$Area left of $t = -0.828\} = 2(0.2126) = 0.4252$

   (e) Fail to reject H_0. There is insufficient evidence at the 5% significance level to reject the claim that the mean annual salary for full-time male workers ages 25 to 34 with a bachelor's degree is $48,718.

6. (a) $H_0: \mu = \$201$ (claim); $H_a: \mu \neq \$201$

(b) "= vs ≠" → Two-tailed

σ is unknown, $n \geq 30$ → z-test.

(c) not applicable

(d) $z = \dfrac{\bar{x} - \mu}{\dfrac{s}{\sqrt{n}}} = \dfrac{216 - 201}{\dfrac{30}{\sqrt{35}}} = \dfrac{15}{5.071} \approx 2.958$

$P\text{-value} = 2\{\text{Area right of } z = 2.958\} = 2\{0.0015\} = 0.0030$

(e) Reject H_0. There is sufficient evidence at the 5% significance level to reject the claim that the mean daily cost of meals and lodging for a family of four traveling in Kansas is $201.

Hypothesis Testing with Two Samples

8.1 Try It Yourself Solutions

Note: Answers may differ due to rounding.

1. (1) Independent

(2) Dependent

2a. $H_0: \mu_1 = \mu_2; H_a: \mu_1 \neq \mu_2$ (claim)

 b. $\alpha = 0.01$

 c. $z_0 = \pm 2.575$; Reject H_0 if $z > 2.575$ or $z < -2.575$.

 d. $z = \dfrac{(\bar{x}_1 - \bar{x}_2) - (\mu_1 - \mu_2)}{\sqrt{\dfrac{s_1^2}{n_1} + \dfrac{s_2^2}{n_2}}} = \dfrac{(3900 - 3500) - (0)}{\sqrt{\dfrac{(900)^2}{50} + \dfrac{(500)^2}{50}}} = \dfrac{400}{\sqrt{21200}} \approx 2.747$

 e. Reject H_0.

 f. There is enough evidence to support the claim.

3a. $z = \dfrac{(\bar{x}_1 - \bar{x}_2) - (\mu_1 - \mu_2)}{\sqrt{\dfrac{s_1^2}{n_1} + \dfrac{s_2^2}{n_2}}} = \dfrac{(293 - 286) - (0)}{\sqrt{\dfrac{(24)^2}{150} + \dfrac{(18)^2}{200}}} = \dfrac{7}{\sqrt{5.46}} \approx 3.00$

$\rightarrow$ P-value = {area right of $z = 3.00$} = 0.0014

 b. Reject H_0. There is enough evidence to support the claim.

8.1 EXERCISE SOLUTIONS

1. Two samples are dependent if each member of one sample corresponds to a member of the other sample. Example: The weights of 22 people before starting an exercise program and the weights of the same 22 people 6 weeks after starting the exercise program.

   Two samples are independent if the sample selected from one population is not related to the sample from the second population. Example: The weights of 25 cats and the weights of 25 dogs.

2. State the hypotheses and identify the claim. Specify the level of significance and find the critical value(s). Identify the rejection regions. Find the standardized test statistic. Make a decision and interpret in the context of the claim.

3. Use P-values.

4. (1) The samples must be randomly selected.

   (2) The samples must be independent.

   (3) {$n_1 \geq 30$ and $n_2 \geq 30$} or {each population must be normally distributed with known standard deviations}

5. Independent because different students were sampled.

6. Dependent because the same students were sampled.

7. Dependent because the same adults were sampled.

8. Independent because different individuals were sampled.

9. Independent because different boats were sampled.

10. Dependent because the same cars were sampled.

11. Dependent because the same tire sets were sampled.

12. Dependent because the same people were sampled.

13. $H_0: \mu_1 = \mu_2$ (claim); $H_a: \mu_1 \neq \mu_2$

Rejection regions: $z_0 < -1.96$ and $z_0 > 1.96$ (Two-tailed test)

(a) $\bar{x}_1 - \bar{x}_2 = 16 - 14 = 2$

(b) $z = \dfrac{(\bar{x}_1 - \bar{x}_2) - (\mu_1 - \mu_2)}{\sqrt{\dfrac{s_1^2}{n_1} + \dfrac{s_2^2}{n_2}}} = \dfrac{(16 - 14) - (0)}{\sqrt{\dfrac{(1.1)^2}{50} + \dfrac{(1.5)^2}{50}}} = \dfrac{2}{\sqrt{0.0692}} \approx 7.60$

(c) z is in the rejection region because $7.60 > 1.96$.

(d) Reject H_0. There is enough evidence to reject the claim.

14. $H_0: \mu_1 \leq \mu_2$; $H_1: \mu_1 > \mu_2$ (claim)

Rejection region: $z_0 > 1.28$ (Right-tailed test)

(a) $\bar{x}_1 - \bar{x}_2 = 500 - 510 = -10$

(b) $z = \dfrac{(\bar{x}_1 - \bar{x}_2) - (\mu_1 - \mu_2)}{\sqrt{\dfrac{s_1^2}{n_1} + \dfrac{s_2^2}{n_2}}} = \dfrac{(500 - 510) - (0)}{\sqrt{\dfrac{(30)^2}{100} + \dfrac{(15)^2}{75}}} = \dfrac{-10}{\sqrt{12}} \approx -2.89$

(c) z is not in the rejection region because $-2.89 < 1.28$.

(d) Fail to reject H_0. There is not enough evidence to support the claim.

15. $H_0: \mu_1 \geq \mu_2$; $H_a: \mu_1 < \mu_2$ (claim)

Rejection region: $z_0 < -2.33$ (Left-tailed test)

(a) $\bar{x}_1 - \bar{x}_2 = 1225 - 1195 = 30$

(b) $z = \dfrac{(\bar{x}_1 - \bar{x}_2) - (\mu_1 - \mu_2)}{\sqrt{\dfrac{s_1^2}{n_1} + \dfrac{s_2^2}{n_2}}} = \dfrac{(1225 - 1195) - (0)}{\sqrt{\dfrac{(75)^2}{35} + \dfrac{(105)^2}{105}}} = \dfrac{30}{\sqrt{265.714}} \approx 1.84$

(c) z is not in the rejection region because $1.84 > -2.330$.

(d) Fail to reject H_0. There is not enough evidence to support the claim.

16. H_0: $\mu_1 \le \mu_2$ (claim); H_1: $\mu_1 > \mu_2$

Rejection region: $z_0 > 1.88$ (Right-tailed test)

(a) $\bar{x}_1 - \bar{x}_2 = 5004 - 4895 = 109$

(b) $z = \dfrac{(\bar{x}_1 - \bar{x}_2) - (\mu_1 - \mu_2)}{\sqrt{\dfrac{s_1^2}{n_1} + \dfrac{s_2^2}{n_2}}} = \dfrac{(5004 - 4895) - (0)}{\sqrt{\dfrac{(136)^2}{144} + \dfrac{(215)^2}{156}}} = \dfrac{109}{\sqrt{424.759}} \approx 5.29$

(c) z is in the rejection region because $5.29 > 1.88$.

(d) Reject H_0. There is enough evidence to reject the claim.

17. H_0: $\mu_1 \le \mu_2$; H_a: $\mu_1 > \mu_2$ (claim)

$z_0 = 2.33$; Reject H_0 if $z > 2.33$.

$z = \dfrac{(\bar{x}_1 - \bar{x}_2) - (\mu_1 - \mu_2)}{\sqrt{\dfrac{s_1^2}{n_1} + \dfrac{s_2^2}{n_2}}} = \dfrac{(5.2 - 5.5) - (0)}{\sqrt{\dfrac{(0.2)^2}{45} + \dfrac{(0.3)^2}{37}}} = \dfrac{-.30}{\sqrt{0.00332}} \approx -5.207$

Fail to reject H_0. There is not enough evidence to support the claim.

18. H_0: $\mu_1 = \mu_2$ (claim); H_a: $\mu_1 \ne \mu_2$

$z_0 = \pm 1.96$

$z = \dfrac{(\bar{x}_1 - \bar{x}_2) - (\mu_1 - \mu_2)}{\sqrt{\dfrac{s_1^2}{n_1} + \dfrac{s_2^2}{n_2}}} = \dfrac{(52 - 45) - (0)}{\sqrt{\dfrac{(2.5)^2}{70} + \dfrac{(5.5)^2}{60}}} = \dfrac{7}{\sqrt{0.59345}} \approx 9.087$

Reject H_0. There is enough evidence to support the claim.

19. (a) H_0: $\mu_1 = \mu_2$; H_a: $\mu_1 \ne \mu_2$ (claim)

(b) $z_0 = \pm 1.645$; Reject H_0 if $z < -1.645$ or $z > 1.645$.

(c) $z = \dfrac{(\bar{x}_1 - \bar{x}_2) - (\mu_1 - \mu_2)}{\sqrt{\dfrac{s_1^2}{n_1} + \dfrac{s_2^2}{n_2}}} = \dfrac{(42 - 45) - (0)}{\sqrt{\dfrac{(4.7)^2}{35} + \dfrac{(4.3)^2}{35}}} = \dfrac{-3}{\sqrt{1.159}} \approx -2.786$

(d) Reject H_0.

(e) There is sufficient evidence at the 10% significance level to support the claim that the mean braking distance is different for both types of tires.

20. (a) H_0: $\mu_1 \le \mu_2$; H_a: $\mu_1 > \mu_2$ (claim)

(b) $z_0 = 1.28$; Reject H_0 if $z > 1.28$.

(c) $z = \dfrac{(\bar{x}_1 - \bar{x}_2) - (\mu_1 - \mu_2)}{\sqrt{\dfrac{s_1^2}{n_1} + \dfrac{s_2^2}{n_2}}} = \dfrac{(55 - 51) - (0)}{\sqrt{\dfrac{(5.3)^2}{50} + \dfrac{(4.9)^2}{50}}} = \dfrac{4}{\sqrt{1.042}} \approx 3.92$

(d) Reject H_0.

(e) There is sufficient evidence at the 10% significance level to support the claim that the mean braking distance for Type C is greater than for Type D.

21. (a) $H_0: \mu_1 \geq \mu_2; H_a: \mu_1 < \mu_2$ (claim)

(b) $z_0 = -2.33$; Reject H_0 if $z < -2.33$.

(c) $z = \dfrac{(\bar{x}_1 - \bar{x}_2) - (\mu_1 - \mu_2)}{\sqrt{\dfrac{s_1^2}{n_1} + \dfrac{s_2^2}{n_2}}} = \dfrac{(75 - 80) - (0)}{\sqrt{\dfrac{(12.50)^2}{47} + \dfrac{(20)^2}{55}}} = \dfrac{-5}{\sqrt{10.597}} \approx -1.54$

(d) Fail to reject H_0.

(e) There is insufficient evidence at the 1% significance level to conclude that the repair costs for Model A are lower than for Model B.

22. (a) $H_0: \mu_1 = \mu_2$ (claim); $H_a: \mu_1 \neq \mu_2$

(b) $z_0 = \pm 2.575$; Reject H_0 if $z < -2.575$ or $z > 2.575$.

(c) $z = \dfrac{(\bar{x}_1 - \bar{x}_2) - (\mu_1 - \mu_2)}{\sqrt{\dfrac{s_1^2}{n_1} + \dfrac{s_2^2}{n_2}}} = \dfrac{(50 - 60) - (0)}{\sqrt{\dfrac{(10)^2}{34} + \dfrac{(18)^2}{46}}} = \dfrac{-10}{\sqrt{9.985}} \approx -3.165$

(d) Reject H_0.

(e) There is sufficient evidence at the 1% significance level to reject the claim that the mean repair costs for Model A and Model B are the same.

23. (a) $H_0: \mu_1 = \mu_2$ (claim); $H_a: \mu_1 \neq \mu_2$

(b) $z_0 = \pm 2.575$; Reject H_0 if $z < -2.575$ or $z > 2.575$.

(c) $z = \dfrac{(\bar{x}_1 - \bar{x}_2) - (\mu_1 - \mu_2)}{\sqrt{\dfrac{s_1^2}{n_1} + \dfrac{s_2^2}{n_2}}} = \dfrac{(21.0 - 20.8) - (0)}{\sqrt{\dfrac{(5.0)^2}{43} + \dfrac{(4.7)^2}{56}}} = \dfrac{0.2}{\sqrt{0.976}} \approx 0.202$

(d) Fail to reject H_0.

(e) There is insufficient evidence at the 1% significance level to reject the claim that the male and female high school students have equal ACT scores.

24. (a) $H_0: \mu_1 \leq \mu_2; H_a: \mu_1 > \mu_2$ (claim)

(b) $z_0 = 1.28$; Reject H_0 if $z > 1.28$.

(c) $z = \dfrac{(\bar{x}_1 - \bar{x}_2) - (\mu_1 - \mu_2)}{\sqrt{\dfrac{s_1^2}{n_1} + \dfrac{s_2^2}{n_2}}} = \dfrac{(22.2 - 20.0) - (0)}{\sqrt{\dfrac{(4.8)^2}{49} + \dfrac{(5.4)^2}{4}}} = \dfrac{2.2}{\sqrt{1.133}} \approx 2.07$

(d) Reject H_0.

(e) There is sufficient evidence at the 10% significance level to support the claim that the ACT scores are higher for high school students in a college prep program.

25. (a) $H_0: \mu_1 = \mu_2$ (claim); $H_a: \mu_1 \neq \mu_2$

(b) $z_0 = \pm 1.645$; Reject H_0 if $z < -1.645$ or $z > 1.645$.

(c) $z = \dfrac{(\bar{x}_1 - \bar{x}_2) - (\mu_1 - \mu_2)}{\sqrt{\dfrac{s_1^2}{n_1} + \dfrac{s_2^2}{n_2}}} = \dfrac{(131 - 136) - (0)}{\sqrt{\dfrac{(26)^2}{35} + \dfrac{(19)^2}{35}}} = \dfrac{-5}{\sqrt{29.629}} \approx -0.919$

(d) Fail to reject H_0.

(e) There is insufficient evidence at the 10% significance level to reject the claim that the lodging cost for a family traveling in North Carolina is the same as South Carolina.

26. (a) H_0: $\mu_1 \geq \mu_2$; H_a: $\mu_1 < \mu_2$ (claim)

(b) $z_0 = -1.645$; Reject H_0 if $z < -1.645$.

(c) $z = \dfrac{(\bar{x}_1 - \bar{x}_2) - (\mu_1 - \mu_2)}{\sqrt{\dfrac{s_1^2}{n_1} + \dfrac{s_2^2}{n_2}}} = \dfrac{(2015 - 2715) - (0)}{\sqrt{\dfrac{(113)^2}{30} + \dfrac{(97)^2}{30}}} = \dfrac{-700}{\sqrt{739.267}} \approx -25.745$

(d) Reject H_0.

(e) There is sufficient evidence at the 5% significance level to support the claim that households in the United States headed by people under the age of 25 spend less on food away from home than households headed by people ages 55–64.

27. (a) H_0: $\mu_1 = \mu_2$ (claim); H_a: $\mu_1 \neq \mu_2$

(b) $z_0 = \pm 1.645$; Reject H_0 if $z < -1.645$ or $z > 1.645$.

(c) $z = \dfrac{(\bar{x}_1 - \bar{x}_2) - (\mu_1 - \mu_2)}{\sqrt{\dfrac{s_1^2}{n_1} + \dfrac{s_2^2}{n_2}}} = \dfrac{(145 - 138) - (0)}{\sqrt{\dfrac{(28)^2}{50} + \dfrac{(24)^2}{50}}} = \dfrac{7}{\sqrt{27.2}} \approx 1.342$

(d) Fail to reject H_0.

(e) There is insufficient evidence at the 10% significance level to reject the claim that the lodging cost for a family traveling in North Carolina is the same as South Carolina. The new samples do not lead to a different conclusion.

28. H_0: $\mu_1 \geq \mu_1$; H_a: $\mu_1 < \mu_2$ (claim)

$z_0 = -1.645$; Reject H_0 if $z < -1.645$.

$z = \dfrac{(\bar{x}_1 - \bar{x}_2) - (\mu_1 - \mu_2)}{\sqrt{\dfrac{s_1^2}{n_1} + \dfrac{s_2^2}{n_2}}} = \dfrac{(2130 - 2655) - (0)}{\sqrt{\dfrac{(124)^2}{40} + \dfrac{(116)^2}{40}}} = \dfrac{-525}{\sqrt{720.8}} = -19.555$

Reject H_0. There is enough evidence to support the claim.

29. (a) H_0: $\mu_1 \leq \mu_2$; H_a: $\mu_1 < \mu_2$ (claim)

(b) $z_0 = 1.96$; Reject H_0 if $z > 1.96$.

(c) $\bar{x}_1 \approx 2.130$, $s_1 \approx 0.490$, $n_1 = 30$
$\bar{x}_2 \approx 1.757$, $s_2 \approx 0.470$, $n_2 = 30$
$z = \dfrac{(\bar{x}_1 - \bar{x}_2) - (\mu_1 - \mu_2)}{\sqrt{\dfrac{s_1^2}{n_1} + \dfrac{s_2^2}{n_2}}} = \dfrac{(2.130 - 1.757) - (0)}{\sqrt{\dfrac{(0.490)^2}{30} + \dfrac{(0.470)^2}{30}}} = \dfrac{0.373}{\sqrt{0.0154}} \approx 3.01$

(d) Reject H_0.

(e) At the 2.5% level of significance, there is sufficient evidence to support the claim.

30. (a) H_0: $\mu_1 \geq \mu_2$; H_a: $\mu_1 < \mu_2$ (claim)

(b) $z_0 = -1.88$; Reject H_0 if $z < -1.88$.

(c) $\bar{x}_1 \approx 32.523, s_1 \approx 4.477, n_1 = 35$

$\bar{x}_2 = 47.989, s_2 \approx 4.651, n_2 = 35$

$$z = \frac{(\bar{x}_1 - \bar{x}_2) - (\mu_1 - \mu_2)}{\sqrt{\frac{s_1^2}{n_1} + \frac{s_2^2}{n_2}}} = \frac{(32.523 - 47.989) - (0)}{\sqrt{\frac{(4.477)^2}{35} + \frac{(4.651)^2}{35}}} = \frac{-15.466}{\sqrt{1.191}} \approx -14.17$$

(d) Reject H_0.

(e) There is sufficient evidence at the 3% significance level to support the sociologist's claim.

31. (a) $H_0: \mu_1 = \mu_2$ (claim); $H_a: \mu_1 \neq \mu_2$

(b) $z_0 = \pm 2.575$; Reject H_0 if $z < -2.575$ or $z > 2.575$.

(c) $\bar{x}_1 \approx 0.875, s_1 \approx 0.011, n_1 = 35$
$\bar{x}_2 \approx 0.701, s_2 \approx 0.011, n_2 = 35$

$$z = \frac{(\bar{x}_1 - \bar{x}_2) - (\mu_1 - \mu_2)}{\sqrt{\frac{s_1^2}{n_1} + \frac{s_2^2}{n_2}}} = \frac{(0.875 - 0.701) - (0)}{\sqrt{\frac{(0.011)^2}{35} + \frac{(0.011)^2}{35}}} = \frac{0.174}{\sqrt{0.0000006914}} \approx 66.172$$

(d) Reject H_0.

(e) At the 1% level of significance, there is sufficient evidence to reject the claim.

32. (a) $H_0: \mu_1 = \mu_2$ (claim); $H_a: \mu_1 \neq \mu_2$

(b) $z_0 \approx \pm 2.05$; Reject H_0 if $z < -2.05$ or $z > 2.05$.

(c) $\bar{x}_1 = 3.337, s_1 = 0.011, n_1 = 40$

$\bar{x}_2 = 3.500, s_2 = 0.010, n_2 = 40$

$$z = \frac{(\bar{x}_1 - \bar{x}_2) - (\mu_1 - \mu_2)}{\sqrt{\frac{s_1^2}{n_1} + \frac{s_2^2}{n_2}}} = \frac{(3.337 - 3.500) - (0)}{\sqrt{\frac{(0.011)^2}{40} + \frac{(0.010)^2}{40}}} = \frac{-0.163}{\sqrt{0.000005525}} \approx -69.346$$

(d) Reject H_0.

(e) There is sufficient evidence at the 4% level of significance to reject the claim.

33. They are equivalent through algebraic manipulation of the equation.

$\mu_1 = \mu_2 \rightarrow \mu_1 - \mu_2 = 0$

34. They are equivalent through algebraic manipulation of the equation.

$\mu_1 \geq \mu_2 \rightarrow \mu_1 - \mu_2 \geq 0$

35. $H_0: \mu_1 - \mu_2 = -9$ (claim); $H_a: \mu_1 - \mu_2 \neq -9$

$z_0 = \pm 2.575$; Reject H_0 if $z < -2.575$ or $z > 2.575$.

$$z = \frac{(\bar{x}_1 - \bar{x}_2) - (\mu_1 - \mu_2)}{\sqrt{\frac{s_1^2}{n_1} + \frac{s_2^2}{n_2}}} = \frac{(11.5 - 20) - (-9)}{\sqrt{\frac{(3.8)^2}{70} + \frac{(6.7)^2}{65}}} = \frac{0.5}{\sqrt{0.897}} \approx 0.528$$

Fail to reject H_0. There is not enough evidence to reject the claim.

36. H_0: $\mu_1 - \mu_2 = -2$ (claim); H_a: $\mu_1 - \mu_2 \neq -2$

$z_0 = \pm1.96$; Reject H_0 if $z < -1.96$ or $z > 1.96$.

$$z = \frac{(\bar{x}_1 - \bar{x}_2) - (\mu_1 - \mu_2)}{\sqrt{\frac{s_1^2}{n_1} + \frac{s_2^2}{n_2}}} = \frac{(12.95 - 15.02) - (-2)}{\sqrt{\frac{(4.31)^2}{48} + \frac{(4.99)^2}{56}}} = \frac{-0.07}{\sqrt{0.8316}} \approx -0.077$$

Fail to reject H_0. There is not enough evidence to reject the claim.

37. H_0: $\mu_1 - \mu_2 \leq 6000$; H_a: $\mu_1 - \mu_2 > 6000$ (claim)

$z_0 = 1.28$; Reject H_0 if $z > 1.28$.

$$z = \frac{(\bar{x}_1 - \bar{x}_2) - (\mu_1 - \mu_2)}{\sqrt{\frac{s_1^2}{n_1} + \frac{s_2^2}{n_2}}} = \frac{(67,900 - 64,000) - (6000)}{\sqrt{\frac{(8875)^2}{45} + \frac{(9175)^2}{42}}} = \frac{-2100}{\sqrt{3,754,647.817}} \approx -1.084$$

Fail to reject H_0. There is not enough evidence to support the claim.

38. H_0: $\mu_1 - \mu_2 \leq 30,000$; H_a: $\mu_1 - \mu_2 > 30,000$ (claim)

$z_0 = 1.645$; Reject H_0 if $z > 1.645$.

$$z = \frac{(\bar{x}_1 - \bar{x}_2) - (\mu_1 - \mu_2)}{\sqrt{\frac{s_1^2}{n_1} + \frac{s_2^2}{n_2}}} = \frac{(54,900 - 27,200) - (30,000)}{\sqrt{\frac{(8250)^2}{31} + \frac{(3200)^2}{33}}} = \frac{-2300}{\sqrt{2,505,867.546}} = -1.453$$

Fail to reject H_0. There is not enough evidence to support the claim.

39. $(\bar{x}_1 - \bar{x}_2) - z_c\sqrt{\frac{s_1^2}{n_1} + \frac{s_2^2}{n_2}} < \mu_1 - \mu_2 < (\bar{x}_1 - \bar{x}_2) + z_c\sqrt{\frac{s_1^2}{n_1} + \frac{s_2^2}{n_2}}$

$(123.1 - 125) - 1.96\sqrt{\frac{(9.9)^2}{269} + \frac{(10.1)^2}{268}} < \mu_1 - \mu_2 < (123.1 - 125) + 1.96\sqrt{\frac{(9.9)^2}{269} + \frac{(10.1)^2}{268}}$

$-1.9 - 1.96\sqrt{0.745} < \mu_1 - \mu_2 < -1.9 + 1.96\sqrt{0.745}$

$-3.6 < \mu_1 - \mu_2 < -0.2$

40. $(\bar{x}_1 - \bar{x}_2) - z_c\sqrt{\frac{s_1^2}{n_1} + \frac{s_2^2}{n_2}} < \mu_1 - \mu_2 < (\bar{x}_1 - \bar{x}_2) + z_c\sqrt{\frac{s_1^2}{n_1} + \frac{s_2^2}{n_2}}$

$(10.3 - 8.5) - 1.96\sqrt{\frac{(1.2)^2}{140} + \frac{(1.5)^2}{127}} < \mu_1 - \mu_2 < (10.3 - 8.5) + 1.96\sqrt{\frac{(1.2)^2}{140} + \frac{(1.5)^2}{127}}$

$1.8 - 1.96\sqrt{0.028} < \mu_1 - \mu_2 < 1.8 + 1.96\sqrt{0.028}$

$1.5 < \mu_1 - \mu_2 < 2.1$

41. H_0: $\mu_1 - \mu_2 \geq 0$; H_a: $\mu_1 - \mu_2 < 0$ (claim)

$z_0 = -1.645$; Reject H_0 if $z < -1.645$.

$$z = \frac{(\bar{x}_1 - \bar{x}_2) - (\mu_1 - \mu_2)}{\sqrt{\frac{s_1^2}{n_1} + \frac{s_2^2}{n_2}}} = \frac{(123.1 - 125) - (0)}{\sqrt{\frac{(9.9)^2}{269} + \frac{(10.1)^2}{268}}} = \frac{-1.9}{\sqrt{0.745}} \approx -2.20$$

Reject H_0. There is enough evidence to support the claim. I would recommend using the DASH diet and exercise program over the traditional diet and exercise program.

42. $H_0: \mu_1 - \mu_2 \leq 0; H_a: \mu_1 - \mu_2 > 0$ (claim)

$z_0 = -1.645$; Reject H_0 if $z > -1.645$.

$$z = \frac{(\bar{x}_1 - \bar{x}_2) - (\mu_1 - \mu_2)}{\sqrt{\dfrac{s_1^2}{n_1} + \dfrac{s_2^2}{n_2}}} = \frac{(10.3 - 8.5) - (0)}{\sqrt{\dfrac{(1.2)^2}{140} + \dfrac{(1.5)^2}{127}}} = \frac{1.8}{\sqrt{0.028}} \approx 10.76$$

Reject H_0. There is enough evidence to support the claim. I would recommend using Irinotecan over Fluorouracil because the average number of months with no reported cancer related pain was significantly higher.

43. $H_0: \mu_1 - \mu_2 \geq 0; H_a: \mu_1 - \mu_2 < 0$ (claim)

The 95% CI for $\mu_1 - \mu_2$ in Exercise 39 contained values greater than or equal to zero and, as found in Exercise 41, there was enough evidence at the 5% level of significance to support the claim. If zero is not contained in the CI for $\mu_1 - \mu_2$, you reject H_0 because the null hypothesis states that $\mu_1 - \mu_2$ is greater than or equal to zero.

44. $H_0: \mu_1 - \mu_2 \leq 0; H_1: \mu_1 - \mu_2 > 0$ (claim)

The 95% CI for $\mu_1 - \mu_2$ in Exercise 40 contained only values greater than zero and, as found in Exercise 42, there was sufficient evidence at the 5% level of significance to support the claim. If the CI for $\mu_1 - \mu_2$ contains only positive numbers, you reject H_0, because the null hypothesis states that $\mu_1 - \mu_2$ is less than or equal to zero.

8.2 TESTING THE DIFFERENCE BETWEEN MEANS (SMALL INDEPENDENT SAMPLES)

8.2 Try It Yourself Solutions

1a. $H_0: \mu_1 = \mu_2; H_a: \mu_1 \neq \mu_2$ (claim)

b. $\alpha = 0.05$

c. d.f. $= \min\{n_1 - 1, n_2 - 1\} = \min\{8 - 1, 10 - 1\} = 7$

d. $t_0 = \pm 2.365$; Reject H_0 if $t < -2.365$ or $t > 2.365$.

e. $t = \dfrac{(\bar{x}_1 - \bar{x}_2) - (\mu_1 - \mu_2)}{\sqrt{\dfrac{s_1^2}{n_1} + \dfrac{s_2^2}{n_2}}} = \dfrac{(141 - 151) - (0)}{\sqrt{\dfrac{(7.0)^2}{8} + \dfrac{(3.1)^2}{10}}} = \dfrac{-10}{\sqrt{7.086}} \approx -3.757$

f. Reject H_0.

g. There is enough evidence to support the claim.

2a. $H_0: \mu_1 \geq \mu_2; H_a: \mu_1 < \mu_2$ (claim)   **b.** $\alpha = 0.10$

c. d.f. $= n_1 = n_2 - 2 = 12 + 15 - 2 = 25$ **d.** $t_0 = -1.316$; Reject H_0 if $t < -1.316$.

e. $t = \dfrac{(\bar{x}_1 - \bar{x}_2) - (\mu_1 - \mu_2)}{\sqrt{\dfrac{(n_1 - 1)s_1^2 + (n_2 - 1)s_2^2}{n_1 + n_2 - 2}}\sqrt{\dfrac{1}{n_1} + \dfrac{1}{n_2}}} = \dfrac{(32 - 35) - (0)}{\sqrt{\dfrac{(12 - 1)(2.1)^2 + (15 - 1)(1.8)^2}{12 + 15 - 2}}\sqrt{\dfrac{1}{12} + \dfrac{1}{15}}}$

$= \dfrac{-3}{\sqrt{3.755}\sqrt{0.15}} \approx -3.997$

f. Reject H_0.

g. There is enough evidence to support the claim.

8.2 EXERCISE SOLUTIONS

1. State hypotheses and identify the claim. Specify the level of significance. Determine the degrees of freedom. Find the critical value(s) and identify the rejection region(s). Find the standardized test statistic. Make a decision and interpret in the context of the original claim.

2. (1) The samples must be randomly selected.

   (2) The samples must be independent.

   (3) Each population must have a normal distribution.

3. (a) d.f. $= n_1 + n_2 - 2 = 23$
      $t_0 = \pm 1.714$

   (b) d.f. $= \min\{n_1 - 1, n_2 - 1\} = 10$
      $t_0 = \pm 1.812$

4. (a) d.f. $= n_1 + n_2 - 2 = 25$
      $t_0 = 2.485$

   (b) d.f. $= \min\{n_1 - 1, n_2 - 1\} = 11$
      $t_0 = 2.718$

5. (a) d.f. $= n_1 + n_2 - 2 = 22$
      $t_0 = -2.074$

   (b) d.f. $= \min\{n_1 - 1, n_2 - 1\} = 8$
      $t_0 = -2.306$

6. (a) d.f. $= n_1 + n_2 - 2 = 39$
      $t_0 = \pm 1.96$

   (b) d.f. $= \min\{n_1 - 1, n_2 - 1\} = 18$
      $t_0 = \pm 2.101$

7. (a) d.f. $= n_1 + n_2 - 2 = 19$
      $t_0 = 1.729$

   (b) d.f. $= \min\{n_1 - 1, n_2 - 1\} = 7$
      $t_0 = 1.895$

8. (a) d.f. $= n_1 + n_2 - 2 = 11$
      $t_0 = -1.363$

   (b) d.f. $= \min\{n_1 - 1, n_2 - 1\} = 3$
      $t_0 = -1.638$

9. (a) d.f. $= n_1 + n_2 - 2 = 27$
      $t_0 = \pm 2.771$

   (b) d.f. $= \min\{n_1 - 1, n_2 - 1\} = 11$
      $t_0 = \pm 3.106$

10. (a) d.f. $= n_1 + n_2 - 2 = 16$
       $t_0 = 2.921$

    (b) d.f. $= \min\{n_1 - 1, n_2 - 1\} = 6$
       $t_0 = 3.707$

11. H_0: $\mu_1 = \mu_2$ (claim); H_a: $\mu_1 \neq \mu_2$
    d.f. $= n_1 + n_2 - 2 = 15$
    $t_0 = \pm 2.947$ (Two-tailed test)

    (a) $\bar{x}_1 - \bar{x}_2 = 33.7 - 35.5 = -1.8$

    (b) $t = \dfrac{(\bar{x}_1 - \bar{x}_2) - (\mu_1 - \mu_2)}{\sqrt{\dfrac{(n_1 - 1)s_1^2 + (n_2 - 1)s_2^2}{n_1 + n_2 - 2}} \sqrt{\dfrac{1}{n_1} + \dfrac{1}{n_2}}} = \dfrac{(33.7 - 35.5) - (0)}{\sqrt{\dfrac{(10 - 1)(3.5)^2 + (7 - 1)(2.2)^2}{10 + 7 - 2}} \sqrt{\dfrac{1}{10} + \dfrac{1}{7}}}$

    $ = \dfrac{-1.8}{\sqrt{9.286}\,\sqrt{0.243}} \approx -1.199$

    (c) t is not in the rejection region.

    (d) Fail to reject H_0.

12. H_0: $\mu_1 \geq \mu_2$ (claim); H_a: $\mu_1 < \mu_2$

d.f. $= n_1 + n_2 - 2 = 18$

$t_0 = -1.330$ (Left-tailed test)

(a) $\bar{x}_1 - \bar{x}_2 = 0.515 - 0.475 = 0.04$

(b) $t = \dfrac{(\bar{x}_1 - \bar{x}_2) - (\mu_1 - \mu_2)}{\sqrt{\dfrac{(n_1 - 1)s_1^2 + (n_2 - 1)s_2^2}{n_1 + n_2 - 2}}\sqrt{\dfrac{1}{n_1} + \dfrac{1}{n_2}}}$

$= \dfrac{(0.515 - 0.475) - (0)}{\sqrt{\dfrac{(11 - 1)(0.305)^2 + (9 - 1)(0.215)^2}{11 + 9 - 2}}\sqrt{\dfrac{1}{11} + \dfrac{1}{9}}} = \dfrac{0.4}{\sqrt{0.0722}\sqrt{0.202}} \approx 0.331$

(c) t is not in the rejection region.

(d) Fail to reject H_0.

13. There is no need to run the test since this is a right-tailed test and the test statistic is negative. It is obvious that the standardized test statistic will also be negative and fall outside of the rejection region. So, the decision is to fail to reject H_0.

14. H_0: $\mu_1 \leq \mu_2$ (claim); H_a: $\mu_1 > \mu_2$

d.f. $= \min\{n_1 - 1, n_2 - 1\} = 13$

(a) $\bar{x}_1 - \bar{x}_2 = 45 - 50 = -5$

(b) $t_0 = 2.650$ (Right-tailed test)

$t = \dfrac{(\bar{x}_1 - \bar{x}_2) - (\mu_1 - \mu_2)}{\sqrt{\dfrac{s_1^2}{n_1} + \dfrac{s_2^2}{n_2}}} = \dfrac{(45 - 50) - (0)}{\sqrt{\dfrac{(4.8)^2}{16} + \dfrac{(1.2)^2}{14}}} = \dfrac{-5}{\sqrt{1.543}} \approx -4.025$

(c) t is not in the rejection region.

(d) Fail to reject H_0.

15. (a) H_0: $\mu_1 = \mu_2$ (claim); H_a: $\mu_1 \neq \mu_2$

(b) d.f. $= n_1 + n_2 - 2 = 12 + 17 - 2 = 27$

$t_0 = \pm 1.703$; Reject H_0 if $t < -1.703$ or $t > 1.703$.

(c) $t = \dfrac{(\bar{x}_1 - \bar{x}_2) - (\mu_1 - \mu_2)}{\sqrt{\dfrac{(n_1 - 1)s_1^2 + (n_2 - 1)s_2^2}{n_1 + n_2 - 2}}\sqrt{\dfrac{1}{n_1} + \dfrac{1}{n_2}}}$

$= \dfrac{(10.1 - 8.3) - (0)}{\sqrt{\dfrac{(12 - 1)(4.11)^2 + (17 - 1)(4.02)^2}{12 + 17 - 2}}\sqrt{\dfrac{1}{12} + \dfrac{1}{17}}} = \dfrac{1.8}{\sqrt{16.459}\sqrt{0.142}} \approx 1.177$

(d) Fail to reject H_0.

(e) There is not enough evidence to reject the claim.

16. (a) H_0: $\mu_1 = \mu_2$ (claim); H_a: $\mu_1 \neq \mu_2$

(b) d.f. $= n_1 + n_2 - 2 = 5 + 8 - 2 = 11$

$t_0 = \pm 2.201$; Reject H_0 if $t < -2.201$ or $t > 2.201$.

(c) $t = \dfrac{(\bar{x}_1 - \bar{x}_2) - (\mu_1 - \mu_2)}{\sqrt{\dfrac{(n_1 - 1)s_1^2 + (n_2 - 1)s_2^2}{n_1 + n_2 - 2}}\sqrt{\dfrac{1}{n_1} + \dfrac{1}{n_2}}}$

$= \dfrac{(11.0 - 10.6) - (0)}{\sqrt{\dfrac{(5 - 1)(4.07)^2 + (8 - 1)(6.62)^2}{5 + 8 - 2}}\sqrt{\dfrac{1}{5} + \dfrac{1}{8}}} = \dfrac{0.4}{\sqrt{33.912}\,\sqrt{0.325}} \approx 0.120$

(d) Fail to reject H_0.

(e) There is not enough evidence to reject the claim.

17. (a) $H_0: \mu_1 \geq \mu_2$; $H_a: \mu_1 < \mu_2$ (claim)

(b) d.f. $= n_1 + n_2 - 2 = 35$

$t_0 = -1.282$; Reject H_0 if $t < -1.282$.

(c) $t = \dfrac{(\bar{x}_1 - \bar{x}_2) - (\mu_1 - \mu_2)}{\sqrt{\dfrac{(n_1 - 1)s_1^2 + (n_2 - 1)s_2^2}{n_1 + n_2 - 2}}\sqrt{\dfrac{1}{n_1} + \dfrac{1}{n_2}}} = \dfrac{(473 - 741) - (0)}{\sqrt{\dfrac{(14 - 1)(190)^2 + (23 - 1)(205)^2}{14 + 23 - 2}}\sqrt{\dfrac{1}{14} + \dfrac{1}{23}}}$

$= \dfrac{-268}{\sqrt{39,824.286}\,\sqrt{0.115}} = -3.960$

(d) Reject H_0.

(e) There is enough evidence to support the claim.

18. (a) $H_0: \mu_1 \leq \mu_2$; $H_a: \mu_1 > \mu_2$ (claim)

(b) d.f. $= n_1 + n_2 - 2 = 11$

$t_0 = 1.363$; Reject H_0 if $t > 1.363$.

(c) $t = \dfrac{(\bar{x}_1 - \bar{x}_2) - (\mu_1 - \mu_2)}{\sqrt{\dfrac{(n_1 - 1)s_1^2 + (n_2 - 1)s_2^2}{n_1 + n_2 - 2}}\sqrt{\dfrac{1}{n_1} + \dfrac{1}{n_2}}}$

$= \dfrac{(1090 - 485) - (0)}{\sqrt{\dfrac{(5 - 1)(403)^2 + (8 - 1)(382)^2}{5 + 8 - 2}}\sqrt{\dfrac{1}{5} + \dfrac{1}{8}}}$

$= \dfrac{605}{\sqrt{151,918.546}\,\sqrt{0.325}} \approx 2.723$

(d) Reject H_0.

(e) There is enough evidence to support the claim.

19. (a) $H_0: \mu_1 \leq \mu_2$; $H_a: \mu_1 > \mu_2$ (claim)

(b) d.f. $= \min\{n_1 - 1, n_2 - 1\} = 14$

$t_0 = 1.345$; Reject H_0 if $t > 1.345$.

(c) $z = \dfrac{(\bar{x}_1 - \bar{x}_2) - (\mu_1 - \mu_2)}{\sqrt{\dfrac{s_1^2}{n_1} + \dfrac{s_2^2}{n_2}}} = \dfrac{(42,200 - 37,900) - (0)}{\sqrt{\dfrac{(8600)^2}{19} + \dfrac{(5500)^2}{15}}} = \dfrac{4300}{\sqrt{5,909,298.246}} \approx 1.769$

(d) Reject H_0.

(e) There is enough evidence to support the claim.

20. (a) H_0: $\mu_1 = \mu_2$; H_a: $\mu_1 \neq \mu_2$ (claim)

(b) d.f. $= n_1 + n_2 - 2 = 33$

$t_0 = \pm 2.576$; Reject H_0 if $t < -2.576$ or $t > 2.576$.

(c) $t = \dfrac{(\bar{x}_1 - \bar{x}_2) - (\mu_1 - \mu_2)}{\sqrt{\dfrac{(n_1 - 1)s_1^2 + (n_2 - 1)s_2^2}{n_1 + n_2 - 2}} \sqrt{\dfrac{1}{n_1} + \dfrac{1}{n_2}}}$

$= \dfrac{(36{,}700 - 34{,}700) - (0)}{\sqrt{\dfrac{(17 - 1)(7800)^2 + (18 - 1)(7375)^2}{17 + 18 - 2}} \sqrt{\dfrac{1}{17} + \dfrac{1}{18}}}$

$= \dfrac{2000}{\sqrt{57{,}517{,}594.70} \sqrt{0.1144}} \approx 0.780$

(d) Fail to reject H_0.

(e) There is not enough evidence to support the claim.

21. (a) H_0: $\mu_1 = \mu_2$; H_a: $\mu_1 \neq \mu_2$ (claim)

(b) d.f. $= n_1 + n_2 - 2 = 21$

$t_0 \pm 2.831$; Reject H_0 if $t < -2.831$ or $t > 2.831$.

(c) $\bar{x}_1 = 340.300$, $s_1 = 22.301$, $n_1 = 10$

$\bar{x}_2 = 389.538$, $s_2 = 14.512$, $n_2 = 13$

$t = \dfrac{(\bar{x}_1 - \bar{x}_2) - (\mu_1 - \mu_2)}{\sqrt{\dfrac{(n_1 - 1)s_1^2 + (n_2 - 1)s_2^2}{n_1 + n_2 - 2}} \sqrt{\dfrac{1}{n_1} + \dfrac{1}{n_2}}}$

$= \dfrac{(340.300 - 389.538) - (0)}{\sqrt{\dfrac{(10 - 1)(22.301)^2 + (13 - 1)(14.512)^2}{10 + 13 - 2}} \sqrt{\dfrac{1}{10} + \dfrac{1}{13}}} = \dfrac{-49.238}{\sqrt{333.485} \sqrt{0.177}} \approx -6.410$

(d) Reject H_0.

(e) There is enough evidence to support the claim.

22. (a) H_0: $\mu_1 \leq \mu_2$; H_a: $\mu_1 > \mu_2$ (claim)

(b) d.f. $= \min\{n_1 - 1, n_2 - 1\} = 13$

$t_0 = 1.350$; Reject H_0 if $t > 1.350$.

(c) $\bar{x}_1 = 402.765$, $s_1 = 11.344$, $n_1 = 17$

$\bar{x}_2 = 384.000$, $s_2 = 17.698$, $n_2 = 14$

$t = \dfrac{(\bar{x}_1 - \bar{x}_2) - (\mu_1 - \mu_2)}{\sqrt{\dfrac{s_1^2}{n_1} + \dfrac{s_2^2}{n_2}}} = \dfrac{(402.765 - 384.000) - (0)}{\sqrt{\dfrac{(11.344)^2}{17} + \dfrac{(17.698)^2}{14}}} = \dfrac{18.765}{\sqrt{29.943}} \approx 3.429$

(d) Reject H_0.

(e) There is enough evidence to support the claim and to recommend using the experimental method.

23. (a) H_0: $\mu_1 \geq \mu_2$; H_a: $\mu_1 < \mu_2$ (claim)

(b) d.f. $= n_1 + n_2 - 2 = 42$

$t_0 = -1.282 \to t < -1.282$

(c) $\bar{x}_1 = 56.684$, $s_1 = 6.961$, $n_1 = 19$

$\bar{x}_2 = 67.400$, $s_2 = 9.014$, $n_2 = 25$

$$t = \frac{(\bar{x}_1 - \bar{x}_2) - (\mu_1 - \mu_2)}{\sqrt{\dfrac{(n_1 - 1)s_1^2 + (n_2 - 1)s_2^2}{n_1 + n_2 - 2}} \sqrt{\dfrac{1}{n_1} + \dfrac{1}{n_2}}}$$

$$= \frac{(56.684 - 67.400) - (0)}{\sqrt{\dfrac{(19 - 1)(6.961)^2 + (25 - 1)(9.014)^2}{19 + 25 - 2}} \sqrt{\dfrac{1}{19} + \dfrac{1}{25}}} = \frac{-10.716}{\sqrt{67.196}\,\sqrt{0.0926}} \approx -4.295$$

(d) Reject H_0.

(e) There is enough evidence to support the claim and to recommend changing to the new method.

24. (a) H_0: $\mu_1 \geq \mu_2$; H_a: $\mu_1 < \mu_2$ (claim)

(b) d.f. $= n_1 + n_2 - 2 = 39$

$t_0 = -1.645$; Reject H_0 if $t < -1.645$.

(c) $\bar{x}_1 = 79.091$, $s_1 = 6.900$, $n_1 = 22$

$\bar{x}_2 = 83.000$, $s_2 = 7.645$, $n_2 = 19$

$$t = \frac{(\bar{x}_1 - \bar{x}_2) - (\mu_1 - \mu_2)}{\sqrt{\dfrac{(n_1 - 1)s_1^2 + (n_2 - 1)s_2^2}{n_1 + n_2 - 2}} \sqrt{\dfrac{1}{n_1} + \dfrac{1}{n_2}}}$$

$$= \frac{(79.091 - 83.000) - (0)}{\sqrt{\dfrac{(22 - 1)(6.900)^2 + (19 - 1)(7.645)^2}{22 + 19 - 2}} \sqrt{\dfrac{1}{22} + \dfrac{1}{19}}} = \frac{-3.909}{\sqrt{52.611}\,\sqrt{0.0981}} \approx -1.721$$

(d) Reject H_0.

(e) There is enough evidence to support the claim.

25. $\hat{\sigma} = \sqrt{\dfrac{(n_1 - 1)s_1^2 + (n_2 - 1)s_2^2}{n_1 + n_2 - 2}} = \sqrt{\dfrac{(15 - 1)(6.2)^2 + (12 - 1)(8.1)^2}{15 + 12 - 2}} \approx 7.099$

$(\bar{x}_1 - \bar{x}_2) \pm t_c \hat{\sigma} \sqrt{\dfrac{1}{n_1} + \dfrac{1}{n_2}} \to (450 - 420) \pm 2.060 \cdot 7.099 \sqrt{\dfrac{1}{15} + \dfrac{1}{12}}$

$\to 30 \pm 5.664 \to 24.336 < \mu_1 - \mu_2 < 35.664 \to 24 < \mu_1 - \mu_2 < 36$

26. $\hat{\sigma} = \sqrt{\dfrac{(n_1 - 1)s_1^2 + (n_2 - 1)s_2^2}{n_1 + n_2 - 2}} = \sqrt{\dfrac{(15 - 1)(2.1)^2 + (12 - 1)(1.8)^2}{15 + 12 - 2}} \approx 1.974$

$(\bar{x}_1 - \bar{x}_2) \pm t_c \hat{\sigma} \sqrt{\dfrac{1}{n_1} + \dfrac{1}{n_2}} \to (37 - 32) \pm 2.060 \cdot 1.974 \sqrt{\dfrac{1}{15} + \dfrac{1}{12}}$

$\to 5 \pm 1.575 \to 3.425 < \mu_1 - \mu_2 < 6.575 \to 3 < \mu_1 - \mu_2 < 7$

27. $(\bar{x}_1 - \bar{x}_2) \pm t_c \sqrt{\dfrac{s_1^2}{n_1} + \dfrac{s_2^2}{n_2}} \rightarrow (75 - 70) \pm 1.771 \sqrt{\dfrac{(3.64)^2}{16} + \dfrac{(2.12)^2}{14}}$

$\rightarrow -5 \pm 1.898 \rightarrow 3.102 < \mu_1 - \mu_2 < 6.898 \rightarrow 3 < \mu_1 - \mu_2 < 7$

28. $(\bar{x}_1 - \bar{x}_2) \pm t_c \sqrt{\dfrac{s_1^2}{n_1} + \dfrac{s_2^2}{n_2}} \rightarrow (39 - 37) \pm 1.345 \sqrt{\dfrac{(2.42)^2}{20} + \dfrac{(1.65)^2}{12}}$

$\rightarrow 2 \pm 0.926 \rightarrow 1.074 < \mu_1 - \mu_2 < 2.926 \rightarrow 1 < \mu_1 - \mu_2 < 3$

8.3 TESTING THE DIFFERENCE BETWEEN MEANS (DEPENDENT SAMPLES)

8.3 Try It Yourself Solutions

1.

Before	After	d	d^2
72	73	−1	1
81	80	1	1
76	79	−3	9
74	76	−2	4
75	76	−1	1
80	80	0	0
68	74	−6	36
75	77	−2	4
78	75	3	9
76	74	2	4
74	76	−2	4
77	78	−1	1
		$\Sigma d = -12$	$\Sigma d^2 = 74$

a. $H_0\colon \mu_d \geq 0$; $H_a\colon \mu_d < 0$ (claim)

b. $\alpha = 0.05$ and d.f. $= n - 1 = 11$

c. $t_0 \approx -1.796$; Reject H_0 if $t < -1.796$.

d. $\bar{d} = \dfrac{\Sigma d}{n} = \dfrac{-12}{12} = -1$

$s_d = \sqrt{\dfrac{n(\Sigma d^2) - (\Sigma d)^2}{n(n-1)}} = \sqrt{\dfrac{12(74) - (-12)^2}{12(11)}} \approx 2.374$

e. $t = \dfrac{\bar{d} - \mu_d}{\dfrac{s_d}{\sqrt{n}}} = \dfrac{-1 - 0}{\dfrac{2.374}{\sqrt{12}}} \approx -1.459$

f. Fail to reject H_0.

g. There is not enough evidence to support the claim.

2.

Before	After	d	d^2
101.8	99.2	2.6	6.76
98.5	98.4	0.1	0.01
98.1	98.2	−0.1	0.01
99.4	99	0.4	0.16
98.9	98.6	0.3	0.09
100.2	99.7	0.5	0.25
97.9	97.8	0.1	0.01
		$\Sigma d = 3.9$	$\Sigma d^2 = 7.29$

a. $H_0\colon \mu_d = 0$; $H_a\colon \mu_d \neq 0$ (claim)

b. $\alpha = 0.05$ and d.f. $= n - 1 = 6$

c. $t_0 = \pm 2.447$; Reject H_0 if $t < -2.447$ or $t > 2.447$.

d. $\bar{d} = \dfrac{\Sigma d}{n} = \dfrac{3.9}{7} \approx 0.557$

$s_d = \sqrt{\dfrac{n(\Sigma d^2) - (\Sigma d)^2}{n(n-1)}} = \sqrt{\dfrac{7(7.29) - (3.9)^2}{7(6)}} \approx 0.924$

e. $t = \dfrac{\bar{d} - \mu_d}{\frac{s_d}{\sqrt{n}}} = \dfrac{0.557 - 0}{\frac{0.924}{\sqrt{7}}} \approx 1.595$

f. Fail to reject H_0.

g. There is not enough evidence at the 5% significance level to conclude that the drug changes the body's temperature.

8.3 EXERCISE SOLUTIONS

1. (1) Each sample must be randomly selected from a normal population.

(2) Each member of the first sample must be paired with a member of the second sample.

2. The symbol $\bar{d}$ represents the mean of the differences between the paired data entries in dependent samples. The symbol S_d represents the standard deviation of the differences between the paired data entries in the dependent samples.

3. $H_0: \mu_d \geq 0; H_a: \mu_d < 0$ (claim)

$\alpha = 0.05$ and d.f. $= n - 1 = 13$

$t_0 = -1.771$ (Left-tailed)

$t = \dfrac{\bar{d} - \mu_d}{\frac{s_d}{\sqrt{n}}} = \dfrac{1.5 - 0}{\frac{3.2}{\sqrt{14}}} = \dfrac{1.5}{855} \approx 1.754$

Fail to reject H_0.

4. $H_0: \mu_d = 0$ (claim); $H_a: \mu_d \neq 0$

$\alpha = 0.01$ and d.f. $= n - 1 = 7$

$t_0 = \pm 3.499$ (Two-tailed)

$t = \dfrac{\bar{d} - \mu_d}{\frac{s_d}{\sqrt{n}}} = \dfrac{3.2 - 0}{\frac{8.45}{\sqrt{8}}} = \dfrac{3.2}{2.988} \approx 1.071$

Fail to reject H_0.

5. $H_0: \mu_d \leq 0$ (claim); $H_a: \mu_d > 0$

$\alpha = 0.10$ and d.f. $= n - 1 = 15$

$t_0 = 1.341$ (Right-tailed)

$t = \dfrac{\bar{d} - \mu_d}{\frac{s_d}{\sqrt{n}}} = \dfrac{6.5 - 0}{\frac{9.54}{\sqrt{16}}} = \dfrac{6.5}{2.385} \approx 2.725$

Reject H_0.

6. $H_0: \mu_d \leq 0; H_a: \mu_d > 0$ (claim)

$\alpha = 0.05$ and d.f. $= n - 1 = 27$

$t_0 = 1.703$ (Right-tailed)

$$t = \frac{\bar{d} - \mu_d}{\frac{s_d}{\sqrt{n}}} = \frac{0.55 - 0}{\frac{0.99}{\sqrt{28}}} = \frac{0.55}{0.187} \approx 2.940$$

Reject H_0.

7. $H_0: \mu_d \geq 0$ (claim); $H_a: \mu_d < 0$

$\alpha = 0.01$ and d.f. $= n - 1 = 14$

$t_0 = -2.624$ (Left-tailed)

$$t = \frac{\bar{d} - \mu_d}{\frac{s_d}{\sqrt{n}}} = \frac{-2.3 - 0}{\frac{1.2}{\sqrt{15}}} = \frac{-2.3}{0.3098} \approx -7.423$$

Reject H_0.

8. $H_0: \mu_d = 0$; $H_a: \mu_d \neq 0$ (claim)

$\alpha = 0.10$ and d.f. $= n - 1 = 19$

$t_0 = \pm 1.729$ (Two-tailed)

$$t = \frac{\bar{d} - \mu_d}{\frac{s_d}{\sqrt{n}}} = \frac{-1 - 0}{\frac{2.75}{\sqrt{20}}} = \frac{-1}{0.615} \approx -1.626$$

Fail to reject H_0.

9. (a) $H_0: \mu_d \geq 0$; $H_a: \mu_d < 0$ (claim)

(b) $t_0 = -2.650$; Reject H_0 if $t < -2.650$.

(c) $\bar{d} \approx -33.714$ and $s_d \approx 42.034$

(d) $t = \frac{\bar{d} - \mu_d}{\frac{s_d}{\sqrt{n}}} = \frac{-33.714 - 0}{\frac{42.034}{\sqrt{14}}} = \frac{-33.714}{11.234} \approx -3.001$

(e) Reject H_0.

(f) There is enough evidence to support the claim that the SAT scores improved.

10. (a) $H_0: \mu_d \geq 0$; $H_a: \mu_d < 0$ (claim)

(b) $t_0 = -2.821$; Reject H_0 if $t < -2.821$.

(c) $\bar{d} \approx -59.9$ and $s_d \approx 26.831$

(d) $t = \frac{\bar{d} - \mu_d}{\frac{s_d}{\sqrt{n}}} = \frac{-59.9 - 0}{\frac{26.831}{\sqrt{10}}} = \frac{-59.9}{8.485} \approx -7.060$

(e) Reject H_0.

(f) There is enough evidence to support the claim that the SAT prep course improves verbal SAT scores.

11. (a) $H_0: \mu_d \geq 0$; $H_a: \mu_d < 0$ (claim)

(b) $t_0 = -1.415$; Reject H_0 if $t < 1.415$.

(c) $\bar{d} \approx -1.575$ and $s_d \approx 0.803$

(d) $t = \dfrac{\bar{d} - \mu_d}{\dfrac{s_d}{\sqrt{n}}} = \dfrac{-1.575 - 0}{\dfrac{0.803}{\sqrt{8}}} = \dfrac{-1.575}{0.284} = -5.548$

(e) Reject H_0.

(f) There is enough evidence to support the fuel additive improved gas mileage.

12. (a) H_0: $\mu_d \geq 0$; H_a: $\mu_d < 0$ (claim)

(b) $t_0 = -1.397$; Reject H_0 if $t < -1.397$.

(c) $\bar{d} \approx -1.567$ and $s_d \approx 0.760$

(d) $t = \dfrac{\bar{d} - \mu_d}{\dfrac{s_d}{\sqrt{n}}} = \dfrac{-1.567 - 0}{\dfrac{0.760}{\sqrt{9}}} = \dfrac{-1.567}{0.253} \approx -6.186$

(e) Reject H_0.

(f) There is enough evidence to support the claim that the fuel additive improved gas mileage.

13. (a) H_0: $\mu_d \leq 0$; H_a: $\mu > 0$ (claim)

(b) $t_0 = 1.363$; Reject H_0 if $t > 1.363$.

(c) $\bar{d} = 3.75$ and $s_d \approx 7.84$

(d) $t = \dfrac{\bar{d} - \mu_d}{\dfrac{s_d}{\sqrt{n}}} = \dfrac{3.75 - 0}{\dfrac{7.84}{\sqrt{12}}} = \dfrac{3.75}{2.26} = 1.657$

(e) Reject H_0.

(f) There is enough evidence to support the claim that the exercise program helps participants lose weight.

14. (a) H_0: $\mu_d \leq 0$; H_a: $\mu_d > 0$ (claim)

(b) $t_0 = 1.350$; Reject H_0 if $t > 1.350$.

(c) $\bar{d} = 1.357$ and $s_d = 3.97$

(d) $t = \dfrac{\bar{d} - \mu_d}{\dfrac{s_d}{\sqrt{n}}} = \dfrac{1.357 - 0}{\dfrac{3.97}{\sqrt{14}}} = \dfrac{1.357}{1.06} = 1.278$

(e) Fail to reject H_0.

(f) There is not enough evidence to support the claim that the program helped adults lose weight.

15. (a) H_0: $\mu_d \leq 0$; H_a: $\mu_d > 0$ (claim)

(b) $t_0 = 2.764$; Reject H_0 if $t > 2.764$.

(c) $\bar{d} \approx 1.255$ and $s_d \approx 0.4414$

(d) $t = \dfrac{\overline{d} - \mu_d}{\dfrac{s_d}{\sqrt{n}}} = \dfrac{1.255 - 0}{\dfrac{0.4414}{\sqrt{11}}} = \dfrac{1.255}{0.1331} \approx 9.438$

(e) Reject H_0.

(f) There is enough evidence to support the claim that soft tissue therapy and spinal manipulation help reduce the length of time patients suffer from headaches.

16. (a) H_0: $\mu_d \geq 0$; H_a: $\mu_d < 0$ (claim)

(b) $t_0 = -1.761$; Reject H_0 if $t < -1.761$.

(c) $\overline{d} \approx 48.467$ and $s_d \approx 239.005$

(d) $t = \dfrac{\overline{d} - \mu_d}{\dfrac{s_d}{\sqrt{n}}} = \dfrac{48.467 - 0}{\dfrac{239.005}{\sqrt{15}}} = \dfrac{48.467}{61.711} \approx 0.785$

(e) Fail to reject H_0.

(f) There is not enough evidence to support the claim that Vitamin C will increase muscular endurance.

17. (a) H_0: $\mu_d \leq 0$; H_a: $\mu_d > 0$ (claim)

(b) $t_0 = 1.895$; Reject H_0 if $t > 1.895$.

(c) $\overline{d} = 14.75$ and $s_d \approx 6.861$

(d) $t = \dfrac{\overline{d} - \mu_d}{\dfrac{s_d}{\sqrt{n}}} = \dfrac{14.75 - 0}{\dfrac{6.86}{\sqrt{8}}} = \dfrac{14.75}{2.4257} \approx 6.081$

(e) Reject H_0.

(f) There is enough evidence to support the claim that the new drug reduces systolic blood pressure.

18. (a) H_0: $\mu_d \leq 0$; H_a: $\mu_d > 0$ (claim)

(b) $t_0 = 1.860$; Reject H_0 if $t > 1.860$.

(c) $\overline{d} \approx 6.333$ and $s_d \approx 5.362$

(d) $t = \dfrac{\overline{d} - \mu_d}{\dfrac{s_d}{\sqrt{n}}} = \dfrac{6.333 - 0}{\dfrac{5.362}{\sqrt{9}}} = \dfrac{6.333}{1.787} \approx 3.544$

(e) Reject H_0.

(f) There is enough evidence to support the claim that the new drug reduces diastolic blood pressure.

19. (a) H_0: $\mu_d = 0$; H_a: $\mu_d \neq 0$ (claim)

(b) $t_0 = \pm 2.365$; Reject H_0 if $t < -2.365$ or $t > 2.365$.

(c) $\overline{d} = -1$ and $s_d = 1.309$

(d) $t = \dfrac{\bar{d} - \mu_d}{\frac{s_d}{\sqrt{n}}} = \dfrac{-1 - 0}{\frac{1.31}{\sqrt{8}}} = \dfrac{-1}{0.463} = -2.160$

(e) Fail to reject H_0.

(f) There is not enough evidence to support the claim that the product ratings have changed.

20. (a) H_0: $\mu_d = 0$; H_a: $\mu_d \neq 0$ (claim)

(b) $t_0 = \pm 2.571$; Reject H_0 if $t < -2.571$ or $t > 2.571$.

(c) $\bar{d} \approx -7.167$ and $s_d \approx 10.245$

(d) $t = \dfrac{\bar{x} - \mu_d}{\frac{s_d}{\sqrt{n}}} = \dfrac{-7.167 - 0}{\frac{10.245}{\sqrt{6}}} = \dfrac{-7.167}{4.183} \approx -1.713$

(e) Fail to reject H_0.

(f) There is not enough evidence to support the claim that the plant performance ratings have changed.

21. $\bar{d} \approx -1.525$ and $s_d \approx 0.542$

$$\bar{d} - t_{\alpha/2}\frac{s_d}{\sqrt{n}} < \mu_d < \bar{d} - t_{\alpha/2}\frac{s_d}{\sqrt{n}}$$

$$-1.525 - 1.753\left(\frac{0.542}{\sqrt{16}}\right) < \mu_d < -1.525 + 1.753\left(\frac{0.542}{\sqrt{16}}\right)$$

$$-1.525 - 0.238 < \mu_d < -1.525 + 0.238$$

$$-1.76 < \mu_d < -1.29$$

22. $\bar{d} = -0.436$ and $s_d \approx 0.677$

$$\bar{d} - t_{\alpha/2}\frac{s_d}{\sqrt{n}} < \mu_d < \bar{d} + t_{\alpha/2}\frac{s_d}{\sqrt{n}}$$

$$-0.436 - 2.160\left(\frac{0.677}{\sqrt{14}}\right) < \mu_d < -0.436 + 2.160\left(\frac{0.677}{\sqrt{14}}\right)$$

$$-0.436 - 0.391 < \mu_d < -0.436 + 0.391$$

$$-0.83 < \mu_d < -0.05$$

8.4 TESTING THE DIFFERENCE BETWEEN PROPORTIONS

8.4 Try It Yourself Solutions

1a. H_0: $p_1 = p_2$; H_a: $p_1 \neq p_2$ (claim)   **b.** $\alpha = 0.05$

c. $z_0 = \pm 1.96$; Reject H_0 if $z < -1.96$ or $z > 1.96$.

d. $\bar{p} = \dfrac{x_1 + x_2}{n_1 + n_2} = \dfrac{1484 + 1497}{6869 + 6869} = 0.217$

$\bar{q} = 0.783$

e. $n_1\bar{p} \approx 1490.573 > 5, n_1\bar{q} \approx 5378.427 > 5, n_2\bar{p} \approx 1490.573 > 5,$ and $n_2\bar{q} \approx 5378.427 > 5.$

f. $z = \dfrac{(\hat{p}_1 - \hat{p}_2) - (p_1 - p_2)}{\sqrt{\bar{p}\bar{q}\left(\dfrac{1}{n_1} + \dfrac{1}{n_2}\right)}} = \dfrac{(0.216 - 0.218) - (0)}{\sqrt{0.217 \cdot 0.783\left(\dfrac{1}{6869} + \dfrac{1}{6869}\right)}} = \dfrac{-0.002}{\sqrt{0.00004947}} \approx -0.284$

g. Fail to reject H_0.

h. There is not enough evidence to support the claim.

2a. $H_0: p_1 \leq p_2; H_a: p_1 > p_2$ (claim)   **b.** $\alpha = 0.05$

c. $z_0 = 1.645$; Reject H_0 if $z > 1.645$.

d. $\bar{p} = \dfrac{x_1 + x_2}{n_1 + n_2} = \dfrac{1264 + 522}{6869 + 6869} = 0.130$

$\bar{q} = 0.870$

e. $n_1\bar{p} \approx 892.97 > 5, n_1\bar{q} \approx 5976.03 > 5, n_2\bar{p} \approx 892.97 > 5,$ and $n_2\bar{q} \approx 5976.03 > 5.$

f. $z = \dfrac{(\hat{p}_1 - \hat{p}_2) - (p_1 - p_2)}{\sqrt{\bar{p}\bar{q}\left(\dfrac{1}{n_1} + \dfrac{1}{n_2}\right)}} = \dfrac{(0.184 - 0.076) - (0)}{\sqrt{0.130 \cdot 0.870\left(\dfrac{1}{6869} + \dfrac{1}{6869}\right)}} = \dfrac{0.108}{\sqrt{0.00003293}} \approx 18.820$

g. Reject H_0.

h. There is enough evidence to support the claim.

8.4 EXERCISE SOLUTIONS

1. State the hypotheses and identify the claim. Specify the level of significance. Find the critical value(s) and rejection region(s). Find $\bar{p}$ and $\bar{q}$. Find the standardized test statistic. Make a decision and interpret in the context of the claim.

2. (1) The samples must be randomly selected.

   (2) The sample must be independent.

   (3) $n_1p_1 \geq 5, n_1q_1 \geq 5, n_2p_2 \geq 5,$ and $n_2q_2 \geq 5$

3. $H_0: p_1 = p_2; H_a: p_1 \neq p_2$ (claim)

   $z_0 = \pm 2.575$ (Two-tailed test)

   $\bar{p} = \dfrac{x_1 + x_2}{n_1 + n_2} = \dfrac{35 + 36}{70 + 60} = 0.546$

   $\bar{q} = 0.454$

   $z = \dfrac{(\hat{p}_1 - \hat{p}_2) - (p_1 - p_2)}{\sqrt{\bar{p}\bar{q}\left(\dfrac{1}{n_1} + \dfrac{1}{n_2}\right)}} = \dfrac{(0.500 - 0.600) - (0)}{\sqrt{0.546 \cdot 0.454\left(\dfrac{1}{70} + \dfrac{1}{60}\right)}} = \dfrac{-0.100}{\sqrt{0.00767}} \approx -1.142$

   Fail to reject H_0.

4. $H_0: p_1 \geq p_2$; $H_a: p_1 < p_2$ (claim)

$z_0 = -1.645$ (Left-tailed test)

$\bar{p} = \dfrac{x_1 + x_2}{n_1 + n_2} = \dfrac{471 + 372}{785 + 465} = 0.674$

$\bar{q} = 0.326$

$z = \dfrac{(\hat{p}_1 - \hat{p}_2) - (p_1 - p_2)}{\sqrt{\bar{p}\,\bar{q}\left(\dfrac{1}{n_1} + \dfrac{1}{n_2}\right)}} = \dfrac{(0.600 - 0.800) - (0)}{\sqrt{0.674 \cdot 0.326\left(\dfrac{1}{785} + \dfrac{1}{465}\right)}} = \dfrac{-0.2}{\sqrt{0.000752428}} \approx -7.291$

Reject H_0.

5. $H_0: p_1 \leq p_2$ (claim); $H_a: p_1 > p_2$

$z_0 = 1.282$ (Right-tailed test)

$\bar{p} = \dfrac{x_1 + x_2}{n_1 + n_2} = \dfrac{344 + 304}{860 + 800} = 0.390$

$\bar{q} = 0.610$

$z = \dfrac{(\hat{p}_1 - \hat{p}_2) - (p_1 - p_2)}{\sqrt{\bar{p}\,\bar{q}\left(\dfrac{1}{n_1} + \dfrac{1}{n_2}\right)}} = \dfrac{(0.400 - 0.380) - (0)}{\sqrt{0.390 \cdot 0.610\left(\dfrac{1}{860} + \dfrac{1}{800}\right)}} = \dfrac{0.020}{\sqrt{0.0000574003}} \approx 0.835$

Fail to reject H_0.

6. $H_0: p_1 = p_2$ (claim); $H_a: p_1 \neq p_2$

$z_0 = \pm 1.96$ (Two-tailed test)

$\bar{p} = \dfrac{x_1 + x_2}{n_1 + n_2} = \dfrac{29 + 25}{45 + 30} = 0.72$

$\bar{q} = 0.28$

$z = \dfrac{(\hat{p}_1 - \hat{p}_2) - (p_1 - p_2)}{\sqrt{\bar{p}\,\bar{q}\left(\dfrac{1}{n_1} + \dfrac{1}{n_2}\right)}} = \dfrac{(0.644 - 0.833) - (0)}{\sqrt{(0.72)(0.28)\left(\dfrac{1}{45} + \dfrac{1}{30}\right)}} = \dfrac{-0.189}{\sqrt{0.0112}} \approx -1.786$

Fail to reject H_0.

7. (a) $H_0: p_1 = p_2$ (claim); $H_a: p_1 \neq p_2$

(b) $z_0 = \pm 1.96$; Reject H_0 if $z < -1.96$ or $z > 1.96$.

$\bar{p} = \dfrac{x_1 + x_2}{n_1 + n_2} = \dfrac{520 + 865}{1539 + 2055} = 0.385$

$\bar{q} = 0.615$

(c) $z = \dfrac{(\hat{p}_1 - \hat{p}_2) - (p_1 - p_2)}{\sqrt{\bar{p}\,\bar{q}\left(\dfrac{1}{n_1} + \dfrac{1}{n_2}\right)}} = \dfrac{(0.338 - 0.421) - (0)}{\sqrt{0.385 \cdot 0.615\left(\dfrac{1}{1539} + \dfrac{1}{2055}\right)}}$

$= \dfrac{-0.083}{\sqrt{0.0000269069}} \approx -5.06$

(d) Reject H_0.

(e) There is sufficient evidence at the 5% level to reject the claim that the proportion of adults using alternative medicines has not changed since 1991.

8. (a) H_0: $p_1 = p_2$ (claim); H_a: $p_1 \neq p_2$

(b) $z_0 = \pm 1.645$; Reject H_0 if $z < -1.645$ or $z > 1.645$.

$$\bar{p} = \frac{x_1 + x_2}{n_1 + n_2} = \frac{5 + 19}{77 + 84} = 0.149$$

$$\bar{q} = 0.851$$

(c) $z = \dfrac{(\hat{p}_1 - \hat{p}_2) - (p_1 - p_2)}{\sqrt{\bar{p}\,\bar{q}\left(\dfrac{1}{n_1} + \dfrac{1}{n_2}\right)}} = \dfrac{(0.06494 - 0.22620) - (0)}{\sqrt{0.149 \cdot 0.851\left(\dfrac{1}{77} + \dfrac{1}{84}\right)}} = \dfrac{-0.16126}{\sqrt{0.0031563}} \approx -2.870$

(d) Reject H_0.

(e) There is sufficient evidence at the 10% significance level to reject the claim that the proportions of patients suffering new bouts of depression are the same for both groups.

9. (a) H_0: $p_1 = p_2$ (claim); H_a: $p_1 \neq p_2$

(b) $z_0 = \pm 1.645$; Reject H_0 if $z < -1.645$ or $z > 1.645$.

$$\bar{p} = \frac{x_1 + x_2}{n_1 + n_2} = \frac{2201 + 2348}{5240 + 6180} = 0.398$$

$$\bar{q} = 0.602$$

(c) $z = \dfrac{(\hat{p}_1 - \hat{p}_2) - (p_1 - p_2)}{\sqrt{\bar{p}\,\bar{q}\left(\dfrac{1}{n_1} + \dfrac{1}{n_2}\right)}} = \dfrac{(0.4200 - 0.37994) - 0}{\sqrt{(0.398)(0.602)\left(\dfrac{1}{5240} + \dfrac{1}{6180}\right)}}$

$\qquad = \dfrac{0.041}{\sqrt{0.000084494}} = 4.362$

(d) Reject H_0.

(e) There is sufficient evidence at the 10% significance level to reject the claim that the proportions of male and female senior citizens that eat the daily recommended number of servings of vegetables are the same.

10. (a) H_0: $p_1 \geq p_2$; H_a: $p_1 < p_2$ (claim)

(b) $z_0 = -2.33$; Reject H_0 if $z < -2.33$.

$$\bar{p} = \frac{x_1 + x_2}{n_1 + n_2} = \frac{361 + 341}{1245 + 1065} = 0.304$$

$$\bar{q} = 0.696$$

(c) $z = \dfrac{(\hat{p}_1 - \hat{p}_2) - (p_1 - p_2)}{\sqrt{\bar{p}\,\bar{q}\left(\dfrac{1}{n_1} + \dfrac{1}{n_2}\right)}} = \dfrac{(0.2900 - 0.3202)}{\sqrt{(0.304)(0.696)\left(\dfrac{1}{1245} + \dfrac{1}{1065}\right)}} = \dfrac{-0.0302}{\sqrt{0.0003686}} = -1.57$

(d) Fail to reject H_0.

(e) There is not enough evidence at the 1% significance level to support the claim that the proportion of male senior citizens eating the daily recommended number of servings of fruit is less than female senior citizens.

11. (a) $H_0: p_1 \leq p_2$; $H_a: p_1 > p_2$ (claim)

(b) $z_0 = 2.33$; Reject H_0 if $z > 2.33$.

$$\bar{p} = \frac{x_1 + x_2}{n_1 + n_2} = \frac{496 + 468}{2000 + 2000} = 0.241$$

$$\bar{q} = 0.759$$

(c) $z = \dfrac{(\hat{p}_1 - \hat{p}_2) - (p_1 - p_2)}{\sqrt{\bar{p}\,\bar{q}\left(\dfrac{1}{n_1} + \dfrac{1}{n_2}\right)}} = \dfrac{(0.248 - 0.234) - (0)}{\sqrt{0.241 \cdot 0.759\left(\dfrac{1}{2000} + \dfrac{1}{2000}\right)}} = \dfrac{0.014}{\sqrt{0.0001829}} \approx 1.04$

(d) Fail to reject H_0.

(e) There is not sufficient evidence at the 1% significance level to support the claim that the proportion of adults who are smokers is greater in Alabama than in Missouri.

12. (a) $H_0: p_1 \geq p_2$; $H_a: p_1 < p_2$ (claim)

(b) $z_0 = -1.645$; Reject H_0 if $z < -1.645$.

$$\bar{p} = \frac{x_1 + x_2}{n_1 + n_2} = \frac{228 + 185}{1500 + 1000} = 0.1652$$

$$\bar{q} = 0.8348$$

(c) $z = \dfrac{(\hat{p}_1 - \hat{p}_2) - (p_1 - p_2)}{\sqrt{\bar{p}\,\bar{q}\left(\dfrac{1}{n_1} + \dfrac{1}{n_2}\right)}} = \dfrac{(0.152 - 0.185) - (0)}{\sqrt{0.1652 \cdot 0.8348\left(\dfrac{1}{1500} + \dfrac{1}{1000}\right)}} = \dfrac{-0.033}{\sqrt{0.0002298}} \approx -2.177$

(d) Reject H_0.

(e) There is sufficient evidence at the 5% level to conclude that the proportion of adults who are smokers is lower in California than in Oregon.

13. (a) $H_0: p_1 \geq p_2$; $H_a: p_1 < p_2$ (claim)

(b) $z_0 = -2.33$; Reject H_0 if $z < -2.33$.

$$\bar{p} = \frac{x_1 + x_2}{n_1 + n_2} = \frac{2083 + 985}{9300 + 4900} = 0.216$$

$$\bar{q} = 0.784$$

(c) $z = \dfrac{(\hat{p}_1 - \hat{p}_2) - (p_1 - p_2)}{\sqrt{\bar{p}\,\bar{q}\left(\dfrac{1}{n_1} + \dfrac{1}{n_2}\right)}} = \dfrac{(0.22398 - 0.20102) - (0)}{\sqrt{0.216 \cdot 0.784\left(\dfrac{1}{9300} + \dfrac{1}{4900}\right)}} = \dfrac{0.02296}{\sqrt{0.00005277}} \approx 3.16$

(d) Fail to reject H_0.

(e) There is sufficient evidence at the 1% significance level to support the claim that the proportion of twelfth grade males who said they had smoked in the last 30 days is less than the proportion of twelfth grade females.

14. (a) $H_0: p_1 = p_2$ (claim); $H_a: p_1 \neq p_2$

(b) $z_0 = \pm1.645$; Reject H_0 if $z < -1.645$ or $z > 1.645$.

$$\bar{p} = \frac{x_1 + x_2}{n_1 + n_2} = \frac{3444 + 3067}{12{,}900 + 14{,}200} = 0.240$$

$$\bar{q} = 0.760$$

(c) $z = \dfrac{(\hat{p}_1 - \hat{p}_2) - (p_1 - p_2)}{\sqrt{\bar{p}\,\bar{q}\left(\dfrac{1}{n_1} + \dfrac{1}{n_2}\right)}} = \dfrac{(0.267 - 0.216) - (0)}{\sqrt{0.240 \cdot 0.760\left(\dfrac{1}{12{,}900} + \dfrac{1}{14{,}200}\right)}}$

$= \dfrac{0.051}{\sqrt{0.00002698}} \approx 9.818$

(d) Reject H_0.

(e) There is sufficient evidence at the 10% significance level to reject the claim that the proportion of college students who said they had smoked in the last 30 days has not changed.

15. (a) $H_0: p_1 = p_2$ (claim); $H_a: p_1 \neq p_2$

(b) $z_0 = \pm 1.96$; Reject H_0 if $z < -1.96$ or $z > 1.96$.

$\bar{p} = \dfrac{x_1 + x_2}{n_1 + n_2} = \dfrac{805 + 746}{1150 + 1050} = 0.705$

$\bar{q} = 0.295$

(c) $z = \dfrac{(\hat{p}_1 - \hat{p}_2) - (p_1 - p_2)}{\sqrt{\bar{p}\,\bar{q}\left(\dfrac{1}{n_1} + \dfrac{1}{n_2}\right)}} = \dfrac{(0.70000 - 0.71047) - 0}{\sqrt{(0.705)(0.295)\left(\dfrac{1}{1150} + \dfrac{1}{1050}\right)}} = \dfrac{-0.01047}{\sqrt{0.0003789}} = -0.54$

(d) Fail to reject H_0.

(e) There is insufficient evidence at the 5% significance level to reject the claim that the proportions of Internet users are the same for both groups.

16. (a) $H_0: p_1 \leq p_2$; $H_a: p_1 > p_2$ (claim);

(b) $z_0 = 1.645$; Reject H_0 if $z > 1.645$.

(c) $\bar{p} = \dfrac{x_1 + x_2}{n_1 + n_2} = \dfrac{354 + 189}{485 + 315} = 0.679$

$\bar{q} = 0.321$

$z = \dfrac{(\hat{p}_1 - \hat{p}_2) - (p_1 - p_2)}{\sqrt{\bar{p}\,\bar{q}\left(\dfrac{1}{n_1} + \dfrac{1}{n_2}\right)}} = \dfrac{(0.730 - 0.600) - 0}{\sqrt{(0.679)(0.321)\left(\dfrac{1}{485} + \dfrac{1}{315}\right)}} = \dfrac{0.13}{\sqrt{0.0011413}}$

$= 3.848$

(d) Reject H_0.

(e) There is sufficient evidence at the 5% significance level to support the claim that the proportion of adults who use the Internet is greater for adults who live in an urban area than for adults who live in a rural area.

17. $H_0: p_1 \geq p_2; H_a: p_1 < p_2$ (claim)

$z_0 = -2.33$

$\bar{p} = \dfrac{x_1 + x_2}{n_1 + n_2} = \dfrac{28 + 35}{700 + 500} = 0.053$

$\bar{q} = 0.947$

$z = \dfrac{(\hat{p}_1 - \hat{p}_2) - (p_1 - p_2)}{\sqrt{\bar{p}\,\bar{q}\left(\dfrac{1}{n_1} + \dfrac{1}{n_2}\right)}} = \dfrac{(0.04 - 0.07) - (0)}{\sqrt{0.053 \cdot 0.947\left(\dfrac{1}{700} + \dfrac{1}{500}\right)}} = \dfrac{-0.03}{\sqrt{0.0001721}} \approx -2.287$

Fail to reject H_0. There is insufficient evidence at the 1% significance level to support the claim.

18. $H_0: p_1 \leq p_2; H_a: p_1 > p_2$ (claim)

$z_0 = 2.33$

$\bar{p} = \dfrac{x_1 + x_2}{n_1 + n_2} = \dfrac{161 + 85}{700 + 500} = 0.205$

$\bar{q} = 0.795$

$z = \dfrac{(\hat{p}_1 - \hat{p}_2) - (p_1 - p_2)}{\sqrt{\bar{p}\,\bar{q}\left(\dfrac{1}{n_1} + \dfrac{1}{n_2}\right)}} = \dfrac{(0.23 - 0.17) - (0)}{\sqrt{0.205 \cdot 0.795\left(\dfrac{1}{700} + \dfrac{1}{500}\right)}} = \dfrac{0.06}{\sqrt{0.0005588}} \approx 2.33$

Reject H_0. There is sufficient evidence at the 1% significance level to support the representatives belief.

19. $H_0: p_1 \geq p_2; H_a: p_1 < p_2$ (claim)

$z_0 = -1.645$

$\bar{p} = \dfrac{x_1 + x_2}{n_1 + n_2} = \dfrac{189 + 185}{700 + 500} = 0.312$

$\bar{q} = 0.688$

$z = \dfrac{(\hat{p}_1 - \hat{p}_2) - (p_1 - p_2)}{\sqrt{\bar{p}\,\bar{q}\left(\dfrac{1}{n_1} + \dfrac{1}{n_2}\right)}} = \dfrac{(0.27 - 0.37) - (0)}{\sqrt{0.312 \cdot 0.688\left(\dfrac{1}{700} + \dfrac{1}{500}\right)}} = \dfrac{-0.10}{\sqrt{0.0007359}} \approx -3.686$

Reject H_0. There is sufficient evidence at the 5% significance level to support the organization's claim.

20. $H_0: p_1 = p_2$ (claim); $H_a: p_1 \neq p_2$

$z_0 = \pm 1.96$

$\bar{p} = \dfrac{x_1 + x_2}{n_1 + n_2} = \dfrac{224 + 145}{700 + 500} = 0.308$

$\bar{q} = 0.692$

$z = \dfrac{(\hat{p}_1 - \hat{p}_2) - (p_1 - p_2)}{\sqrt{\bar{p}\,\bar{q}\left(\dfrac{1}{n_1} + \dfrac{1}{n_2}\right)}} = \dfrac{(0.32 - 0.29) - (0)}{\sqrt{0.308 \cdot 0.692\left(\dfrac{1}{700} + \dfrac{1}{500}\right)}} = \dfrac{0.03}{\sqrt{0.00073075}} \approx 1.110$

Fail to reject H_0. There is insufficient evidence at the 5% significance level to reject the claim.

21. $H_0: p_1 \leq p_2$; $H_a: p_1 > p_2$ (claim)

$z_0 = 1.645$

$\bar{p} = \dfrac{x_1 + x_2}{n_1 + n_2} = \dfrac{7501 + 7501}{13,300 + 14,100} = 0.548$

$\bar{q} = 0.452$

$z = \dfrac{(\hat{p}_1 - \hat{p}_2) - (p_1 - p_2)}{\sqrt{\bar{p}\,\bar{q}\left(\dfrac{1}{n_1} + \dfrac{1}{n_2}\right)}} = \dfrac{(0.564 - 0.532) - (0)}{\sqrt{(0.548)(0.452)\left(\dfrac{1}{13,300} + \dfrac{1}{14,100}\right)}} = \dfrac{0.032}{\sqrt{0.00003619}} \approx 5.319$

Reject H_0. There is sufficient evidence at the 5% significance level to support the claim.

22. $H_0: p_1 \leq p_2$; $H_a: p_1 > p_2$ (claim)

$z_0 = 1.645$

$\bar{p} = \dfrac{x_1 + x_2}{n_1 + n_2} = \dfrac{5610 + 5934}{13,200 + 13,800} = 0.428$

$\bar{q} = 0.572$

$z = \dfrac{(\hat{p}_1 - \hat{p}_2) - (p_1 - p_2)}{\sqrt{\bar{p}\,\bar{q}\left(\dfrac{1}{n_1} + \dfrac{1}{n_2}\right)}} = \dfrac{(0.425 - 0.430) - (0)}{\sqrt{(0.428)(0.572)\left(\dfrac{1}{13,200} + \dfrac{1}{13,800}\right)}} = \dfrac{-0.005}{\sqrt{0.00003629}} = -0.830$

Fail to reject H_0. There is not enough evidence at the 5% significance level to support the claim.

23. $H_0: p_1 = p_2$ (claim); $H_a: p_1 \neq p_2$

$z_0 = \pm 2.576$

$\bar{p} = \dfrac{x_1 + x_2}{n_1 + n_2} = \dfrac{7501 + 5610}{13,300 + 13,200} = 0.495$

$\bar{q} = 0.505$

$z = \dfrac{(\hat{p}_1 - \hat{p}_2) - (p_1 - p_2)}{\sqrt{\bar{p}\,\bar{q}\left(\dfrac{1}{n_1} + \dfrac{1}{n_2}\right)}} = \dfrac{(0.564 - 0.425) - (0)}{\sqrt{(0.495)(0.505)\left(\dfrac{1}{13,300} + \dfrac{1}{13,200}\right)}} = \dfrac{0.139}{\sqrt{0.00003773}} \approx 22,629$

Reject H_0. There is sufficient evidence at the 1% significance level to reject the claim.

24. $H_0: p_1 = p_2$ (claim); $H_a: p_1 \neq p_2$

$z_0 = \pm 2.576$

$\bar{p} = \dfrac{x_1 + x_2}{n_1 + n_2} = \dfrac{7501 + 5934}{14,100 + 13,800} = 0.482$

$\bar{q} = 0.518$

$z = \dfrac{(\hat{p}_1 - \hat{p}_2) - (p_1 - p_2)}{\sqrt{\bar{p}\,\bar{q}\left(\dfrac{1}{n_1} + \dfrac{1}{n_2}\right)}} = \dfrac{(0.532 - 0.430) - (0)}{\sqrt{(0.482)(0.518)\left(\dfrac{1}{14,100} + \dfrac{1}{13,800}\right)}} = \dfrac{0.102}{\sqrt{0.00003580}} = 17.047$

Reject H_0. There is sufficient evidence at the 1% significance level to reject the claim.

25. $(\hat{p}_1 - \hat{p}_2) \pm z_c \sqrt{\dfrac{\hat{p}_1 \hat{q}_1}{n_1} + \dfrac{\hat{p}_2 \hat{q}_2}{n_2}} \rightarrow (0.088 - 0.083) \pm 1.96 \sqrt{\dfrac{0.088 \cdot 0.912}{1,068,000} + \dfrac{0.083 \cdot 0.917}{1,476,000}}$

$$\rightarrow 0.005 \pm 1.96 \sqrt{0.00000012671}$$

$$\rightarrow 0.004 < p_1 - p_2 < 0.006$$

26. $(\hat{p}_1 - \hat{p}_2) \pm z_c \sqrt{\dfrac{\hat{p}_1 \hat{q}_1}{n_1} + \dfrac{\hat{p}_2 \hat{q}_2}{n_2}} \rightarrow (0.116 - 0.092) \pm 1.96 \sqrt{\dfrac{0.116 \cdot 0.884}{1,068,000} + \dfrac{0.092 \cdot 0.908}{1,476,000}}$

$$\rightarrow 0.024 \pm 1.96 \sqrt{0.0000001526}$$

$$\rightarrow 0.023 < p_1 - p_2 < -0.025$$

CHAPTER 8 REVIEW EXERCISE SOLUTIONS

1. Independent because the two samples of laboratory mice are different.

2. Dependent, because the same mice were used for both experiments.

3. $H_0: \mu_1 \geq \mu_2$ (claim); $H_1: \mu_1 < \mu_2$

$z_0 = -1.645$

$z = \dfrac{(\bar{x}_1 - \bar{x}_2) - (\mu_1 - \mu_2)}{\sqrt{\dfrac{s_1^2}{n_1} + \dfrac{s_2^2}{n_2}}} = \dfrac{(1.28 - 1.34) - (0)}{\sqrt{\dfrac{(0.30)^2}{96} + \dfrac{(0.23)^2}{85}}} = \dfrac{-0.06}{\sqrt{0.001560}} \approx -1.519$

Fail to reject H_0. There is not enough evidence to reject the claim.

4. $H_0: \mu_1 = \mu_2$ (claim); $H_a: \mu_1 \neq \mu_2$

$z_0 = \pm 2.575$

$z = \dfrac{(\bar{x}_1 - \bar{x}_2) - (\mu_1 - \mu_2)}{\sqrt{\dfrac{s_1^2}{n_1} + \dfrac{s_2^2}{n_2}}} = \dfrac{(5595 - 5575) - (0)}{\sqrt{\dfrac{(52)^2}{156} + \dfrac{(68)^2}{216}}} = \dfrac{20}{\sqrt{38.741}} \approx 3.213$

Reject H_0. There is enough evidence to reject the claim.

5. $H_0: \mu_1 \geq \mu_2$; $H_1: \mu_1 < \mu_2$ (claim)

$z_0 = -1.282$

$z = \dfrac{(\bar{x}_1 - \bar{x}_2) - (\mu_1 - \mu_2)}{\sqrt{\dfrac{s_1^2}{n_1} + \dfrac{s_2^2}{n_2}}} = \dfrac{(0.28 - 0.33) - (0)}{\sqrt{\dfrac{(0.11)^2}{41} + \dfrac{(0.10)^2}{34}}} = \dfrac{-0.50}{\sqrt{0.00058924}} \approx -2.060$

Reject H_0. There is enough evidence to support the claim.

6. $H_0: \mu_1 = \mu_2$; $H_a: \mu_1 \neq \mu_2$ (claim)

$z_0 = \pm 1.96$

$z = \dfrac{(\bar{x}_1 - \bar{x}_2) - (\mu_1 - \mu_2)}{\sqrt{\dfrac{s_1^2}{n_1} + \dfrac{s_2^2}{n_2}}} = \dfrac{(87 - 85) - (0)}{\sqrt{\dfrac{(14)^2}{410} + \dfrac{(15)^2}{340}}} = \dfrac{2}{\sqrt{1.13981}} \approx 1.87$

Fail to reject H_0. There is not enough evidence to support the claim.

7. (a) $H_0: \mu_1 \le \mu_2; H_1: \mu_1 > \mu_2$ (claim)

(b) $z_0 = 1.645$; Reject H_0 if $z > 1.645$.

(c) $z = \dfrac{(\bar{x}_1 - \bar{x}_2) - (\mu_1 - \mu_2)}{\sqrt{\dfrac{s_1^2}{n_1} + \dfrac{s_2^2}{n_2}}} = \dfrac{(480 - 470) - (0)}{\sqrt{\dfrac{(32)^2}{36} + \dfrac{(54)^2}{41}}} = \dfrac{10}{\sqrt{99.566}} \approx 1.002$

(d) Fail to reject H_0.

(e) There is not enough evidence to support the claim.

8. (a) $H_0: \mu_1 = \mu_2; H_a: \mu_1 \ne \mu_2$ (claim)

(b) $z_0 = \pm 1.645$; Reject H_0 if $z < -1.645$ or $z > 1.645$.

(c) $z = \dfrac{(\bar{x}_1 - \bar{x}_2) - (\mu_1 - \mu_2)}{\sqrt{\dfrac{s_1^2}{n_1} + \dfrac{s_2^2}{n_2}}} = \dfrac{(360 - 380) - (0)}{\sqrt{\dfrac{(50)^2}{38} + \dfrac{(45)^2}{35}}} = \dfrac{-20}{\sqrt{123.647}} \approx -1.799$

(d) Reject H_0.

(e) There is sufficient evidence at the 10% significance level to support the claim that the caloric content of the two types of french fries is different.

9. $H_0: \mu_1 = \mu_2$ (claim); $H_a: \mu_1 \ne \mu_2$

d.f. $= n_1 + n_2 - 2 = 29$

$t_0 = \pm 2.045$

$t = \dfrac{(\bar{x}_1 - \bar{x}_2) - (\mu_1 - \mu_2)}{\sqrt{\dfrac{(n_1-1)s_1^2 + (n_2-1)s_2^2}{n_1 + n_2 - 2}}\sqrt{\dfrac{1}{n_1} + \dfrac{1}{n_2}}} = \dfrac{(300 - 290) - (0)}{\sqrt{\dfrac{(19-1)(26)^2 + (12-1)(22)^2}{19 + 12 - 2}}\sqrt{\dfrac{1}{19} + \dfrac{1}{12}}}$

$= \dfrac{10}{\sqrt{603.172}\,\sqrt{0.1360}} \approx 1.104$

Fail to reject H_0. There is not enough evidence to reject the claim.

10. $H_0: \mu_1 = \mu_2$ (claim); $H_a: \mu_1 \ne \mu_2$

d.f. $= \min\{n_1 - 1, n_2 - 1\} = 5$

$t_0 = \pm 2.015$

$t = \dfrac{(\bar{x}_1 - \bar{x}_2) - (\mu_1 - \mu_2)}{\sqrt{\dfrac{s_1^2}{n_1} + \dfrac{s_2^2}{n_2}}} = \dfrac{(0.015 - 0.019) - (0)}{\sqrt{\dfrac{(0.011)^2}{8} + \dfrac{(0.004)^2}{6}}} = \dfrac{-.004}{\sqrt{0.0000178}} \approx -0.948$

Fail to reject H_0.

There is not enough evidence to reject the claim.

11. $H_0: \mu_1 \le \mu_2$ (claim); $H_a: \mu_1 > \mu_2$

d.f. $= \min\{n_1 - 1, n_2 - 1\} = 24$

$t_0 = 1.711$

$$t = \frac{(\bar{x}_1 - \bar{x}_2) - (\mu_1 - \mu_2)}{\sqrt{\dfrac{s_1^2}{n_1} + \dfrac{s_2^2}{n_2}}} = \frac{(183.5 - 184.7) - (0)}{\sqrt{\dfrac{(1.3)^2}{25} + \dfrac{(3.9)^2}{25}}} = \frac{-1.2}{\sqrt{0.676}} \approx -1.460$$

Fail to reject H_0. There is not enough evidence to reject the claim.

12. $H_0: \mu_1 \ge \mu_2$ (claim); $H_a: \mu_1 < \mu_2$

d.f. $= n_1 + n_2 - 2 = 39$

$t_0 = -2.326$

$$t = \frac{(\bar{x}_1 - \bar{x}_2) - (\mu_1 - \mu_2)}{\sqrt{\dfrac{(n_1 - 1)s_1^2 + (n_2 - 1)s_2^2}{n_1 + n_2 - 2}} \sqrt{\dfrac{1}{n_1} + \dfrac{1}{n_2}}}$$

$$= \frac{(24.5 - 26.4) - (0)}{\sqrt{\dfrac{(19 - 1)(2.95)^2 + (20 - 1)(2.15)^2}{19 + 20 - 2}} \sqrt{\dfrac{1}{19} + \dfrac{1}{20}}} = \frac{-1.9}{\sqrt{6.607}\,\sqrt{0.1026}} \approx -2.308$$

Fail to reject H_0. There is not enough evidence to reject the claim.

13. $H_0: \mu_1 = \mu_2$; $H_a: \mu_1 \ne \mu_2$ (claim)

d.f. $= n_1 + n_2 - 2 = 10$

$t_0 = \pm 3.169$

$$t = \frac{(\bar{x}_1 - \bar{x}_2) - (\mu_1 - \mu_2)}{\sqrt{\dfrac{(n_1 - 1)s_1^2 + (n_2 - 1)s_2^2}{n_1 + n_2 - 2}} \cdot \sqrt{\dfrac{1}{n_1} + \dfrac{1}{n_2}}} = \frac{(61 - 55) - (0)}{\sqrt{\dfrac{(5 - 1)\,3.3^2 + (7 - 1)\,1.2^2}{5 + 7 - 2}} \cdot \sqrt{\dfrac{1}{5} + \dfrac{1}{7}}}$$

$$= \frac{6}{\sqrt{5.22}\,\sqrt{0.343}} \approx 4.484$$

Reject H_0. There is enough evidence to support the claim.

14. $H_0: \mu_1 \ge \mu_2$ (claim); $H_a: \mu_1 < \mu_2$

d.f. $= \min\{n_1 - 1, n_2 - 1\} = 5$

$t_0 = -1.476$

$$t = \frac{(\bar{x}_1 - \bar{x}_2) - (\mu_1 - \mu_2)}{\sqrt{\dfrac{s_1^2}{n_1} + \dfrac{s_2^2}{n_2}}} = \frac{(520 - 500) - (0)}{\sqrt{\dfrac{(25)^2}{7} + \dfrac{(55)^2}{6}}} = \frac{20}{\sqrt{593.452}} \approx 0.821$$

Fail to reject H_0. There is not enough evidence to reject the claim.

15. (a) $H_0: \mu_1 \le \mu_2$; $H_a: \mu_1 > \mu_2$ (claim)

(b) d.f. $= n_1 + n_2 - 2 = 42$

  $t_0 = 1.645$; Reject H_0 if $t > 1.645$.

(c) $\bar{x}_1 = 51.476$, $s_1 = 11.007$, $n_1 = 21$

$\bar{x}_2 = 41.522$, $s_2 = 17.149$, $n_2 = 23$

$$t = \frac{(\bar{x}_1 - \bar{x}_2) - (\mu_1 - \mu_2)}{\sqrt{\dfrac{(n_1 - 1)s_1^2 + (n_2 - 1)s_2^2}{n_1 + n_2 - 2}}\sqrt{\dfrac{1}{n_1} + \dfrac{1}{n_2}}} = \frac{(51.476 - 41.522) - (0)}{\sqrt{\dfrac{(21 - 1)(11.007)^2 + (23 - 1)(17.149)^2}{21 + 23 - 2}}\sqrt{\dfrac{1}{21} + \dfrac{1}{23}}}$$

$$= \frac{9.954}{\sqrt{211.7386}\,\sqrt{0.0911}} \approx 2.266$$

(d) Reject H_0.

(e) There is sufficient evidence at the 5% significance level to support the claim that the third graders taught with the directed reading activities scored higher than those taught without the activities.

16. (a) H_0: $\mu_1 = \mu_2$ (claim); H_a: $\mu_1 \neq \mu_2$

(b) d.f. $= n_1 + n_2 - 2 = 20$

$t_0 = \pm 2.086$; Reject H_0 if $t < -2.086$ or $t > 2.086$.

(c) $t = \dfrac{(\bar{x}_1 - \bar{x}_2) - (\mu_1 - \mu_2)}{\sqrt{\dfrac{(n_1 - 1)s_1^2 + (n_2 - 1)s_2^2}{n_1 + n_2 - 2}}\sqrt{\dfrac{1}{n_1} + \dfrac{1}{n_2}}}$

$= \dfrac{(32{,}750 - 31{,}200) - (0)}{\sqrt{\dfrac{(12 - 1)(1900)^2 + (10 - 1)(1825)^2}{12 + 10 - 2}}\sqrt{\dfrac{1}{12} + \dfrac{1}{10}}}$

$= \dfrac{750}{\sqrt{3{,}484{,}281.25}\,\sqrt{0.183}} \approx 1.939$

(d) Fail to reject H_0. There is not enough evidence to reject the claim.

17. H_0: $\mu_d = 0$ (claim); H_a: $\mu_d \neq 0$

$\alpha = 0.05$ and d.f. $= n - 1 = 99$

$t_0 = \pm 1.96$ (Two-tailed test)

$t = \dfrac{\bar{d} - \mu_d}{\dfrac{s_d}{\sqrt{n}}} = \dfrac{10 - 0}{\dfrac{12.4}{\sqrt{100}}} = \dfrac{10}{1.24} \approx 8.065$

Reject H_0.

18. H_0: $\mu_d \geq 0$; H_a: $\mu_d < 0$ (claim)

$\alpha = 0.01$ and d.f. $= n - 1 = 24$

$t_0 = -2.492$ (Left-tailed test)

$t = \dfrac{\bar{d} - \mu_d}{\dfrac{s_d}{\sqrt{n}}} = \dfrac{3.2 - 0}{\dfrac{1.38}{\sqrt{25}}} = \dfrac{3.2}{0.276} \approx 11.594$

Fail to reject H_0.

19. $H_0: \mu_d \leq 6$ (claim); $H_a: \mu_d > 6$

$\alpha = 0.10$ and d.f. $= n - 1 = 32$

$t_0 = 1.282$ (Right-tailed test)

$t = \dfrac{\overline{d} - \mu_d}{\dfrac{s_d}{\sqrt{n}}} = \dfrac{10.3 - 6}{\dfrac{1.24}{\sqrt{33}}} = \dfrac{4.3}{0.21586} \approx 19.921$

Reject H_0.

20. $H_0: \mu_d = 0$; $H_a: \mu_d \neq 15$ (claim)

$\alpha = 0.05$ and d.f. $= n - 1 = 36$

$t_0 = \pm 1.96$ (Two-tailed test)

$t = \dfrac{\overline{d} - \mu_d}{\dfrac{s_d}{\sqrt{n}}} = \dfrac{17.5 - 15}{\dfrac{4.05}{\sqrt{37}}} = \dfrac{2.5}{0.6691} \approx 3.755$

Reject H_0.

21. (a) $H_0: \mu_d \leq 0$; $H_a: \mu_d > 0$ (claim)

(b) $t_0 = 1.383$; Reject H_0 if $t > 1.383$.

(c) $\overline{d} = 5$ and $s_d \approx 8.743$

(d) $t = \dfrac{\overline{d} - \mu_d}{\dfrac{s_d}{\sqrt{n}}} = \dfrac{5 - 0}{\dfrac{8.743}{\sqrt{10}}} = \dfrac{5}{2.765} \approx 1.808$

(e) Reject H_0.

(f) There is enough evidence to support the claim.

22. (a) $H_0: \mu_d \leq 0$; $H_a: \mu_d > 0$ (claim)

(b) $t_0 = 1.372$; Reject H_0 if $t > 1.372$.

(c) $\overline{d} \approx -0.636$ and $s_d \approx 5.870$

(d) $t = \dfrac{\overline{d} - \mu_d}{\dfrac{s_d}{\sqrt{n}}} = \dfrac{-0.636 - 0}{\dfrac{5.870}{\sqrt{11}}} = \dfrac{-0.636}{1.7699} \approx -0.359$

(e) Fail to reject H_0.

(f) There is not enough evidence to support the claim.

23. $H_0: p_1 = p_2$; $H_a: p_1 \neq p_2$ (claim)

$z_0 = \pm 1.96$ (Two-tailed test)

$$\overline{p} = \frac{x_1 + x_2}{n_1 + n_2} = \frac{425 + 410}{840 + 760} = 0.522$$

$\overline{q} = 0.478$

$$z = \frac{(\hat{p}_1 - \hat{p}_2) - (p_1 - p_2)}{\sqrt{\overline{p}\,\overline{q}\left(\frac{1}{n_1} + \frac{1}{n_2}\right)}} = \frac{(0.506 - 0.539) - (0)}{\sqrt{0.522 \cdot 0.478\left(\frac{1}{840} + \frac{1}{760}\right)}} = \frac{-0.033}{\sqrt{0.0006254}} \approx -1.320$$

Fail to reject H_0.

24. $H_0: p_1 \leq p_2$ (claim); $H_a: p_1 > p_2$

$z_0 = 2.33$ (Right-tailed test)

$$\overline{p} = \frac{x_1 + x_2}{n_1 + n_2} = \frac{36 + 46}{100 + 200} = 0.273$$

$\overline{q} = 0.727$

$$z = \frac{(\hat{p}_1 - \hat{p}_2) - (p_1 - p_2)}{\sqrt{\overline{p}\,\overline{q}\left(\frac{1}{n_1} + \frac{1}{n_2}\right)}} = \frac{(0.360 - 0.230) - (0)}{\sqrt{0.273 \cdot 0.727\left(\frac{1}{100} + \frac{1}{200}\right)}} = \frac{-0.130}{\sqrt{0.002977}} \approx -2.383$$

Reject H_0.

25. $H_0: p_1 \leq p_2$; $H_a: p_1 > p_2$ (claim)

$z_0 = 1.282$ (Right-tailed test)

$$\overline{p} = \frac{x_1 + x_2}{n_1 + n_2} = \frac{261 + 207}{556 + 483} = 0.450$$

$\overline{q} = 0.550$

$$z = \frac{(\hat{p}_1 - \hat{p}_2) - (p_1 - p_2)}{\sqrt{\overline{p}\,\overline{q}\left(\frac{1}{n_1} + \frac{1}{n_2}\right)}} = \frac{(0.469 - 0.429) - (0)}{\sqrt{0.450 \cdot 0.550\left(\frac{1}{556} + \frac{1}{483}\right)}} = \frac{0.04}{\sqrt{0.0009576}} \approx 1.293$$

Reject H_0.

26. $H_0: p_1 \geq p_2$; $H_a: p_1 < p_2$ (claim)

$z_0 = -1.645$ (Left-tailed test)

$$\overline{p} = \frac{x_1 + x_2}{n_1 + n_2} = \frac{86 + 107}{900 + 1200} \approx 0.092$$

$\overline{q} = 0.908$

$$z = \frac{(\hat{p}_1 - \hat{p}_2) - (p_1 - p_2)}{\sqrt{\overline{p}\,\overline{q}\left(\frac{1}{n_1} + \frac{1}{n_2}\right)}} = \frac{(0.096 - 0.089) - (0)}{\sqrt{0.092 \cdot 0.908\left(\frac{1}{900} + \frac{1}{1200}\right)}} = \frac{0.007}{\sqrt{0.00016243}} \approx 0.549$$

Fail to reject H_0.

27. (a) $H_0: p_1 = p_2$ (claim); $H_a: p_1 \neq p_2$

(b) $z_0 = \pm 1.645$; Reject H_0 if $z < -1.645$ or $z > 1.645$.

$$\bar{p} = \frac{x_1 + x_2}{n_1 + n_2} = \frac{398 + 530}{800 + 1000} = 0.516$$

$$\bar{q} = 0.484$$

(c) $z = \dfrac{(\hat{p}_1 - \hat{p}_2) - (p_1 - p_2)}{\sqrt{\bar{p}\bar{q}\left(\dfrac{1}{n_1} + \dfrac{1}{n_2}\right)}} = \dfrac{(0.4975 - 0.5300) - (0)}{\sqrt{0.516 \cdot 0.484\left(\dfrac{1}{800} + \dfrac{1}{1000}\right)}} = \dfrac{-0.0325}{\sqrt{0.0005619}} \approx -1.371$

(d) Fail to reject H_0.

(e) There is not enough evidence to reject the claim.

28. (a) $H_0: p_1 \leq p_2$; $H_a: p_1 > p_2$ (claim)

(b) $z_0 = 1.645$; Reject H_0 if $z > 1.645$.

$$\bar{p} = \frac{x_1 + x_2}{n_1 + n_2} = \frac{986 + 5576}{6164 + 42{,}890} \approx 0.134$$

$$\bar{q} = 0.866$$

(c) $z = \dfrac{(\hat{p}_1 - \hat{p}_2) - (p_1 - p_2)}{\sqrt{\bar{p}\bar{q}\left(\dfrac{1}{n_1} + \dfrac{1}{n_2}\right)}} = \dfrac{(0.160 - 0.130) - (0)}{\sqrt{0.134 \cdot 0.866\left(\dfrac{1}{6164} + \dfrac{1}{42{,}890}\right)}}$

$= \dfrac{0.030}{\sqrt{0.00002153}} \approx 6.465$

(d) Reject H_0.

(e) There is enough evidence to support the claim.

CHAPTER 8 QUIZ SOLUTIONS

1. (a) $H_0: \mu_1 \leq \mu_2$; $H_a: \mu_1 > \mu_2$ (claim)

(b) n_1 and $n_2 > 30$ and the samples are independent $\rightarrow$ Right tailed z-test

(c) $z_0 = 1.645$; Reject H_0 if $z > 1.645$.

(d) $z = \dfrac{(\bar{x}_1 - \bar{x}_2) - (\mu_1 - \mu_2)}{\sqrt{\dfrac{s_1^2}{n_1} + \dfrac{s_2^2}{n_2}}} = \dfrac{(149 - 145) - (0)}{\sqrt{\dfrac{(35)^2}{49} + \dfrac{(33)^2}{50}}} = \dfrac{4}{\sqrt{46.780}} \approx 0.585$

(e) Fail to reject H_0.

(f) There is not enough evidence at the 5% significance level to support the claim that the mean score on the science assessment for male high school students was higher than for the female high school students.

2. (a) $H_0: \mu_1 = \mu_2$ (claim); $H_a: \mu_1 \neq \mu_2$

(b) n_1 and $n_2 < 30$, the samples are independent, and the populations are normally distributed. $\rightarrow$ Two-tailed t-test (assume variances are equal)

(c) d.f. $= n_1 + n_2 - 2 = 26$

$t_0 = \pm 2.779$; Reject H_0 if $t < -2.779$ or $t > 2.779$.

(d) $t = \dfrac{(\bar{x}_1 - \bar{x}_2) - (\mu_1 - \mu_2)}{\sqrt{\dfrac{(n_1 - 1)s_1^2 + (n_2 - 1)s_2^2}{n_1 + n_2 - 2}}\sqrt{\dfrac{1}{n_1} + \dfrac{1}{n_2}}} = \dfrac{(153 - 149) - (0)}{\sqrt{\dfrac{(13 - 1)(32)^2 + (15 - 1)(30)^2}{13 + 15 - 2}}\sqrt{\dfrac{1}{13} + \dfrac{1}{15}}}$

$= \dfrac{4.0}{\sqrt{957.23}\sqrt{0.14359}} \approx 0.341$

(e) Fail to reject H_0.

(f) There is not enough evidence at the 1% significance level to reject the teacher's claim that the mean scores on the science assessment test are the same for fourth grade boys and girls.

3. (a) $H_0: p_1 \leq p_2$; $H_a: p_1 > p_2$ (claim)

(b) Testing 2 proportions, $n_1\bar{p}$, $n_1\bar{q}$, $n_2\bar{p}$, and $n_2\bar{q} \geq 5$, and the samples are independent $\rightarrow$ Right-tailed z-test

(c) $z_0 = 1.28$; Reject H_0 if $z > 1.28$.

(d) $\bar{p} = \dfrac{x_1 + x_2}{n_1 + n_2} = \dfrac{2043 + 3018}{6382 + 11{,}179} = 0.288$

$\bar{q} = 0.712$

$z = \dfrac{(\hat{p}_1 - \hat{p}_2) - (p_1 - p_2)}{\sqrt{\bar{p}\,\bar{q}\left(\dfrac{1}{n_1} + \dfrac{1}{n_2}\right)}} = \dfrac{(0.32 - 0.27) - (0)}{\sqrt{0.288 \cdot 0.712\left(\dfrac{1}{6382} + \dfrac{1}{11{,}179}\right)}}$

$= \dfrac{0.05}{\sqrt{0.000050473}} \approx 7.04$

(e) Reject H_0.

(f) There is sufficient evidence at the 10% significance level to support the claim that the proportion of fatal crashes involving alcohol is higher for drivers in the 21 to 24 age group than for drivers ages 25 to 35.

4. (a) $H_0: \mu_d \geq 0$; $H_a: \mu_d < 0$ (claim)

(b) Dependent samples and both populations are normally distributed. $\rightarrow$ one-tailed t-test

(c) $t_0 = -1.796$; Reject H_0 if $t < -1.796$.

(d) $t = \dfrac{\bar{d} - \mu_d}{\dfrac{s_d}{\sqrt{n}}} = \dfrac{0 - 68.5}{\dfrac{26.318}{\sqrt{12}}} = \dfrac{-68.5}{7.597} \approx -9.016$

(e) Reject H_0.

(f) There is sufficient evidence at the 5% significance level to conclude that the students' SAT scores improved on the second test.

1. (a) $\hat{p} = \dfrac{x}{n} = \dfrac{570}{1000} = 0.570$

$\hat{q} = 0.430$

$n\hat{p} = 570 \geq 5$

$n\hat{q} = 430 \geq 5$

Use normal distribution

$\hat{p} \pm z_c \sqrt{\dfrac{\hat{p}\hat{q}}{n}} = 0.570 \pm 1.96 \sqrt{\dfrac{(0.570)(0.430)}{1000}} = 0.570 \pm 0.031 \Longrightarrow (0.539, 0.601)$

(b) Given that the 95% CI is (0.539, 0.601), it is unlikely that more than 60% of adults believe it is somewhat or very likely that life exists on other planets.

2. H_0: $\mu_d \leq 0$

H_a: $\mu_d > 0$ (claim)

$\overline{d} = 6.833$

$s_d = 3.713$

d.f. $= n - 1 = 11$

$t_0 = 1.796$

$t = \dfrac{\overline{d} - 0}{\dfrac{sd}{\sqrt{n}}} = \dfrac{6.833 - 0}{\dfrac{3.713}{\sqrt{12}}} = 6.375$

Reject H_0. There is enough evidence to support the claim.

3. $\overline{x} \pm z_c \dfrac{s}{\sqrt{n}} = 29.97 \pm 1.96 \dfrac{3.4}{\sqrt{42}} = 26.97 \pm 1.03 \Longrightarrow (25.94, 28.00)$; z-distribution

4. $\overline{x} \pm t_c \dfrac{s}{\sqrt{n}} = 3.46 \pm 1.753 \dfrac{1.63}{\sqrt{16}} = 3.46 \pm 0.71 \Longrightarrow (2.75, 4.17)$; t-distribution

5. $\overline{x} \pm z_c \dfrac{s}{\sqrt{n}} = 12.1 \pm 2.787 \dfrac{2.64}{\sqrt{26}} = 12.1 \pm 1.4 \Longrightarrow (10.7, 13.5)$; t-distribution

6. $\overline{x} \pm t_c \dfrac{s}{\sqrt{n}} = 8.21 \pm 2.365 \dfrac{0.62}{\sqrt{8}} = 8.21 \pm 0.52 \Longrightarrow (7.69, 8.73)$; t-distribution

7. H_0: $\mu_1 \leq \mu_2$

H_a: $\mu_1 > \mu_2$ (claim)

$z = 1.282$

$z = \dfrac{(\overline{x}_1 - \overline{x}_2) - 0}{\sqrt{\dfrac{s_1^2}{n_1} + \dfrac{s_2^2}{n_2}}} = \dfrac{3086 - 2263}{\sqrt{\dfrac{(563)^2}{85} + \dfrac{(624)^2}{68}}} = \dfrac{823}{\sqrt{9790.282}} = 8.318$

Reject H_0. There is enough evidence to support the claim.

8. H_0: $\mu \geq 33$

H_a: $\mu < 33$ (claim)

9. H_0: $p \geq 0.19$ (claim)

H_a: $p < 0.19$

10. H_0: $\sigma = 0.63$ (claim)

H_a: $\sigma \neq 0.63$

11. H_0: $\mu = 2.28$

H_a: $\mu \neq 2.28$ (claim)

12. (a) $\dfrac{(n-1)s^2}{\chi_R^2} < \sigma^2 < \dfrac{(n-1)s^2}{\chi_L^2} \Rightarrow \dfrac{(26-1)(3.1)^2}{46.928} < \sigma^2 < \dfrac{(26-1)(3.1)^2}{10.520}$

$\Rightarrow 5.1 < \sigma < 22.8$

(b) $\dfrac{(n-1)s^2}{\chi_R^2} < \sigma < \dfrac{(n-1)s^2}{\chi_L^2} \Rightarrow \sqrt{5.1} < \sigma < \sqrt{22.8}$

$\Rightarrow 2.3 < \sigma < 4.8$

(c) Because the 99% CI of σ is (2.3, 4.8) and contains 2.5, there is not enough evidence to support the pharmacist's claim.

13. H_0: $\mu_1 \geq \mu_2$

H_a: $\mu_1 < \mu_2$ (claim)

d.f. $= n_1 + n_2 - 2 = 15 + 15 - 2 = 28$

$t_0 = 1.701$

$$t = \frac{(\bar{x}_1 - \bar{x}_2) - 0}{\sqrt{\dfrac{(n_1-1)s_1^2 + (n_2-1)s_2^2}{n_1 + n_2 - 2}}\sqrt{\dfrac{1}{n_1} + \dfrac{1}{n_2}}}$$

$$= \frac{57.9 - 61.1}{\sqrt{\dfrac{(15-1)(0.8)^2 + (15-1)(0.6)^2}{15 + 15 - 2}}\sqrt{\dfrac{1}{15} + \dfrac{1}{15}}}$$

$$= \frac{-3.2}{\sqrt{0.5}\sqrt{0.133}} = -12.394$$

Reject H_0. There is enough evidence to support the claim.

14. (a) $\bar{x} = 296.231$

$s = 111.533$

$\bar{x} + t_c \dfrac{s}{\sqrt{n}} \Rightarrow 296.231 \pm 2.060 \dfrac{111.533}{\sqrt{26}}$

$\Rightarrow 296.231 \pm 45.059$

$\Rightarrow (251.2, 341.3)$

(b) H_0: $\mu \geq 280$

H_a: $\mu < 280$ (claim)

$t_0 = -1.316$

$$t = \frac{\bar{x} - 280}{\dfrac{s}{\sqrt{n}}} = \frac{296.231 - 280}{\dfrac{111.533}{\sqrt{26}}} = 0.742$$

Fail to reject H_0. There is not enough evidence to support the claim.

15. (a) H_0: $p_1 = p_2$ (claim)

H_a: $p_1 \neq p_2$

$$\bar{p} = \frac{x_1 + x_2}{n_1 + n_2} = \frac{195 + 204}{319 + 323} = 0.621$$

$$\bar{q} = 0.379$$

$$z_0 = \pm 1.645$$

$$z = \frac{(\hat{p}_1 - \hat{p}_2) - 0}{\sqrt{\bar{p}\,\bar{q}\left(\dfrac{1}{n_1} + \dfrac{1}{n_2}\right)}}$$

$$= \frac{(0.611 - 0.632)}{\sqrt{(0.621)(0.379)\left(\dfrac{1}{319} + \dfrac{1}{323}\right)}}$$

$$= \frac{-0.021}{\sqrt{0.001466}} = -0.548$$

Fail to reject H_0. There is not enough evidence to reject the claim.

16. $\bar{x} \pm z\dfrac{s}{\sqrt{n}} \Rightarrow 150 \pm 1.645 \dfrac{34}{\sqrt{120}} \Rightarrow 150 \pm 5 \Rightarrow (145, 155)$

(b) H_0: $\mu \geq 145$ (claim)

H_a: $\mu < 145$

$$z_0 = -1.645$$

$$z = \frac{\bar{x} - 145}{\dfrac{s}{\sqrt{n}}} = \frac{150 - 145}{\dfrac{34}{\sqrt{120}}} = 1.611$$

Fail to reject H_0. There is not enough evidence to reject the claim.

Correlation and Regression

9.1 Try It Yourself Solutions

1ab.

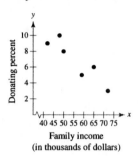

c. Yes, it appears that there is a negative linear correlation. As family income increases, the percent of income donated to charity decreases.

2ab.

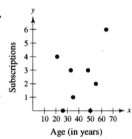

c. No, it appears that there is no linear correlation between age and subscriptions.

3ab.

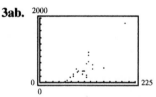

c. Yes, there appears to be a positive linear relationship between budget and worldwide gross.

4a. $n = 6$

b.

x	y	xy	x^2	y^2
42	9	378	1764	81
48	10	480	2304	100
50	8	400	2500	64
59	5	295	3481	25
65	6	390	4225	36
72	3	216	5184	9
$\Sigma x = 336$	$\Sigma y = 41$	$\Sigma xy = 2159$	$\Sigma x^2 = 19{,}458$	$\Sigma y^2 = 315$

c. $r = \dfrac{n\Sigma xy - (\Sigma x)(\Sigma y)}{\sqrt{n\Sigma x^2 - (\Sigma x)^2}\sqrt{n\Sigma y^2 - (\Sigma y)^2}} = \dfrac{6(2159) - (336)(41)}{\sqrt{6(19{,}458) - (336)^2}\sqrt{6(315) - (41)^2}}$

$= \dfrac{-822}{\sqrt{3852}\,\sqrt{209}} \approx -0.916$

d. Because r is close to -1, there appears to be a strong negative linear correlation between income level and donating percent.

5a. Enter the data.

b. $r \approx 0.838$

c. Because r is close to 1, there appears to be a strong positive linear correlation between budgets and worldwide grosses.

6a. $n = 6$

b. $\alpha = 0.01$

c. 0.917

d. Because $|r| \approx 0.916 < 0.917$, the correlation is not significant.

e. There is not enough evidence to conclude that there is a significant linear correlation between income level and the donating percent.

7a. $H_0: \rho = 0; H_a: \rho \neq 0$

b. $\alpha = 0.01$

c. d.f. $= n - 2 = 23$

d. ± 2.807; Reject H_0 if $t < -2.807$ or $t > 2.807$.

e. $t = \dfrac{r}{\sqrt{\dfrac{1 - r^2}{n - 2}}} = \dfrac{0.83773}{\sqrt{\dfrac{1 - (0.83773)^2}{25 - 2}}} = \dfrac{0.83773}{\sqrt{0.012966}} \approx 7.357$

f. Reject H_0.

g. There is enough evidence in the sample to conclude that a significant linear correlation exists.

9.1 EXERCISE SOLUTIONS

1. $r = -0.925$ represents a stronger correlation because it is closer to -1 than $r = 0.834$ is to $+1$.

2. (c) Correlation coefficients must be contained in the interval $[-1, 1]$.

3. A table can be used to compare r with a critical value or a hypotheses test can be performed using a t-test.

4. r = sample correlation coefficient
   ρ = population correlation coefficient

5. Negative linear correlation   6. No linear correlation

7. No linear correlation   8. Positive linear correlation

9. (c), You would expect a positive linear correlation between age and income.

10. (d), You would not expect age and height to be correlated.

11. (b), You would expect a negative linear correlation between age and balance on student loans.

12. (a), You would expect the relationship between age and body temperature to be fairly constant.

13. Explanatory variable: Amount of water consumed

Response variable: Weight loss

14. Explanatory variable: Hours of safety classes

Response variable: Number of accidents

15. (a)

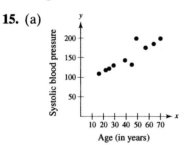

(b)

x	y	xy	x²	y²
16	109	1744	256	11,881
25	122	3050	625	14,884
39	143	5577	1521	20,449
45	132	5940	2025	17,424
49	199	9751	2401	39,601
64	185	11,840	4096	34,225
70	199	13,930	4900	39,601
29	130	3770	841	16,900
57	175	9975	3249	30,625
20	118	2360	400	13,924
$\Sigma x = 414$	$\Sigma y = 1512$	$\Sigma xy = 67{,}937$	$\Sigma x^2 = 20{,}314$	$\Sigma y^2 = 239{,}514$

$$r = \frac{n\Sigma xy - (\Sigma x)(\Sigma y)}{\sqrt{n\Sigma x^2 - (\Sigma x)^2}\sqrt{n\Sigma y^2 - (\Sigma y)^2}} = \frac{10(67{,}937) - (414)(1512)}{\sqrt{10(20{,}314) - (414)^2}\sqrt{10(239{,}514) - (1512)^2}}$$

$$= \frac{53{,}402}{\sqrt{31{,}744}\sqrt{108{,}996}} \approx 0.908$$

(c) Strong positive linear correlation

16. (a)

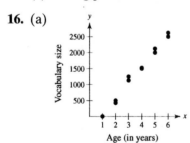

(b)

x	y	xy	x²	y²
1	3	3	1	9
2	400	880	4	193,600
3	1200	3600	9	1,440,000
4	1500	6000	16	2,250,000
5	2100	10,500	25	4,410,000
6	2600	15,600	36	6,760,000
3	1100	3300	9	1,210,000
5	2000	10,000	25	4,000,000
2	500	1000	4	250,000
4	1525	6100	16	2,325,625
6	2500	15,000	36	6,250,000
$\Sigma x = 41$	$\Sigma y = 15{,}468$	$\Sigma xy = 71{,}983$	$\Sigma x^2 = 181$	$\Sigma y^2 = 29{,}089{,}234$

$$r = \frac{n\Sigma xy(\Sigma x)(\Sigma y)}{\sqrt{n\Sigma x^2 - (\Sigma x)^2}\sqrt{n\Sigma y^2 - (\Sigma y)^2}}$$

$$= \frac{11(71{,}983) - (41)(15{,}468)}{\sqrt{11(181) - (41)^2}\sqrt{11(29{,}089{,}234) - (15{,}468)^2}} = \frac{157{,}625}{\sqrt{310}\sqrt{80{,}722{,}550}} \approx 0.996$$

(c) Strong positive linear correlation

17. (a)

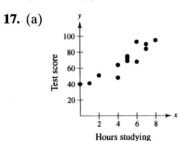

Hours studying

(b)

x	y	xy	x²	y²
0	40	0	0	1600
1	41	41	1	1681
2	51	102	4	2601
4	48	192	16	2304
4	64	256	16	4096
5	69	345	25	4761
5	73	365	25	5329
5	75	375	25	5625
6	68	408	36	4624
6	93	558	36	8649
7	84	588	49	7056
7	90	630	49	8100
8	95	760	64	9025
$\Sigma x = 60$	$\Sigma y = 891$	$\Sigma xy = 4620$	$\Sigma x^2 = 346$	$\Sigma y^2 = 65{,}451$

$$r = \frac{n\Sigma xy - (\Sigma x)(\Sigma y)}{\sqrt{n\Sigma x^2 - (\Sigma x)^2}\sqrt{n\Sigma y^2 - (\Sigma y)^2}} = \frac{13(4620) - (60)(891)}{\sqrt{13(346) - (60)^2}\sqrt{13(65{,}451) - (891)^2}}$$

$$= \frac{6600}{\sqrt{898}\sqrt{56{,}982}} \approx 0.923$$

(c) Strong positive linear correlation

18. (a)

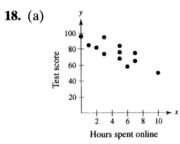

Hours spent online

(b)

x	y	xy	x²	y²
0	96	0	0	9216
1	85	85	1	7225
2	82	164	4	6724
3	74	222	9	5476
3	95	285	9	9025
5	68	340	25	4624
5	76	380	25	5776
5	84	420	25	7056
6	58	348	36	3364
7	65	455	49	4225
7	75	525	49	5625
10	50	500	100	2500
$\Sigma x = 54$	$\Sigma y = 908$	$\Sigma xy = 3724$	$\Sigma x^2 = 332$	$\Sigma y^2 = 70{,}836$

$$r = \frac{n\Sigma xy - (\Sigma x)(\Sigma y)}{\sqrt{n\Sigma x^2 - (\Sigma x)^2}\sqrt{n\Sigma y^2 - (\Sigma y)^2}}$$

$$= \frac{12(3724) - (54)(908)}{\sqrt{12(332) - (54)^2}\sqrt{12(70836) - (908)^2}} = \frac{-4344}{\sqrt{1068}\sqrt{25{,}568}} \approx -0.831$$

(c) Strong negative linear correlation

19. (a)

(b)

x	y	xy	x²	y²
207	553	114,471	42,849	305,809
204	391	79,764	41,616	152,881
200	1835	367,000	40,000	3.37×10^6
200	784	156,800	40,000	614,656
180	749	134,820	32,400	561,001
175	218	38,150	30,625	47,524
175	255	44,625	30,625	65,025
170	433	73,610	28,900	187,489
$\Sigma x = 1511$	$\Sigma y = 5218$	$\Sigma xy = 1{,}009{,}240$	$\Sigma x^2 = 287{,}015$	$\Sigma y^2 = 5{,}301{,}610$

$$r = \frac{n\Sigma xy - (\Sigma x)(\Sigma y)}{\sqrt{n\Sigma x^2 - (\Sigma x)^2}\sqrt{n\Sigma y^2 - (\Sigma y)^2}} = \frac{8(1{,}009{,}240) - (1511)(5218)}{\sqrt{8(287{,}015) - (1511)^2}\sqrt{8(5{,}301{,}610) - (5218)^2}}$$

$$= \frac{189{,}522}{\sqrt{12{,}999}\sqrt{15{,}185{,}356}} \approx 0.427$$

(c) Weak positive linear correlation

20. (a)

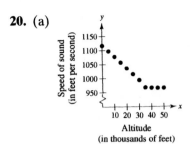

(b)

x	y	xy	x²	y²
0	1116.3	0.000	0	1.25×10^6
5	1096.9	5484.5	25	1.2×10^6
10	1077.3	10,773	100	1.16×10^6
15	1057.2	15,858	225	1.12×10^6
20	1036.8	20,736	400	1.07×10^6
25	1015.8	25,395	625	1.03×10^6
30	994.5	29,835	900	989,030
35	969.0	33,915	1225	938,961
40	967.7	38,708	1600	936,443
45	967.7	43,547	2025	936,443
50	967.7	48,385	2500	936,443
$\Sigma x = 275$	$\Sigma y = 11,266.9$	$\Sigma xy = 272,636$	$\Sigma x^2 = 9625$	$\Sigma y^2 = 11,571,687.43$

$$r = \frac{n\Sigma xy - (\Sigma x)(\Sigma y)}{\sqrt{n\Sigma x^2 - (\Sigma x)^2}\sqrt{n\Sigma y^2 - (\Sigma y)^2}} = \frac{11(272,636) - (275)(11,266.9)}{\sqrt{11(9625) - (275)^2}\sqrt{11(11,571,687.43) - (11,266.9)^2}}$$

$$= \frac{-99,401.5}{\sqrt{30,250}\sqrt{345,526.12}} \approx -0.972$$

(c) Strong negative linear correlation

21. (a)

(b)

x	y	xy	x²	y²
2.34	1.33	3.112	5.476	1.769
1.96	1.07	2.097	3.842	1.145
1.39	1.15	1.599	1.932	1.323
3.07	0.25	0.768	9.425	0.063
0.65	1.00	0.650	0.423	1.000
5.21	1.00	5.210	27.144	1.000
0.88	1.59	1.399	0.774	2.528
3.23	1.20	3.876	10.433	1.440
2.54	1.62	4.115	6.452	2.624
1.03	0.20	0.206	1.061	0.040
$\Sigma x = 22.3$	$\Sigma y = 10.41$	$\Sigma xy = 23.031$	$\Sigma x^2 = 66.961$	$\Sigma y^2 = 12.931$

$$r = \frac{n\Sigma xy - (\Sigma x)(\Sigma y)}{\sqrt{n\Sigma x^2 - (\Sigma x)^2}\sqrt{n\Sigma y^2 - (\Sigma y)^2}} = \frac{10(23.031) - (22.3)(10.41)}{\sqrt{10(66.961) - (22.3)^2}\sqrt{10(12.931) - (10.41)^2}}$$

$$= \frac{-1.833}{\sqrt{172.32}\sqrt{20.942}} = -0.031$$

(c) No linear correlation

22. (a)

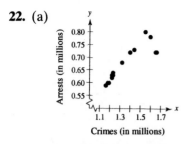

(b)

x	y	xy	x^2	y^2
1.66	0.72	1.195	2.756	0.518
1.65	0.72	1.188	2.723	0.518
1.60	0.78	1.248	2.560	0.608
1.55	0.80	1.240	2.403	0.640
1.44	0.73	1.051	2.074	0.533
1.40	0.72	1.008	1.960	0.518
1.32	0.68	0.898	1.742	0.462
1.23	0.64	0.787	1.513	0.410
1.22	0.63	0.769	1.488	0.397
1.23	0.63	0.775	1.513	0.397
1.22	0.62	0.756	1.488	0.384
1.18	0.60	0.708	1.392	0.360
1.16	0.59	0.684	1.346	0.348
1.19	0.60	0.714	1.416	0.360
$\Sigma x = 19.05$	$\Sigma y = 9.46$	$\Sigma xy = 13.021$	$\Sigma x^2 = 26.373$	$\Sigma y^2 = 6.455$

$$r = \frac{n\Sigma xy - (\Sigma x)(\Sigma y)}{\sqrt{n\Sigma x^2 - (\Sigma x)^2}\ \sqrt{n\Sigma y^2 - (\Sigma y)^2}}$$

$$= \frac{14(13.021) - (19.05)(9.46)}{\sqrt{14(26.373) - (19.05)^2}\ \sqrt{14(6.455) - (9.46)^2}} = \frac{2.081}{\sqrt{6.320}\ \sqrt{0.878}}$$

$$= 0.883$$

(c) Strong positive linear correlation

23. $r \approx 0.623$

$n = 8$ and $\alpha = 0.01$

$cv = 0.834$

$|r| \approx 0.623 < 0.834 \Rightarrow$ The correlation is not significant.

or

$H_0: \rho = 0$; and $H_a: \rho \neq 0$

$\alpha = 0.01$

d.f. $= n - 2 = 6$

$cu = \pm 3.703$; Reject H_0 if $t < -3.707$ or $t > 3.707$.

$$t = \frac{r}{\sqrt{\dfrac{1 - r^2}{n - 2}}} = \frac{0.623}{\sqrt{\dfrac{1 - (0.623)^2}{8 - 2}}} = \frac{0.623}{\sqrt{0.10198}} = 1.951$$

Fail to reject H_0. There is not enough evidence to conclude that there is a significant linear correlation between vehicle weight and the variability in braking distance.

24. $r \approx 0.955$

$n = 8$ and $\alpha = 0.05$

$cv = 0.707$

$|r| = 0.955 > 0.707 \Rightarrow$ The correlation is significant.

or

$H_0: p = 0; H_a: p \neq 0$

$\alpha = 0.05$

d.f. $= n - 2 = 6$

$cu = \pm 2.447$; Reject H_0 if $t < -2.447$ or $t > 2.447$.

$$t = \frac{r}{\sqrt{\dfrac{1-r^2}{n-2}}} = \frac{0.955}{\sqrt{\dfrac{1-(0.955)^2}{8-2}}} = \frac{0.955}{\sqrt{0.01466}} = 7.887$$

Reject H_0. There is enough evidence to conclude that there is a significant linear correlation between vehicle weight and variability in braking distance.

25. $r \approx 0.923$

$n = 8$ and $\alpha = 0.01$

$cv = 0.834$

$|r| \approx 0.923 > 0.834 \Rightarrow$ The correlation is significant.

or

$H_0: \rho = 0; H_a: \rho \neq 0$

$\alpha = 0.01$

d.f. $= n - 2 = 11$

$cu = \pm 3.106$; Reject H_0 if $t < -3.106$ or $t > 3.106$.

$r \approx 0.923$

$$t = \frac{r}{\sqrt{\dfrac{1-r^2}{n-2}}} = \frac{0.923}{\sqrt{\dfrac{1-(0.923)^2}{13-2}}} = \frac{0.923}{\sqrt{0.01346}} \approx 7.955$$

Reject H_0. There is enough evidence to conclude that a significant linear correlation exists.

26. $r \approx 0.831$

$n = 12$ and $\alpha = 0.05$

$cv = 0.576$

$|r| = 0.831 > 0.576 \Rightarrow$ The correlation is significant.

or

$H_0: \rho = 0; H_a: \rho \neq 0$

$\alpha = 0.05$

d.f. $= n - 2 = 10$

$cu = \pm 2.228$; Reject H_0 if $t < -2.228$ or $t > 2.228$.

$r \approx -0.831$

$$t = \frac{r}{\sqrt{\dfrac{1-r^2}{n-2}}} = \frac{-0.831}{\sqrt{\dfrac{1-(-0.831)^2}{12-2}}} \approx -4.724$$

Reject H_0. There is enough evidence to conclude that there is a significant linear correlation between the data.

27. $r \approx -0.030$

$n = 10$ and $\alpha = 0.01$

$cv = 0.765$

$|r| \approx 0.030 < 0.765 \Rightarrow$ The correlation is not significant.

or

$H_0: \rho = 0; H_a: \rho \neq 0$

$\alpha = 0.01$

d.f. $= n - 2 = 8$

$cu = \pm 3.355$; Reject H_0 if $t < -3.355$ or $t > 3.355$.

$r \approx -0.030$

$$t = \frac{r}{\sqrt{\dfrac{1 - r^2}{n - 2}}} = \frac{-0.030}{\sqrt{\dfrac{1 - (-0.030)^2}{10 - 2}}}$$

$$= \frac{-0.030}{\sqrt{0.125}} = -0.085$$

Fail to reject H_0. There is not enough evidence at the 1% significance level to conclude there is a significant linear correlation between earnings per share and dividends per share.

28. $r \approx 0.842$

$n = 11$ and $\alpha = 0.05$

$cv = 0.602$

$|r| = 0.842 > 0.602 \Rightarrow$ The correlation is significant.

or

$H_0: \rho = 0; H_a: \rho \neq 0$

$\alpha = 0.05$

d.f. $= n - 2 = 9$

$cu = \pm 2.262$; Reject H_0 if $t < -2.262$ or $t > 2.262$.

$r \approx 0.842$

$$t = \frac{r}{\sqrt{\dfrac{1 - r^2}{n - 2}}} = \frac{0.832}{\sqrt{\dfrac{1 - (-0.842)^2}{11 - 2}}} = 4.682$$

Reject H_0. There is enough evidence to conclude that there is a significant correlation between the data.

29. The correlation coefficient remains unchanged when the *x*-values and *y*-values are switched.

30. The correlation coefficient remains unchanged when the *x*-values and *y*-values are switched.

31. Answers will vary.

9.2 Try It Yourself Solutions

1a. $n = 6$

x	y	xy	x^2
42	9	378	1764
48	10	480	2304
50	8	400	2500
59	5	295	3481
65	6	390	4225
72	3	216	5184
$\Sigma x = 336$	$\Sigma y = 41$	$\Sigma xy = 2159$	$\Sigma x^2 = 19,458$

b. $m = \dfrac{n\Sigma xy - (\Sigma x)(\Sigma y)}{n\Sigma x^2 - (\Sigma x)^2} = \dfrac{6(2159) - (336)(41)}{6(19,458) - (336)^2} = \dfrac{-822}{3852} \approx -0.2133956$

c. $b = \bar{y} - m\bar{x} = \left(\dfrac{41}{6}\right) - (-0.2133956)\left(\dfrac{336}{6}\right) \approx 18.7837$

d. $\hat{y} = -0.213x + 18.783$

2a. Enter the data.

b. $m \approx 10.93477; b \approx -710.61551$

c. $\hat{y} = 10.935x - 710.616$

3a. (1) $\hat{y} = 12.481(2) + 33.683$ (2) $\hat{y} = 12.481(3.32) + 33.683$

b. (1) 58.645   (2) 75.120

c. (1) 58.645 minutes   (2) 75.120 minutes

9.2 EXERCISE SOLUTIONS

1. c   **2.** a   **3.** d   **4.** b   **5.** g   **6.** e

7. h   **8.** f   **9.** c   **10.** b   **11.** a   **12.** d

13.

x	y	xy	x^2
764	55	42,020	583,696
625	47	29,375	390,625
520	51	26,520	270,400
510	28	14,280	260,100
492	39	19,188	242,064
484	34	16,456	234,256
450	33	14,850	202,500
430	31	13,330	184,900
410	40	16,400	168,100
$\Sigma x = 4685$	$\Sigma y = 358$	$\Sigma xy = 192,419$	$\Sigma x^2 = 2,536,641$

$m = \dfrac{n\Sigma xy - (\Sigma x)(\Sigma y)}{n\Sigma x^2 - (\Sigma x)^2} = \dfrac{9(192,419) - (4685)(358)}{9(2,536,641) - (4685)^2}$

$= \dfrac{54,541}{880,544} \approx 0.06194 \approx 0.062$

$b = \bar{y} - m\bar{x} = \left(\dfrac{358}{9}\right) - (0.06194)\left(\dfrac{4685}{9}\right) \approx 7.535$

$\hat{y} = 0.062x + 7.535$

(a) $\hat{y} = 0.062(500) + 7.535 \approx 38.535 \approx 39$ stories

(b) $\hat{y} = 0.062(650) + 7.535 \approx 47.835 \approx 48$ stories

(c) It is not meaningful to predict the value of y for $x = 310$ because $x = 310$ is outside the range of the original data.

(d) $\hat{y} = 0.062(725) + 7.535 \approx 52.485 \approx 52$ stories

14.

x	y	xy	x²
3	1100	3,300	9
4	1300	5,200	16
4	1500	6,000	16
5	2100	10,500	25
6	2600	15,600	36
2	460	920	4
3	1200	3,600	9
$\Sigma x = 27$	$\Sigma y = 10{,}260$	$\Sigma xy = 45{,}120$	$\Sigma x^2 = 115$

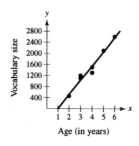

$m = \dfrac{n\Sigma xy - (\Sigma x)(\Sigma y)}{n\Sigma x^2 - (\Sigma x)^2} = \dfrac{7(45{,}120) - (27)(10{,}260)}{7(115) - (27)^2} = \dfrac{38{,}820}{76} = 510.789$

$b = \bar{y} - m\bar{x} = \left(\dfrac{10{,}260}{7}\right) - (510.789)\left(\dfrac{27}{7}\right) = -504.474$

$\hat{y} = 510.789x - 504.474$

(a) $\hat{y} = 510.789(2) - 504.474 \approx 517$ \qquad (b) $\hat{y} = 510.789(3) - 504.474 \approx 1028$

(c) $\hat{y} = 510.789(6) - 504.474 \approx 2560$

(d) It is not meaningful to predict the value of y for $x = 12$ because $x = 12$ is outside the range of the original data.

15.

x	y	xy	x²
0	40	0	0
1	41	41	1
2	51	102	4
4	48	192	16
4	64	256	16
5	69	345	25
5	73	365	25
5	75	375	25
6	68	408	36
6	93	558	36
7	84	588	49
7	90	630	49
8	95	760	64
$\Sigma x = 60$	$\Sigma y = 891$	$\Sigma xy = 4620$	$\Sigma x^2 = 346$

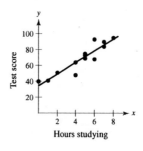

$m = \dfrac{n\Sigma xy - (\Sigma x)(\Sigma y)}{n\Sigma x^2 - (\Sigma x)^2} = \dfrac{13(4620) - (60)(891)}{13(346) - (60)^2} = \dfrac{6600}{898} \approx 7.350$

$b = \bar{y} - m\bar{x} = \left(\dfrac{60}{13}\right) - (7.350)\left(\dfrac{891}{13}\right) \approx 34.617$

$\hat{y} = 7.350x + 34.617$

(a) $\hat{y} = 7.350(3) + 34.617 \approx 56.7$

(b) $\hat{y} = 7.350(6.5) + 34.617 \approx 82.4$

(c) It is not meaningful to predict the value of y for $x = 13$ because $x = 13$ is outside the range of the original data.

(d) $\hat{y} = 7.350(4.5) + 34.617 \approx 67.7$

16.

x	y	xy	x^2
0	96	0	0
1	85	85	1
2	82	164	4
3	74	222	9
3	95	285	9
5	68	340	25
5	76	380	25
5	84	420	25
6	58	348	36
7	65	455	49
7	75	525	49
10	50	500	100
$\Sigma x = 54$	$\Sigma y = 908$	$\Sigma xy = 3724$	$\Sigma x^2 = 332$

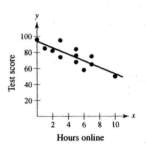

$$m = \frac{n\Sigma xy - (\Sigma x)(\Sigma y)}{n\Sigma x^2 - (\Sigma x)^2} = \frac{12(3724) - (54)(908)}{12(332) - (54)^2} = \frac{-4344}{1068} \approx -4.067$$

$$b = \bar{y} - m\bar{x} = \left(\frac{908}{12}\right) - (-4.067)\left(\frac{54}{12}\right) \approx 93.970$$

$$\hat{y} = -4.067x + 93.970$$

(a) $\hat{y} = -4.067(4) + 93.970 \approx 77.7$ (b) $\hat{y} = -4.067(8) + 93.970 \approx 61.4$

(c) $\hat{y} = -4.067(9) + 93.970 \approx 57.4$

(d) It is not meaningful to predict the values of y for $x = 15$ because $x = 15$ is outside the range of the original data.

17.

x	y	xy	x^2
150	420	63,000	22,500
170	470	79,900	28,900
120	350	42,000	14,400
120	360	43,200	14,400
90	270	24,300	8,100
180	550	99,000	32,400
170	530	90,100	28,900
140	460	64,400	19,600
90	380	34,200	8,100
110	330	36,300	12,100
$\Sigma x = 1340$	$\Sigma y = 4120$	$\Sigma xy = 576,400$	$\Sigma x^2 = 189,400$

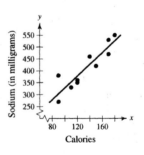

$$m = \frac{n\Sigma xy - (\Sigma x)(\Sigma y)}{n\Sigma x^2 - (\Sigma x)^2} = \frac{10(576,400) - (1340)(4120)}{10(189,400) - (1340)^2} = \frac{243,200}{98,400} \approx 2.471545 \approx 2.472$$

$$b = \bar{y} - m\bar{x} = \left(\frac{4120}{10}\right) - 2.471545\left(\frac{1340}{10}\right) = 80.813$$

$$\hat{y} = 2.472x + 80.813$$

(a) $\hat{y} = 2.472(170) + 80.813 = 501.053$ milligrams

(b) $\hat{y} = 2.472(100) + 80.813 = 328.213$ milligrams

(c) $\hat{y} = 2.472(140) + 80.813 = 426.893$ milligrams

(d) It is not meaningful to predict the value of y for $x = 210$ because $x = 210$ is outside the range of the original data.

18.

x	y	xy	x^2
70	8.3	581	4900
72	10.5	756	5184
75	11.0	825	5625
76	11.4	866.4	5776
71	9.2	653.2	5041
73	10.9	795.7	5329
85	14.9	1266.5	7225
78	14.0	1092	6084
77	16.3	1255.1	5929
80	18.0	1440	6400
82	15.8	1295.6	6724
$\Sigma x = 839$	$\Sigma y = 140.3$	$\Sigma xy = 10{,}826.5$	$\Sigma x^2 = 64{,}217$

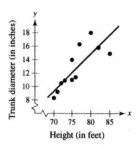

$$m = \frac{n\Sigma xy - (\Sigma x)(\Sigma y)}{n\Sigma x^2 - (\Sigma x)^2} = \frac{11(10{,}826.5) - (839)(140.3)}{11(64{,}217) - (839)^2} = \frac{1379.8}{2466} = 0.55953 \approx 0.560$$

$$b = \bar{y} - m\bar{x} = \left(\frac{140.3}{11}\right) - (0.55953)\left(\frac{839}{11}\right) = -29.922$$

$$\hat{y} = 0.560x - 29.922$$

(a) $\hat{y} = 0.560(74) - 29.922 = 11.52$ inches

(b) $\hat{y} = 0.560(81) - 29.922 = 15.44$ inches

(c) It is not meaningful to predict y for $x = 95$ because $x = 95$ is outside the range of the original data.

(d) $\hat{y} = 0.560(79) - 29.922 = 14.32$ inches

19.

x	y	xy	x^2
8.5	66.0	561.00	72.25
9.0	68.5	616.50	81.00
9.0	67.5	607.50	81.00
9.5	70.0	665.00	90.25
10.0	70.0	700.00	100.00
10.0	72.0	720.00	100.00
10.5	71.5	750.75	110.25
10.5	69.5	729.75	110.25
11.0	71.5	786.50	121.00
11.0	72.0	792.00	121.00
11.0	73.0	803.00	121.00
12.0	73.5	882.00	144.00
12.0	74.0	888.00	144.00
12.5	74.0	925.00	156.25
$\Sigma x = 146.5$	$\Sigma y = 993.0$	$\Sigma xy = 10{,}427.0$	$\Sigma x^2 = 1552.3$

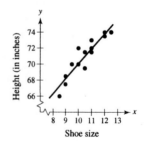

$$m = \frac{n\Sigma xy - (\Sigma x)(\Sigma y)}{n\Sigma x^2 - (\Sigma x)^2} = \frac{14(10{,}427.0) - (146.5)(993.0)}{14(1552.3) - (146.5)^2} = \frac{503.5}{269.95} \approx 1.870$$

$$b = \bar{y} - m\bar{x} = \left(\frac{993.0}{14}\right) - (1.870)\left(\frac{146.5}{14}\right) \approx 51.360$$

$$\hat{y} = 1.870x + 51.360$$

(a) $\hat{y} = 1.870(11.5) + 51.360 \approx 72.865$ inches

(b) $\hat{y} = 1.870(8.0) + 51.360 \approx 66.32$ inches

(c) It is not meaningful to predict the value of y for $x = 15.5$ because $x = 15.5$ is outside the range of the original data.

(d) $\hat{y} = 1.870(10.0) + 51.360 \approx 70.06$ inches

20.

x	y	xy	x²
0.1	14.9	1.49	0.01
0.2	14.5	2.90	0.04
0.4	13.9	xxx	xxx
0.7	14.1	9.87	0.49
0.6	13.9	xxx	xxx
0.9	13.7	12.33	0.81
2.0	14.3	xxx	xxx
0.6	13.9	8.34	0.36
0.5	14.0	7.00	0.25
0.1	14.1	xxx	xxx
$\Sigma x = 4.5$	$\Sigma y = 141.3$	$\Sigma xy = 62.92$	$\Sigma x^2 = 2.61$

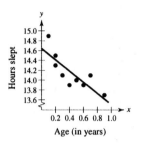

$$m = \frac{n\Sigma xy - (\Sigma x)(\Sigma y)}{n\Sigma x^2 - (\Sigma x)^2} = \frac{10(62.92) - (4.5)(141.3)}{10(2.61) - (4.5)^2} = \frac{-6.65}{5.85} \approx -1.137$$

$$b = \bar{y} - m\bar{x} = \left(\frac{141.3}{10}\right) - (-1.137)\left(\frac{4.5}{10}\right) \approx 14.642$$

$$\hat{y} = -1.137x + 14.642$$

(a) $\hat{y} = -1.137(0.3) + 14.642 \approx 14.30$ hours

(b) It is not meaningful to predict the value of y for $x = 3.9$ because $x = 3.9$ is outside the range of the original data.

(c) $\hat{y} = -1.137(0.6) + 14.642 \approx 13.96$ hours

(d) $\hat{y} = -1.137(0.8) + 14.642 \approx 13.73$ hours

21.

x	y	xy	x²
5720	2.19	12,527	32,718,400
4050	1.36	5508	16,402,500
6130	2.58	15,815	37,576,900
5000	1.74	8700	25,000,000
5010	1.78	8917.8	25,100,100
4270	1.69	7216.3	18,232,900
5500	1.80	9900	30,250,000
5550	1.87	10,379	30,802,500
$\Sigma x = 41,230$	$\Sigma y = 15.01$	$\Sigma xy = 78,962.8$	$\Sigma x^2 = 216,083,300$

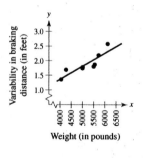

$$m = \frac{n\Sigma xy - (\Sigma x)(\Sigma y)}{n\Sigma x^2 - (\Sigma x)^2} = \frac{8(78,962.8) - (41,230)(15.01)}{8(216,083,300) - (41,230)^2} = \frac{12,833.7}{28,753,500} = 0.000447$$

$$b = \bar{y} - m\bar{x} = \left(\frac{15.01}{8}\right) - (0.000447)\left(\frac{41,230}{8}\right) = -0.425$$

$$\hat{y} = 0.000447x - 0.425$$

(a) $\hat{y} = 0.000447(4500) - 0.425 = 1.587$ feet

(b) $\hat{y} = 0.000447(6000) - 0.425 = 2.257$ feet

(c) It is not meaningful to predict the value of y for $x = 7500$ because $x = 7500$ is outside the range of the original data.

(d) $\hat{y} = 0.000447(5750) - 0.425 = 2.145$ feet

22.

x	y	xy	x^2
5720	3.78	21,622	32,718,400
4050	2.43	9841.5	16,402,500
6130	4.63	28,381.9	37,576,900
5000	2.88	14,400	25,000,000
5010	3.25	16,282.5	25,100,100
4270	2.76	11,785.2	18,232,900
5500	3.42	18,810	30,250,000
5550	3.51	19,480.5	30,802,500
$\Sigma x = 41{,}230$	$\Sigma y = 26.66$	$\Sigma xy = 140{,}603.2$	$\Sigma x^2 = 216{,}083{,}300$

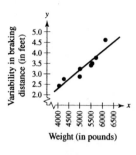

Weight (in pounds)

$$m = \frac{n\Sigma xy - (\Sigma x)(\Sigma y)}{n\Sigma x^2 - (\Sigma x)^2} = \frac{8(140{,}603.2) - (41{,}230)(26.66)}{8(216{,}083{,}300) - (41{,}230)^2} = \frac{25{,}632.2}{28{,}753{,}500} \approx 0.000891502$$

$$b = \bar{y} - m\bar{x} = \left(\frac{26.66}{8}\right) - (-0.000891502)\left(\frac{41{,}230}{8}\right) = -1.26208$$

$$\hat{y} = 0.000892x - 1.262$$

(a) $\hat{y} = 0.000892(4800) - 1.262 = 3.020$ feet   (b) $\hat{y} = 0.000892(5850) - 1.262 = 3.956$ feet

(c) $\hat{y} = 0.000892(4075) - 1.262 = 2.373$ feet

(d) It is not meaningful to predict the value of y for $x = 3000$ because $x = 3000$ is outside the range of the original data.

23. Substitute a value x into the equation of a regression line and solve for y.

24. Prediction values are meaningful only for x-values in (or close to) the range of the data.

25. Strong positive linear correlation; As the ages of the engineers increase, the salaries of the engineers tend to increase.

26. $m = 0.510$

$b = 45.642$

$\hat{y} = mx + b = 0.510x + 45.642$

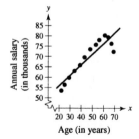

Age (in years)

27. It is not meaningful to predict y for $x = 74$ because $x = 74$ is outside the range of the original data.

28. $r = 0.917$

$cv = 0.684 \Rightarrow$ Reject H_0: $\rho = 0$

29. In general, as age increases; salary increases until age 61 when salary decreases. (Answers will vary.)

30. (a) $\hat{y} = 1.724x + 79.733$

(b) $\hat{y} = 0.453x - 26.448$

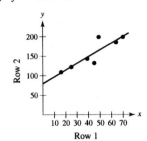

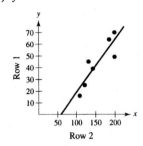

(c) The slope of the line keeps the same sign, but the values of *m* and *b* change.

31. (a)

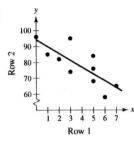

(b)

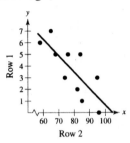

$\hat{y} = -4.297x + 94.200$

$\hat{y} = -0.1413x + 14.763$

(c) The slope of the line keeps the same sign, but the values of *m* and *b* change.

32. (a) $m = 1.711$

$b = 3.912$

$\hat{y} = 1.711x + 3.912$

(b)

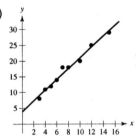

(c)

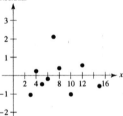

x	y	$\hat{y} = 1.711x + 3.912$	$y - \hat{y}$
8	18	17.600	0.400
4	11	10.756	0.244
15	29	29.577	-0.577
7	18	15.889	2.111
6	14	14.178	-0.178
3	8	9.045	-1.045
12	25	24.444	0.556
10	20	21.022	-1.022
5	12	12.467	-0.467

(d) The residual plot shows no pattern in the residuals because the residuals fluctuate about 0. This suggests that the regression line is a good representation of the data.

33. (a) $m = 0.139$

$b = 21.024$

$\hat{y} = 0.139x + 21.024$

(b)

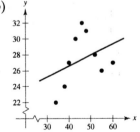

(c)

x	y	$\hat{y} = 0.139x + 21.024$	$y - \hat{y}$
38	24	26.306	−2.306
34	22	25.750	−3.750
40	27	26.584	0.416
46	32	27.418	4.582
43	30	27.001	2.999
48	31	27.696	3.304
60	27	29.364	−2.364
55	26	28.669	−2.669
52	28	28.252	−0.252

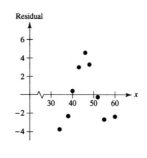

(d) The regression line may not be a good model for the data because the data do not fluctuate about 0.

34. (a)

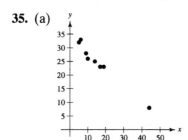

(b) The point (14, 3) may be an outlier.

(c) The point (14, 3) is influential because using all 6 points $\Rightarrow \hat{y} = 0.212x + 6.445$.

However, excluding the point (14, 3) $\Rightarrow \hat{y} = 0.905 + 3.568$.

35. (a)

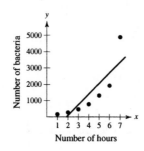

(b) The point (44, 8) may be an outlier.

(c) Excluding the point (44, 8)$\Rightarrow \hat{y} = -0.711x + 35.263$. The point (44, 8) is not influential because using all 8 points $\Rightarrow \hat{y} = -0.607x + 34.160$.

36. The point (200, 1835.4) is an outlier, but is not an influential point because the regression line shows no significant change.

37. $m = 654.536$

$b = -1214.857$

$\hat{y} = 654.536x - 1214.857$

38.

x	y	log y
1	165	2.218
2	280	2.447
3	468	2.670
4	780	2.892
5	1310	3.117
6	1920	3.283
7	4900	3.690

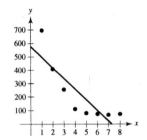

$$\log y = 0.233x + 1.969$$

39. $\log y = 0.233x + 1.969 \Rightarrow y = 10^{0.233x + 1.969}$

$$\Rightarrow y = 93.111(10)^{0.233x}$$

40. The exponential equation is a better model for the data. The exponential equation's error is smaller.

41. $m = -78.929$

$b = 576.179$

$\hat{y} = -78.929x + 576.179$

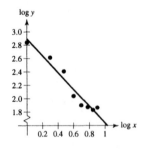

42.

x	y	log x	log y
1	695	0	2.842
2	410	0.301	2.613
3	256	0.477	2.408
4	110	0.602	2.041
5	80	0.699	1.903
6	75	0.778	1.875
7	68	0.845	1.833
8	74	0.903	1.869

$$\log y = m(\log x) + b \Rightarrow \log y = -1.251 \log x + 2.893$$

A linear model is more appropriate for the transformed data.

43. $\log y = m(\log x) + b \Rightarrow y = 10^{m \log x + b} = 10^{b} 10^{m \log x}$

$$= 10^{2.893} 10^{-1.251 \log x} = 781.628 \cdot x^{-1.251}$$

44. The power equation is a much better model for the data. The power equation's error is smaller.

45. $y = a + b \ln x = 25.035 + 19.599 \ln x$

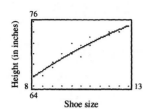

Shoe size

46. $\hat{y} = a + b \ln x = 13.667 - 0.471 \ln x$

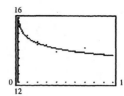

47. The logarithmic equation is a better model for the data. The logarithmic equation's error is smaller.

48. The logarithmic equation is a better model for the data. The logarithmic equation's error is smaller.

9.3 MEASURES OF REGRESSION AND PREDICTION INTERVALS

9.3 Try It Yourself Solutions

1a. $r = 0.979$ **b.** $r^2 = (0.979)^2 = 0.958$

c. 95.8% of the variation in the times is explained.
4.2% of the variation is unexplained.

2a.

x_i	y_i	$\hat{y}_i$	$(y_i - \hat{y}_i)^2$
15	26	28.386	5.693
20	32	35.411	11.635
20	38	35.411	6.703
30	56	49.461	42.759
40	54	63.511	90.459
45	78	70.536	55.711
50	80	77.561	5.949
60	88	91.611	13.039
			$\Sigma = 231.948$

b. $n = 8$

c. $s_e = \sqrt{\dfrac{\Sigma(y_i - \hat{y}_i)^2}{n - 2}} = \sqrt{\dfrac{231.948}{6}} \approx 6.218$

d. The standard error of estimate of the weekly sales for a specific radio ad time is about $621.80.

3a. $n = 8$, d.f. $= 6$, $t_c = 2.447$, $s_e \approx 10.290$

b. $\hat{y} = 50.729x + 104.061 = 50.729(2.5) + 104.061 \approx 230.884$

c. $E = t_c s_e \sqrt{1 + \dfrac{1}{n} + \dfrac{n(x - \bar{x})^2}{n(\Sigma x^2) - (\Sigma x)^2}} = (2.447)(10.290)\sqrt{1 + \dfrac{1}{8} + \dfrac{8(2.5 - 1.975)^2}{8(32.44) - (15.8)^2}}$

$= (2.447)(10.290)\sqrt{1.34818} \approx 29.236$

d. $\hat{y} \pm E \rightarrow (201.648, 260.120)$

e. You can be 95% confident that the company sales will be between $201,648 and $260,120 when advertising expenditures are $2500.

9.3 EXERCISE SOLUTIONS

1. Total variation $= \Sigma(y_i - \bar{y})^2$; the sum of the squares of the differences between the y-values of each ordered pair and the mean of the y-values of the ordered pairs.

2. Explained variation $= \Sigma(\hat{y}_i - \bar{y})^2$; the sum of the squares of the differences between the predicted y-values and the mean of the y-values of the ordered pairs.

3. Unexplained variation $= \Sigma(y_i - \hat{y}_i)^2$; the sum of the squares of the differences between the observed y-values and the predicted y-values.

4. Coefficient of determination: $r^2 = \dfrac{\Sigma(\hat{y}_i - \bar{y})^2}{\Sigma(y_i - \bar{y})^2}$

   r^2 is the ratio of the explained variation to the total variation and is the percent of variation of y that is explained by the relationship between x and y.

5. $r^2 = (0.350)^2 \approx 0.123$

   12.3% of the variation is explained. 87.7% of the variation is unexplained.

6. $r^2 = (-0.275)^2 \approx 0.076$

   7.6% of the variation is explained. 92.4% of the variation is unexplained.

7. $r = (-0.891)^2 \approx 0.794$

   79.4% of the variation is explained. 20.6% of the variation is unexplained.

8. $r^2 = (0.964)^2 \approx 0.929$

   92.9% of the variation is explained. 7.1% of the variation is unexplained.

9. (a) $r^2 = \dfrac{\Sigma(\hat{y}_i - \bar{y})^2}{\Sigma(y_i - \bar{y})^2} \approx 0.233$

   23.3% of the variation in proceeds can be explained by the variation in the number of issues and 76.7% of the variation is unexplained.

   (b) $s_e = \sqrt{\dfrac{\Sigma(y_i - \hat{y}_i)^2}{n - 2}} = \sqrt{\dfrac{2{,}356{,}140{,}670}{10}} \approx 15{,}349.725$

   The standard error of estimate of the proceeds for a specific number of issues is about $15,349,725,000.

10. (a) $r^2 = \dfrac{\Sigma(\hat{y}_i - \bar{y})^2}{\Sigma(y_i - \bar{y})^2} \approx 0.787$

    78.7% of the variation in cigarettes exported from the United States can be explained by the variation in the number of cigarettes consumed in the United States and 21.3% of the variation is unexplained.

    (b) $s_e = \sqrt{\dfrac{\Sigma(y_i - \hat{y}_i)^2}{n - 2}} = \sqrt{\dfrac{367.604}{5}} \approx 8.574$

    The standard error of estimate of the number of cigarettes exported for a specific number of cigarettes consumed is about 8,574,000,000.

11. (a) $r^2 = \dfrac{\Sigma(\hat{y}_i - \bar{y})^2}{\Sigma(y_i - \bar{y})^2} \approx 0.981$

    98.1% of the variation in sales can be explained by the variation in the total square footage and 1.9% of the variation is unexplained.

(b) $s_e = \sqrt{\dfrac{\Sigma(y_i - \hat{y}_i)^2}{n-2}} = \sqrt{\dfrac{8413.958}{9}} \approx 30.576$

The standard error of estimate of the sales for a specific total square footage is about 30,576,000,000.

12. (a) $r^2 = \dfrac{\Sigma(\hat{y}_i - \bar{y})^2}{\Sigma(y_i - \bar{y})^2} \approx 0.578$

57.8% of the variation in the median number of leisure hours per week can be explained by the variation in the median number of work hours per week and 42.2% of the variation is unexplained.

(b) $s_e = \sqrt{\dfrac{\Sigma(y_i - \hat{y}_i)^2}{n-2}} = \sqrt{\dfrac{33.218}{8}} \approx 2.013$

The standard error of estimate of the median number of leisure hours/week for a specific median number of hours/week is about 2.013 hours.

13. (a) $r^2 = \dfrac{\Sigma(\hat{y}_i - \bar{y})^2}{\Sigma(y_i - \bar{y})^2} \approx 0.994$

99.4% of the variation in the median weekly earnings of female workers can be explained by the variation in the median weekly earnings of male workers and 0.6% of the variation is unexplained.

(b) $s_e = \sqrt{\dfrac{\Sigma(y_i - \hat{y}_i)^2}{n-2}} = \sqrt{\dfrac{22.178}{3}} \approx 2.719$

The standard error of estimate of the median weekly earnings of female workers for a specific median weekly earnings of male workers is about $2.72.

14. (a) $r^2 = \dfrac{\Sigma(\hat{y}_i - \bar{y})^2}{\Sigma(y_i - \bar{y})^2} \approx 0.931$

93.1% of the variation in the turnout for federal elections can be explained by the variation in the voting age population and 6.9% of the variation is unexplained.

(b) $s_e = \sqrt{\dfrac{\Sigma(y_i - \hat{y}_i)^2}{n-2}} = \sqrt{\dfrac{31.619}{6}} \approx 2.296$

The standard error of estimate of the turnout in a federal election for a specified voting age population is about 2,296,000.

15. (a) $r^2 = \dfrac{\Sigma(\hat{y}_i - \bar{y})^2}{\Sigma(y_i - \bar{y})^2} \approx 0.994$

99.4% of the variation in the money spent can be explained by the variation in the money raised and 0.6% of the variation is unexplained.

(b) $s_e = \sqrt{\dfrac{\Sigma(y_i - \hat{y}_i)^2}{n-2}} = \sqrt{\dfrac{2136.181}{6}} \approx 18.869$

The standard error of estimate of the money spent for a specified amount of money raised is about $18,869,000.

16. (a) $r^2 = \dfrac{\Sigma(\hat{y}_i - \bar{y})^2}{\Sigma(y_i - \bar{y})^2} \approx 0.905$

90.5% of the variation in federal pension plans can be explained by the variation in IRA's and 9.5% of the variation is unexplained.

274 CHAPTER 9 | CORRELATION AND REGRESSION

(b) $s_e = \sqrt{\dfrac{\Sigma(y_i - \hat{y}_i)^2}{n-2}} = \sqrt{\dfrac{14{,}166.620}{7}} \approx 44.987$

The standard error of estimate of assets in federal pension plans for a specified total of IRA assets is about \$44,987,000,000.

17. $n = 12$, d.f. $= 10$, $t_c = 2.228$, $s_e = 15{,}349.725$

$\hat{y} = 40.049x + 21{,}843.09 = 40.049(712) + 21{,}843.09 \approx 50{,}357.978$

$E = t_c s_e \sqrt{1 + \dfrac{1}{n} + \dfrac{n(x-\bar{x})^2}{n(\Sigma x^2) - (\Sigma x)^2}} = (2.228)(15{,}349.725)\sqrt{1 + \dfrac{1}{12} + \dfrac{12(712 - 3806/12)^2}{12(1{,}652{,}648) - (3806)^2}}$

$= (2.228)(15{,}349.725)\sqrt{1.43325} \approx 40{,}942.727$

$\hat{y} \pm E \rightarrow (9415.251, 91{,}300.705) \rightarrow (\$9{,}415{,}251{,}000, \$91{,}300{,}705{,}000)$

You can be 95% confident that the proceeds will be between \$9,415,251,000 and \$91,300,705,000 when the number of initial offerings is 712.

18. $n = 7$, d.f. $= 5$, $t_c = 2.571$, $s_e = 8.574$

$\hat{y} = 0.671x + 136.771 = 0.671(450) - 136.771 \approx 165.179$

$E = t_c s_e \sqrt{1 + \dfrac{1}{n} + \dfrac{n(x-\bar{x})^2}{n(\Sigma x^2) - (\Sigma x)^2}}$

$= (2.571)(8.574)\sqrt{1 + \dfrac{1}{7} + \dfrac{7[450 - (2869/7)]^2}{7(1{,}178{,}895) - (2869)^2}}$

$= (2.571)(8.574)\sqrt{1.67736} \approx 28.550$

$\hat{y} \pm E \rightarrow (136.629, 193.729) \rightarrow (136{,}629{,}000{,}000, 193{,}729{,}000{,}000)$

You can be 95% confident that the number of cigarettes exported will be between 136,629,000,000 and 193,729,000,000 when the number of cigarettes consumed is 450 billion.

19. $n = 11$, d.f. $= 9$, $t_c = 1.833$, $s_e \approx 30.576$

$\hat{y} = 548.448x - 1881.694 = 548.448(4.5) - 1881.694 \approx 590.822$

$E = t_c s_e \sqrt{1 + \dfrac{1}{n} + \dfrac{n(x-\bar{x})^2}{n(\Sigma x^2) - (\Sigma x)^2}} = (1.833)(30.576)\sqrt{1 + \dfrac{1}{11} + \dfrac{11[4.5 - (61.2/11)]^2}{11(341.9) - (61.2)^2}}$

$= (1.833)(30.576)\sqrt{1.89586} \approx 77.170$

$\hat{y} \pm E \rightarrow (513.652, 677.992) \rightarrow (\$513{,}652{,}000{,}000, \$667{,}992{,}000{,}000)$

You can be 90% confident that the sales will be between \$513,652,000,000 and \$667,992,000,000 when the total square footage is 4.5 billion.

© 2009 Pearson Education, Inc., Upper Saddle River, NJ. All rights reserved. This material is protected under all copyright laws as they currently exist. No portion of this material may be reproduced, in any form or by any means, without permission in writing from the publisher.

20. $n = 10$, d.f. $= 8$, $t_c = 1.860$, $s_e \approx 2.013$

$$\hat{y} = -0.646x + 50.734 = -0.646(45.1) + 50.734 \approx 21.599$$

$$E = t_c s_e \sqrt{1 + \frac{1}{n} + \frac{n(x - \bar{x})^2}{n(\Sigma x^2) - (\Sigma x)^2}}$$

$$= (1.860)(2.013) \sqrt{1 + \frac{1}{10} + \frac{10[45.1 - (475.5/10)]^2}{10(22,716.29) - (475.5)^2}}$$

$$= (1.860)(2.013) \sqrt{1.15649} \approx 4.026$$

$\hat{y} \pm E \rightarrow (17.573, 25.625)$

You can be 90% confident that the median number of leisure hours/week will be between 17.573 and 25.625 when the median number of work hours/week is 45.1.

21. $n = 5$, d.f. $= 3$, $t_c = 5.841$, $s_e \approx 2.719$

$$\hat{y} = 1.369x - 402.687 = 1.369(650) - 402.687 \approx 487.163$$

$$E = t_c s_e \sqrt{1 + \frac{1}{n} + \frac{n(x - \bar{x})^2}{n(\Sigma x^2) - (\Sigma x)^2}} = (5.841)(2.719) \sqrt{1 + \frac{1}{5} + \frac{5[650 - (3479/5)]^2}{5(2,422,619) - (3479)^2}}$$

$$= (5.841)(2.719) \sqrt{2.28641} \approx 24.014$$

$\hat{y} \pm E \rightarrow (463.149, 511.177)$

You can be 99% confident that the median earnings of female workers will be between $463.15 and $511.18 when the median weekly earnings of male workers is $650.

22. $n = 8$, d.f. $= 6$, $t_c = 3.707$, $s_e \approx 2.296$

$$\hat{y} = 0.342x + 5.990 = 0.342(210) + 5.990 \approx 77.81$$

$$E = t_c s_e \sqrt{1 + \frac{1}{n} + \frac{n(x - \bar{x})^2}{n(\Sigma x^2) - (\Sigma x)^2}}$$

$$= (3.707)(2.296) \sqrt{1 + \frac{1}{8} + \frac{8[210 - (1449.1/8)]^2}{8(266,100.61) - (1449.1)^2}}$$

$$= (3.707)(2.296) \sqrt{1.35549} \approx 9.909$$

$\hat{y} \pm E \rightarrow (67.901, 87.719)$

You can be 99% confident that the voter turnout in federal elections will be between 67.901 million and 87.719 million when the voting age population is 210 million.

23. $n = 8$, d.f. $= 6$, $t_c = 2.447$, $s_e \approx 18.869$

$$\hat{y} = 0.943x - 21.541 = 0.943(775.8) + 21.541 \approx 753.120$$

$$E = t_c s_e \sqrt{1 + \frac{1}{n} + \frac{n(x - \bar{x})^2}{n(\Sigma x^2) - (\Sigma x)^2}} = (2.447)(18.869) \sqrt{1 + \frac{1}{8} + \frac{8[775.8 - (6666.2/8)]^2}{8(5,932,282.32) - (6666.2)^2}}$$

$$= (2.447)(18.869) \sqrt{1.13375} \approx 49.163$$

$\hat{y} \pm E \rightarrow (\$703.957 \text{ million}, \$802.283 \text{ million})$

You can be 95% confident that the money spent in congressional campaigns will be between $703.957 million and $802.283 million when the money raised is $775.8 million.

24. $n = 9$, d.f. $= 7$, $t_c = 1.895$, $s_e \approx 44.987$

$\hat{y} = 0.21x + 288.18 = 0.21(2500) + 288.18 \approx 813.18$

$$E = t_c s_e \sqrt{1 + \frac{1}{n} + \frac{n(x - \bar{x})^2}{n(\Sigma x^2) - (\Sigma x)^2}}$$

$$= (1.895)(44.987) \sqrt{1 + \frac{1}{9} + \frac{9[2500 - (22,332/9)]^2}{9(58,537,290) - (22,332)^2}}$$

$$= (1.895)(44.987) \sqrt{1.11122} \approx 89.866$$

$\hat{y} \pm E \rightarrow (723.314, 903.046)$

You can be 90% confident that the total assets in federal pension plans will be between $723.314 billion and $903.046 billion when the total assets in IRA's is $2500 billion.

25.

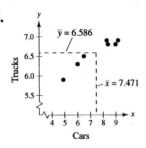

26. $\hat{y} = 0.217x + 4.966$

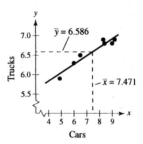

27.

x_i	y_i	$\hat{y}_i$	$\hat{y}_i - \bar{y}$	$y_i - \hat{y}_i$	$y_i - \bar{y}$
9.2	6.9	6.962	0.376	−0.062	0.314
9.0	6.8	6.919	0.333	−0.119	0.214
8.4	6.8	6.789	0.203	0.011	0.214
8.3	6.9	6.767	0.181	0.133	0.314
6.5	6.5	6.377	−0.210	0.124	−0.086
6.0	6.3	6.268	−0.318	0.032	−0.286
4.9	5.9	6.029	−0.557	−0.129	−0.686

28. (a) Explained variation $= \Sigma(\hat{y}_i - \bar{y})^2 \approx 0.781$

(b) Unexplained variation $= \Sigma(y_i - \hat{y}_i)^2 \approx 0.069$

(c) Total variation $= \Sigma(y_i - \bar{y})^2 \approx 0.849$

29. $r^2 \approx 0.919$; About 91.9% of the variation in the median age of trucks can be explained by the variation in the median age of cars, and 8.1% of the variation is unexplained.

30. $s_e \approx 0.117$; The standard error of estimate of the median age of trucks for a specific median age of cars is about 0.117 year.

31. $\hat{y} = 0.217x + 4.966 = 0.217(8.6) + 4.966 \approx 6.832$

$$E = t_c s_e \sqrt{1 + \frac{1}{n} + \frac{n(x - \bar{x})^2}{n(\Sigma x^2) - (\Sigma x)^2}} = (2.571)(0.117) \sqrt{1 + \frac{1}{7} + \frac{7[8.6 - (52.3/7)]^2}{7(407.35) - (52.3)^2}}$$

$$= (2.571)(0.117) \sqrt{1.21961} \approx 0.332$$

$\hat{y} \pm E \rightarrow (6.500, 7.164)$

32. The slope and correlation coefficient will have the same sign because the denominators of both formulas are always positive while the numerators are always equal.

33. $cv = \pm 3.707$

$m = -0.205$

$se = 0.554$

$$t = \frac{m}{s_e}\sqrt{\Sigma x^2 - \frac{(\Sigma x)^2}{n}} = \frac{-0.205}{0.554}\sqrt{838.55 - \frac{(79.5)^2}{8}} = -2.578$$

Fail to reject H_0: $M = 0$

34. $cv = \pm 2.306$

$m = 1.328$

$se = 5.086$

$$t = \frac{m}{s_e}\sqrt{\Sigma x^2 - \frac{(\Sigma x)^2}{n}} = \frac{1.328}{5.086}\sqrt{14,660 - \frac{(374)^2}{10}} = 6.771$$

Reject H_0: $M = 0$

35. $E = t_c s_e \sqrt{\dfrac{1}{n} + \dfrac{\bar{x}^2}{\Sigma x^2 - [(\Sigma x)^2/n]}} = 2.447(10.290)\sqrt{\dfrac{1}{8} + \dfrac{\left(\frac{15.8}{8}\right)^2}{(32.44) - \frac{(15.8)^2}{8}}} = 45.626$

$b \pm E \Rightarrow 104.061 \pm 45.626 = (58.435, 149.687)$

$E = \dfrac{t_c s_e}{\sqrt{\Sigma x^2 - \dfrac{(\Sigma x)^2}{n}}} = \dfrac{2.447(10.290)}{\sqrt{32.44 - \dfrac{(15.8)^2}{8}}} = 22.658$

$m \pm E \Rightarrow 50.729 \pm 22.658 \Rightarrow (28.071, 73.387)$

36. $E = t_c s_e \sqrt{\dfrac{1}{n} + \dfrac{\bar{x}^2}{\Sigma x^2 - (\Sigma x)^2}} = (3.707)(10.290)\sqrt{\dfrac{1}{8} + \dfrac{\left(\frac{15.8}{8}\right)^2}{(32.44) - \frac{(15.8)^2}{8}}} = 69.119$

$b \pm E \Rightarrow 104.061 \pm 69.119 \Rightarrow (34.942, \ 173.18)$

$E = \dfrac{t_c s_e}{\sqrt{\Sigma x^2 - \dfrac{(\Sigma x)^2}{n}}} = \dfrac{(3.707)(10.290)}{\sqrt{32.44 - \dfrac{(15.8)^2}{8}}} = 34.325$

$m \pm E \Rightarrow 50.729 \pm 34.325 \Rightarrow (16.404, \ 85.054)$

9.4 MULTIPLE REGRESSION

9.4 Try It Yourself Solutions

1a. Enter the data.   **b.** $\hat{y} = 46.385 + 0.540x_1 - 4.897x_2$

2ab. (1) $\hat{y} = 46.385 + 0.540(89) - 4.897(1)$

  (2) $\hat{y} = 46.385 + 0.540(78) - 4.897(3)$

  (3) $\hat{y} = 46.385 + 0.540(83) - 4.897(2)$

c. (1) $\hat{y} = 89.548$ (2) $\hat{y} = 73.814$ (3) $\hat{y} = 81.411$

d. (1) 90   (2) 74   (3) 81

9.4 EXERCISE SOLUTIONS

1. $\hat{y} = 640 - 0.105x_1 + 0.124x_2$

(a) $\hat{y} = 640 - 0.105(13,500) + 0.124(12,000) = 710.5$ pounds

(b) $\hat{y} = 640 - 0.105(15,000) + 0.124(13,500) = 739$ pounds

(c) $\hat{y} = 640 - 0.105(14,000) + 0.124(13,500) = 844$ pounds

(d) $\hat{y} = 640 - 0.105(14,500) + 0.124(13,000) = 729.5$ pounds

2. $\hat{y} = 17 + -0.000642x_1 + 0.00125x_2$

(a) $\hat{y} = 17 + -0.000642(60,000) + 0.00125(45,000) = 34.73$ bushels per acre

(b) $\hat{y} = 17 + -0.000642(62,500) + 0.00125(53,000) = 43.125$ bushels per acre

(c) $\hat{y} = 17 + -0.000642(57,500) + 0.00125(50,000) = 42.585$ bushels per acre

(d) $\hat{y} = 17 + -0.000642(59,000) + 0.00125(48,500) = 39.747$ bushels per acre

3. $\hat{y} = -52.2 + 0.3x_1 + 4.5x_2$

(a) $\hat{y} = -52.2 + 0.3(70) + 4.5(8.6) = 7.5$ cubic feet

(b) $\hat{y} = -52.2 + 0.3(65) + 4.5(11.0) = 16.8$ cubic feet

(c) $\hat{y} = -52.2 + 0.3(83) + 4.5(17.6) = 51.9$ cubic feet

(d) $\hat{y} = -52.2 + 0.3(87) + 4.5(19.6) = 62.1$ cubic feet

4. $\hat{y} = -0.415 + 0.0009x_1 + 0.168x_2$

(a) $\hat{y} = -0.415 + 0.0009(15.2) + 0.168(10.3) = \$1.33$

(b) $\hat{y} = -0.415 + 0.0009(19.8) + 0.168(14.5) = \$2.04$

(c) $\hat{y} = -0.415 + 0.0009(12.1) + 0.168(9.1) = \$1.12$

(d) $\hat{y} = -0.415 + 0.0009(21.5) + 0.168(15.8) = \$2.26$

5. $\hat{y} = -2518.4 + 126.8x_1 + 66.4x_2$

(a) $s = 28.489$

(b) $r^2 = 0.985$

(c) The standard error of estimate of the predicted sales given a specific total square footage and number of shopping centers is \$28.489 billion. The multiple regression model explains 98.5% of the variation in y.

6. $\hat{y} = -4.312 + 0.2366x_1 - 0.1093x_2$

(a) $s = 0.828$ (b) $r^2 = 0.996$

(c) The standard error of estimate of the predicted equity given a specific net sales and total assets is \$0.828 billion. The multiple regression model explains 99.6% of the variation in y.

7. $n = 11, k = 2, r^2 = 0.985$

$$r_{adj}^2 = 1 - \left[\frac{(1 - r^2)(n - 1)}{n - k - 1} \right] = 0.981$$

98.1% of the variation in y can be explained by the relationships between variables.

$r_{adj}^2 < r^2$

8. $n = 6, k = 2, r^2 = 0.996$

$$r_{adj}^2 = 1 - \left[\frac{(1 - r^2)(n - 1)}{n - k - 1} \right] = 0.993$$

99.3% of the variation in y can be explained by the relationships between variables.

$r_{adj}^2 < r^2$

CHAPTER 9 REVIEW EXERCISE SOLUTIONS

1.
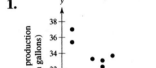

$r \approx -0.939$; strong negative linear correlation; milk production decreases as age increases

2.

$r \approx 0.944$; strong positive linear correlation; The average monthly temperatures increase with convertibles sold.

3.
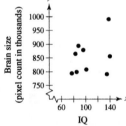

$r \approx 0.361$; weak positive linear correlation; brain size increases as IQ increases.

4.
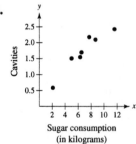

$r \approx 0.953$; strong positive linear correlation; The number of cavities increases as the annual per capita sugar consumption increases.

5. $H_0: \rho = 0; H_a: \rho \neq 0$

$\alpha = 0.01$, d.f. $= n - 2 = 24$

$t_0 = 2.797$

$$t = \frac{r}{\sqrt{\dfrac{1 - r^2}{n - 2}}} = \frac{0.24}{\sqrt{\dfrac{1 - (0.24)^2}{26 - 2}}} = \frac{0.24}{\sqrt{0.03927}} = 1.211$$

Fail to reject H_0. There is not enough evidence to conclude that a significant linear correlation exists.

6. $H_0: \rho = 0;\ H_a: \rho \neq 0$

$\alpha = 0.05,\ d.f. = n - 2 = 20$

$t_0 = \pm 2.086$

$$t = \frac{r}{\sqrt{\dfrac{1 - r^2}{n - 2}}} = \frac{-0.55}{\sqrt{\dfrac{1 - (-0.55)^2}{22 - 2}}} = \frac{-0.55}{\sqrt{0.03488}} \approx -2.945$$

Reject H_0. There is enough evidence to conclude that a significant linear correlation exists.

7. $H_0: \rho = 0;\ H_a: \rho \neq 0$

$\alpha = 0.05,\ \text{d.f.} = n - 2 = 6$

$t_0 = \pm 2.447$

$$t = \frac{r}{\sqrt{\dfrac{1 - r^2}{n - 2}}} = \frac{-0.939}{\sqrt{\dfrac{1 - (-0.939)^2}{8 - 2}}} = \frac{-0.939}{\sqrt{0.01971}} = -6.688$$

Reject H_0. There is enough evidence to conclude that there is a significant linear correlation between the age of a cow and its milk production.

8. $H_0: \rho = 0;\ H_a: \rho \neq 0$

$\alpha = 0.05,\ \text{d.f.} = n - 2 = 4$

$t_0 = \pm 2.776$

$$t = \frac{r}{\sqrt{\dfrac{1 - r^2}{n - 2}}} = \frac{0.944}{\sqrt{\dfrac{1 - (0.944)^2}{4}}} = 5.722$$

Reject H_0. There is enough evidence to conclude that a significant linear correlation exists between average monthly temperatures and convertible sales.

9. $H_0: \rho = 0;\ H_a: \rho \neq 0$

$\alpha = 0.01,\ \text{d.f.} = n - 2 = 7$

$t_0 = \pm 3.499$

$$t = \frac{r}{\sqrt{\dfrac{1 - r^2}{n - 2}}} = \frac{0.361}{\sqrt{\dfrac{1 - (0.361)^2}{7}}} = 1.024$$

Fail to reject H_0. There is not enough evidence to conclude a significant linear correlation exists between brain size and IQ.

10. $H_0: \rho = 0;\ H_a: \rho \neq 0$

$\alpha = 0.01,\ \text{d.f.} = n - 2 = 5$

$t_0 = \pm 4.032$

$$t = \frac{r}{\sqrt{\dfrac{1 - r^2}{n - 2}}} = \frac{0.953}{\sqrt{\dfrac{1 - (0.953)^2}{7 - 2}}} = \frac{0.953}{\sqrt{0.01836}} \approx 7.034$$

Reject H_0. There is enough evidence to conclude that there is a linear correlation between sugar consumption and tooth decay.

11. $\hat{y} = 0.757x + 21.525$

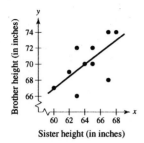

$r \approx 0.688$ (Moderate positive linear correlation)

12. $\hat{y} = 3.431x + 193.845$

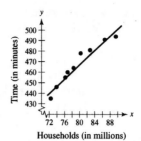

$r \approx 0.970$ (Strong positive linear correlation)

13. $\hat{y} = -0.086x + 10.450$

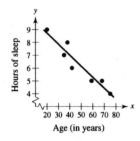

$r \approx -0.949$ (Strong negative linear correlation)

14. $\hat{y} = -0.090x + 44.675$

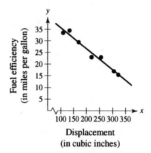

$r \approx -0.984$ (Strong negative linear correlation)

15. (a) $\hat{y} = 0.757(61) + 21.525 = 67.702 \approx 68$ inches

(b) $\hat{y} = 0.757(66) + 21.525 = 71.487 \approx 71$ inches

(c) Not meaningful because $x = 75$ inches is outside range of data.

(d) Not meaningful because $x = 50$ inches is outside range of data.

16. (a) Not meaningful because $x = 60$ is outside the range of the original data.

(b) $\hat{y} = 3.431(85) + 193.845 = 485.48$ minutes

(c) $\hat{y} = 3.431(75) + 193.845 = 451.17$ minutes

(d) Not meaningful because $x = 97$ is outside the range of the original data.

17. (a) 8.902 hours

(b) $\hat{y} = 10.450 = 0.086x = 10.450 - 0.086(25) = 8.3$ hours

(c) Not meaningful because $x = 85$ years is outside range of data.

(d) $\hat{y} = 10.450 - 0.086x = 10.450 - 0.086(50) = 6.15$ hours

18. (a) Not meaningful because $x = 86$ is outside the range of the original data.

(b) $\hat{y} = -0.090(198) + 44.675 = 26.86$ miles per gallon

(c) $\hat{y} = -0.090(289) + 44.675 = 18.67$ miles per gallon

(d) Not meaningful because $x = 407$ is outside the range of the original data.

19. $r^2 = (-0.553)^2 = 0.306$

30.6% of the variation in y is explained.
69.4% of the variation in y is unexplained.

20. $r^2 = (-0.962)^2 = 0.925$

92.5% of the variation in y is explained.
7.5% of the variation in y is unexplained.

21. $r^2 = (0.181)^2 = 0.033$

3.3% of the variation in y is explained.
96.7% of the variation in y is unexplained.

22. $r^2 = (0.740)^2 = 0.548$

54.8% of the variation in y is explained.
45.2% of the variation in y is unexplained.

23. (a) $r^2 = 0.897$

89.7% of the variation in y is explained.
10.3% of the variation in y is unexplained.

(b) $s_e = 568.011$

The standard error of estimate of the cooling capacity for a specific living area is 568.0 Btu per hour.

24. (a) $r^2 \approx 0.446$

44.6% of the variation in y is explained.
55.4% of the variation in y is unexplained.

(b) $s_e = 235.079$

The standard error of the price for a specific area is $235.08.

25. $\hat{y} = 0.757(64) + 21.525 = 69.973$

$$E = t_c s_e \sqrt{1 + \frac{1}{n} + \frac{n(x - \bar{x})^2}{n\Sigma x^2 - (\Sigma x)^2}} \approx (1.833)(2.206)\sqrt{1 + \frac{1}{11} + \frac{11(64 - 64.727)^2}{11(46,154) - 712^2}}$$

$$= (1.833)(2.202)\sqrt{1.09866} \approx 4.231$$

$$\hat{y} - E < y < \hat{y} + E$$
$$69.973 - 4.231 < y < 69.973 + 4.231$$
$$65.742 < y < 74.204$$

You can be 90% confident that the height of a male will be between 65.742 inches and 74.204 inches when his sister is 64 inches tall.

26. $\hat{y} = 193.845 + 3.431(74) = 447.739$

$$E = t_c s_e \sqrt{1 + \frac{1}{n} + \frac{n(x - \bar{x})^2}{n\Sigma x^2 - (\Sigma x)^2}}$$

$$= (1.895)(5.278)\sqrt{1 + \frac{1}{9} + \frac{9[74 - (716.9/9)]^2}{9(57,367.95) - (716.9)^2}} = (1.895)(5.278)\sqrt{1.233} = 11.105$$

$$\hat{y} - E < y < \hat{y} + E$$

$$447.739 - 11.105 < y < 447.739 + 11.105$$

$$436.634 < y < 458.844$$

You can be 90% confident that the average time spent watching television per household will be between 436.634 minutes and 458.844 minutes when 74 million households have multiple television sets.

27. $\hat{y} = -0.086(45) + 10.450 = 6.580$

$$E = t_c s_e \sqrt{1 + \frac{1}{n} + \frac{n(x - \bar{x})^2}{n\Sigma x^2 - (\Sigma x)^2}} \approx (2.571)(0.623)\sqrt{1 + \frac{1}{7} + \frac{7[45 - (337/7)]^2}{7(18,563) - (337)^2}}$$

$$= (2.571)(0.623)\sqrt{1.14708} = 1.715$$

$$\hat{y} - E < y < \hat{y} + E$$
$$6.580 - 1.715 < y < 6.58 + 1.715$$
$$4.865 < y < 8.295$$

You can be 95% confident that the hours slept will be between 4.865 and 8.295 hours for a person who is 45 years old.

28. $\hat{y} = -0.090(265) + 44.675 = 20.825$

$$E = t_c s_e \sqrt{1 + \frac{1}{n} + \frac{n(x - \bar{x})^2}{n\Sigma x^2 - (\Sigma x)^2}}$$

$$= (2.571)(1.476)\sqrt{1 + \frac{1}{7} + \frac{7(265 - 216.6)^2}{7(369,382) - 1516^2}}$$

$$= (2.571)(1.476)\sqrt{1.20133} \approx 4.159$$

$$\hat{y} - E < y < \hat{y} + E$$
$$20.825 - 4.159 < y < 20.825 + 4.159$$
$$16.668 < y < 24.982$$

You can be 95% confident that the fuel efficiency will be between 16.668 miles per gallon and 24.982 miles per gallon when the engine displacement is 265 cubic inches.

29. $\hat{y} = 3002.991 + 9.468(720) = 9819.95$

$$E = t_c s_e \sqrt{1 + \frac{1}{n} + \frac{n(x - \bar{x})^2}{n\Sigma x^2 - (\Sigma x)^2}} = (3.707)(568)\sqrt{1 + \frac{1}{8} + \frac{8(720 - 499.4)^2}{8(2,182,275) - 3995^2}}$$

$$= (3.707)(568)\sqrt{1.38486} \approx 2477.94$$

$$\hat{y} - E < y < \hat{y} + E$$
$$9819.95 - 2477.94 < y < 9819.95 + 2477.94$$
$$7342.01 < y < 12,297.89$$

You can be 99% confident that the cooling capacity will be between 7342.01 Btu per hour and 12,297.89 Btu per hour when the living area is 720 square feet.

30. $\hat{y} = -532.053 + 1.454x = -532.053 + 1.454(900) = 776.547$

$$E = t_c s_e \sqrt{1 + \frac{1}{n} + \frac{n(x - \bar{x})^2}{n\Sigma x^2 - (\Sigma x)^2}}$$

$$= 2.921(235.079)\sqrt{1 + \frac{1}{18} + \frac{18(900 - 733.5)^2}{18(2,314,270) - 6012^2}} = (2.921)(235.079)\sqrt{1.13789} \approx 732.480$$

$$\hat{y} - E < y < \hat{y} + E$$
$$776.547 - 732.480 < y < 776.547 + 732.480$$
$$44.067 < y < 1509.027$$

You can be 99% confident that the price will be between $44.07 and $1509.03 when the cooking area is 900 square inches.

31. $\hat{y} = 3.674 + 1.287x_1 + -7.531x_2$

32. $s_e = 0.710; r^2 = 0.943$

94.3% of the variation in y can be explained by the model.

33. (a) 21.705 (b) 25.21 (c) 30.1 (d) 25.86

34. (a) 11.272 (b) 14.695 (c) 14.380 (d) 9.232

CHAPTER 9 QUIZ SOLUTIONS

1.

The data appear to have a positive linear correlation. The outlays increase as the incomes increase.

2. $r \approx 0.996 \rightarrow$ Strong positive linear correlation

3. $H_0: \rho = 0; H_a: \rho \neq 0$

$\alpha = 0.05$, d.f. $= n - 2 = 9$

$t_0 = \pm 2.262$

$$t = \frac{r}{\sqrt{\dfrac{1 - r^2}{n - 2}}} = \frac{0.996}{\sqrt{\dfrac{1 - (0.996)^2}{11 - 2}}} = \frac{0.996}{\sqrt{0.00088711}} \approx 33.44$$

Reject H_0. There is enough evidence to conclude that a significant correlation exists.

4. $\Sigma y^2 = 616.090$

$\Sigma y = 81.100$

$\Sigma xy = 715.940$

$b = -1.018$

$m = 0.976$

$\hat{y} = 0.976x - 1.018$

5. $\hat{y} = 0.976(6.2) - 1.018 = \$5.033$ trillion

6. $r^2 = 0.993$

99.3% of the variation in y is explained.

0.7% of the variation in y is unexplained.

7. $s_e = 0.121$

The standard deviation of personal outlays for a specified personal income is $0.121 trillion.

8. $\hat{y} = 0.976(7.6) - 1.018 = 6.400$

$$E = t_c s_e \sqrt{1 + \frac{1}{n} + \frac{n(x - \bar{x})^2}{n\Sigma x^2 - (\Sigma x)^2}}$$

$$= 2.262(0.120) \sqrt{1 + \frac{1}{11} + \frac{11[7.6 - (94.6/11)]^2}{11(832.5) - 94.6^2}}$$

$$\approx (2.262)(0.120) \sqrt{1.144} \approx 0.290$$

$$\hat{y} - E < y < \hat{y} + E$$

$$6.400 - 0.290 < y < 6.400 + 0.290$$

$$6.110 < y < 6.690 \rightarrow (\$6.110 \text{ trillion}, \$6.690 \text{ trillion})$$

You can be 95% confident that the personal outlays will be between $6.110 trillion and $6.690 trillion when personal income is $7.6 trillion.

9. (a) 1374.762 pounds

   (b) 1556.755 pounds

   (c) 1183.262 pounds

   (d) 1294.385

   x_2 has the greater influence on y.

Chi-Square Tests and the *F*-Distribution

10.1 GOODNESS OF FIT

10.1 Try It Yourself Solutions

1a.

Music	% of Listeners	Expected Frequency
Classical	4%	12
Country	36%	108
Gospel	11%	33
Oldies	2%	6
Pop	18%	54
Rock	29%	87

2a. The expected frequencies are 64, 80, 32, 56, 60, 48, 40, and 20, all of which are at least 5.

b. Claimed Distribution:

Ages	Distribution
0–9	16%
10–19	20%
20–29	8%
30–39	14%
40–49	15%
50–59	12%
60–69	10%
70+	5%

H_0: Distribution of ages is as shown in table above.

H_a: Distribution of ages differs from the claimed distribution.

c. $\alpha = 0.05$ **d.** d.f. $= n - 1 = 7$

e. $\chi_0^2 = 14.067$; Reject H_0 if $\chi^2 > 14.067$.

f.

Ages	Distribution	Observed	Expected	$\frac{(O-E)^2}{E}$
0–9	16%	76	64	2.250
10–19	20%	84	80	0.200
20–29	8%	30	32	0.125
30–39	14%	60	56	0.286
40–49	15%	54	60	0.600
50–59	12%	40	48	1.333
60–69	10%	42	40	0.100
70+	5%	14	20	1.800
				6.694

$\chi^2 \approx 6.694$

g. Fail to reject H_0.

h. There is not enough evidence to suppport the claim that the distribution of ages differs from the claimed distribution.

3a. The expected frequencies are 123, 138, and 39, all of which are at least 5.

b. Claimed Distribution:

Response	Distribution
Retirement	41%
Children's college education	46%
Not sure	13%

H_0: Distribution of responses is as shown in table above. (claim)

H_a: Distribution of responses differs from the claimed distribution.

c. $\alpha = 0.01$ **d.** d.f. $= n - 1 = 2$ **e.** $\chi_0^2 = 9.210$; Reject H_0 if $\chi^2 > 9.210$.

f.

Response	Distribution	Observed	Expected	$\frac{(O - E)^2}{E}$
Retirement	41%	129	123	0.293
Children's college education	46%	149	138	0.877
Not sure	13%	22	39	7.410
				8.580

$\chi^2 \approx 8.580$

g. Fail to reject H_0.

h. There is not enough evidence to dispute the claimed distribution of responses.

4a. The expected frequencies are 30 for each color.

b. H_0: The distribution of different-colored candies in bags of peanut M&M's is uniform. (claim)

H_1: The distribution of different-colored candies in bags of peanut M&M's is not uniform.

c. $\alpha = 0.05$ **d.** d.f. $= k - 1 = 6 - 1 = 5$ **e.** $\chi_0^2 = 11.071$; Reject H_0 if $\chi^2 > 11.071$.

f.

Color	Observed	Expected	$\frac{(O - E)^2}{E}$
Brown	22	30	2.133
Yellow	27	30	0.300
Red	22	30	2.133
Blue	41	30	4.033
Orange	41	30	4.033
Green	27	30	0.300
			$\chi^2 = \sum \frac{(O - E)^2}{E} = 12.933$

g. Reject H_0.

h. There is enough evidence to dispute the claimed distribution.

10.1 EXERCISE SOLUTIONS

1. A multinomial experiment is a probability experiment consisting of a fixed number of trials in which there are more than two possible outcomes for each independent trial.

2. The observed frequencies must be obtained using a random sample, and each expected frequency must be greater than or equal to 5.

3. $E_i = np_i = (150)(0.3) = 45$ **4.** $E_i = np_i = (500)(0.9) = 450$

5. $E_i = np_i = (230)(0.25) = 57.5$ **6.** $E_i = np_i = (415)(0.08) = 33.2$

7. (a) Claimed distribution:

Response	Distribution
Reward Program	28%
Low Interest Rate	24%
Cash Back	22%
Special Store Discounts	8%
Other	16%

H_0: Distribution of responses is as shown in the table above.

H_a: Distribution of responses differs from the claimed or expected distribution. (claim)

(b) $\chi_0^2 = 9.488$; Reject H_0 if $\chi^2 > 9.488$.

(c)

Response	Distribution	Observed	Expected	$\frac{(O-E)^2}{E}$
Reward Program	28%	112	119	0.412
Low Interest Rate	24%	98	102	0.157
Cash Back	22%	107	93.5	1.949
Special Store Discounts	8%	46	34	4.235
Other	18%	62	76.5	2.748
				9.501

$\chi^2 = 9.501$

(d) Reject H_0. There is enough evidence at the 5% significance level to conclude that the distribution of responses differs from the claimed or expected distribution.

8. (a) Claimed distribution:

Response	Distribution
Home	70%
Work	17%
Commuting	8%
Other	5%

H_0: Distribution of responses is as shown in table above.

H_a: Distribution of responses differs from the claimed or expected distribution. (claim)

(b) $\chi_0^2 = 7.815$; Reject H_0 if $\chi^2 > 7.815$.

(c)

Response	Distribution	Observed	Expected	$\frac{(O-E)^2}{E}$
Home	70%	389	406.70	0.770
Work	17%	110	98.77	1.277
Commuting	8%	55	46.48	1.562
Other	5%	27	29.05	0.145
				3.754

$\chi^2 \approx 3.754$

(d) Fail to reject H_0. There is not enough evidence at the 5% significance level to conclude that the distribution of the responses differs from the claimed or expected distribution.

9. (a) Claimed distribution:

Response	Distribution
Keeping Medicare, Social Security and Medicaid	77%
Tax cuts	15%
Not sure	8%

H_0: Distribution of responses is as shown in the table above.

H_a: Distribution of responses differs from the claimed or expected distribution. (claim)

(b) $\chi_0^2 = 9.210$; Reject H_0 if $\chi^2 > 9.210$.

(c)

Response	Distribution	Observed	Expected	$\frac{(O - E)^2}{E}$
Keeping Medicare, Social Security, and Medicaid	77%	431	462	2.080
Tax cuts	15%	107	90	3.211
Not sure	8%	62	48	4.083
				9.374

$\chi^2 = 9.374$

(d) Reject H_0. There is enough evidence at the 1% significance level to conclude that the distribution of responses differs from the claimed or expected distribution.

10. (a) Claimed distribution:

Response	Distribution
Limited advancement	41%
Lack of recognition	25%
Low salary	15%
Unhappy with mgmt.	10%
Bored/don't know	9%

H_0: Distribution of responses is as shown in table above.

H_a: Distribution of responses differs from the claimed or expected distribution. (claim)

(b) $\chi_0^2 = 13.277$; Reject H_0 if $\chi^2 > 13.277$.

(c)

Response	Distribution	Observed	Expected	$\frac{(O - E)^2}{E}$
Limited advancement	41%	78	82.00	0.195
Lack of recognition	25%	52	50.00	0.080
Low salary	15%	30	30.00	0.000
Unhappy with mgmt.	10%	25	20.00	1.250
Bored/don't know	9%	15	18.00	0.500
				2.025

$\chi^2 \approx 2.025$

(d) Fail to reject H_0. There is not enough evidence at the 1% significance level to conclude that the distribution of the responses differs from the claimed or expected distribution.

11. (a) Claimed distribution:

Day	Distribution
Sunday	14.286%
Monday	14.286%
Tuesday	14.286%
Wednesday	14.286%
Thursday	14.286%
Friday	14.286%
Saturday	14.286%

H_0: The distribution of fatal bicycle accidents throughout the week is uniform as shown in the table above. (claim)

H_a: The distribution of fatal bicycle accidents throughout the week is not uniform.

(b) $\chi_0^2 = 10.645$; Reject H_0 if $\chi^2 > 10.645$.

(c)

Day	Distribution	Observed	Expected	$\dfrac{(O - E)^2}{E}$
Sunday	14.286%	108	111.717	0.124
Monday	14.286%	112	111.717	0.0007
Tuesday	14.286%	105	111.717	0.404
Wednesday	14.286%	111	111.717	0.005
Thursday	14.286%	123	111.717	1.140
Friday	14.286%	105	111.717	0.404
Saturday	14.286%	118	111.717	0.353
				2.430

$\chi^2 \approx 2.430$

(d) Fail to reject H_0. There is not enough evidence at the 10% significance level to reject the claim that the distribution of fatal bicycle accidents throughout the week is uniform.

12. (a) Claimed distribution:

Month	Distribution
January	8.333%
February	8.333%
March	8.333%
April	8.333%
May	8.333%
June	8.333%
July	8.333%
August	8.333%
September	8.333%
October	8.333%
November	8.333%
December	8.333%

H_0: The distribution of fatal bicycle accidents is uniform by month as shown in table above.

H_a: The distribution of fatal bicycle accidents is not uniform by month.

(b) $\chi_0^2 = 17.275$; Reject H_0 if $\chi^2 > 17.275$.

(c)

Month	Distribution	Observed	Expected	$\frac{(O-E)^2}{E}$
January	8.333%	37	65.247	12.229
February	8.333%	42	65.247	8.283
March	8.333%	43	65.247	7.585
April	8.333%	62	65.247	0.162
May	8.333%	79	65.247	2.899
June	8.333%	93	65.247	11.805
July	8.333%	66	65.247	0.009
August	8.333%	86	65.247	6.6001
September	8.333%	81	65.247	3.803
October	8.333%	94	65.247	12.671
November	8.333%	57	65.247	1.042
December	8.333%	43	65.247	7.585
				74.673

$\chi^2 = 74.673$

(d) Reject H_0. There is enough evidence at the 10% significance level to conclude that the distribution of fatal bicycle accidents is not uniform by month.

13. (a) Claimed distribution:

Object struck	Distribution
Tree	50%
Utility pole	14%
Embankment	6%
Guardrail	5%
Ditch	3%
Culvert	2%
Other	20%

H_0: Distribution of objects struck is as shown in table above.

H_a: Distribution of objects struck differs from the claimed or expected distribution. (claim)

(b) $\chi_0^2 = 16.812$; Reject H_0 if $\chi^2 > 16.812$.

(c)

Object Struck	Distribution	Observed	Expected	$\frac{(O-E)^2}{E}$
Tree	50%	375	345.50	2.519
Utility pole	14%	86	96.74	1.192
Embankment	6%	53	41.46	3.212
Guardrail	5%	42	34.55	1.606
Ditch	3%	27	20.73	1.896
Culvert	2%	18	13.82	1.264
Other	20%	90	138.2	16.811
			691	28.501

$\chi^2 \approx 28.501$

(d) Reject H_0. There is enough evidence at the 1% significance level to conclude that the distribution of objects struck changed from the claimed or expected distribution.

14. (a) Claimed distribution:

Time of day	Distribution
Midnight–6 A.M.	32%
6 A.M.–Noon	15%
Noon–6 P.M.	24%
6 P.M.–Midnight	29%

H_0: Distribution of the time of day of roadside hazard crash deaths is as shown in table above.

H_a: Distribution of the time of day of roadside hazard crash deaths differs from the claimed or expected distribution. (claim)

(b) $\chi_0^2 = 11.345$; Reject H_0 if $\chi^2 > 11.345$.

(c)

Time of day	Distribution	Observed	Expected	$\frac{(O-E)^2}{E}$
Midnight–6 A.M.	32%	224	200.64	2.720
6 A.M.–Noon	15%	128	94.05	12.255
Noon–6 P.M.	24%	115	150.48	8.365
6 P.M.–Midnight	29%	160	181.83	2.621
				25.961

$\chi^2 = 25.961$

(d) Reject H_0. There is enough evidence at the 1% significance level to conclude that the distribution of the time of day of roadside hazard crash deaths has changed from the claimed or expected distribution.

15. (a) Claimed distribution:

Response	Distribution
Not a HS grad	33.333%
HS graduate	33.333%
College (1yr+)	33.333%

H_0: Distribution of the responses is uniform as shown in table above. (claim)

H_a: Distribution of the responses is not uniform.

(b) $\chi_0^2 = 7.378$; Reject H_0 if $\chi^2 > 7.378$.

(c)

Response	Distribution	Observed	Expected	$\frac{(O-E)^2}{E}$
Not a HS grad	33.333%	37	33	0.485
HS graduate	33.333%	40	33	1.485
College (1yr+)	33.333%	22	33	3.667
			99	5.637

$\chi^2 \approx 5.637$

(d) Fail to reject H_0. There is not enough evidence at the 2.5% significance level to reject the claim that the distribution of the responses is uniform.

16. (a) Claimed distribution:

Season	Distribution
Spring	25%
Summer	25%
Fall	25%
Winter	25%

H_0: Distribution of births is uniform as shown in table above. (claim)

H_a: Distribution of births is not uniform.

(b) $\chi_0^2 = 11.345$; Reject H_0 if $\chi^2 > 11.345$.

(c)

Season	Distribution	Observed	Expected	$\dfrac{(O-E)^2}{E}$
Spring	25%	564	561.75	0.009
Summer	25%	603	561.75	3.029
Fall	25%	555	561.75	0.081
Winter	25%	525	561.75	2.404
		2247		5.523

$\chi^2 = 5.523$

(d) Fail to reject H_0. There is not enough evidence at the 1% significance level to reject the doctor's claim.

17. (a) Claimed distribution:

Cause	Distribution
Trans. Accidents	43%
Objects/equipment	18%
Assaults	14%
Falls	13%
Exposure	9%
Fire/explosions	3%

H_0: Distribution of the causes is as shown in table above.

H_a: Distribution of the causes differs from the claimed or expected distribution. (claim)

(b) $\chi_0^2 = 11.071$; Reject H_0 if $\chi^2 > 11.071$.

(c)

Cause	Distribution	Observed	Expected	$\dfrac{(O-E)^2}{E}$
Trans. Accidents	43%	2891	2679.33	16.722
Objects/equipment	18%	1159	1121.58	1.248
Assaults	14%	804	872.34	5.354
Falls	13%	754	810.03	3.876
Exposure	9%	531	560.79	1.582
Fires/explosions	3%	92	186.93	48.209
		6231		76.992

$\chi^2 \approx 76.992$

(d) Reject H_0. There is enough evidence at the 5% significance level to conclude that the distributions of the causes in the Western U.S. differs from the national distribution.

18. (a) Claimed distribution:

Response	Distribution
Did not work	22%
Work 1–20 hours	26%
Work 21–34 hours	18%
Work 35+ hours	34%

H_0: Distribution of the responses is as shown in the table above. (claim)

H_a: Distribution of the responses differs from the claimed or expected distribution.

(b) $\chi_0^2 = 11.345$; Reject H_0 if $\chi^2 > 11.345$.

(c)

Response	Distribution	Observed	Expected	$\frac{(O-E)^2}{E}$
Did not work	22%	29	26.4	0.256
Work 1–20 hours	26%	26	31.2	0.867
Work 25–34 hours	18%	25	21.6	0.535
Work 35+ hours	34%	40	40.8	0.016
		120		1.674

$\chi^2 = 1.674$

(d) Fail to reject H_0. There is not enough evidence at the 1% significance level to reject the council's claim.

19. (a) Frequency distribution: $\mu = 69.435$; $\sigma \approx 8.337$

Lower Boundary	Upper Boundary	Lower z-score	Upper z-score	Area
49.5	58.5	−2.39	−1.31	0.0867
58.5	67.5	−1.31	−0.23	0.3139
67.5	76.5	−0.23	0.85	0.3933
76.5	85.5	0.85	1.93	0.1709
85.5	94.5	1.93	3.01	0.0255

Class Boundaries	Distribution	Frequency	Expected	$\frac{(O-E)^2}{E}$
49.5–58.5	8.67%	19	17	0.235
58.5–67.5	31.39%	61	63	0.063
67.5–76.5	39.33%	82	79	0.114
76.5–85.5	17.09%	34	34	0
85.5–94.5	2.55%	4	5	0.2
		200		0.612

H_0: Test scores have a normal distribution. (claim)

H_a: Test scores do not have a normal distribution.

(b) $\chi_0^2 = 13.277$; Reject H_0 if $\chi^2 > 13.277$.

(c) $\chi^2 = 0.612$

(d) Fail to reject H_0. There is not enough evidence at the 1% significance level to reject the claim that the distribution of test scores is normal.

20. (a) Frequency distribution: $\mu = 74.775$, $\sigma = 9.822$

Lower Boundary	Upper Boundary	Lower z-score	Upper z-score	Area
50.5	60.5	−2.47	−1.45	0.0668
60.5	70.5	−1.45	−0.44	0.2564
70.5	80.5	−0.44	0.58	0.3891
80.5	90.5	0.58	1.60	0.2262
90.5	100.5	1.60	2.62	0.0504

Class Boundaries	Distribution	Frequency	Expected	$\dfrac{(O - E)^2}{E}$
50.5–60.5	6.68%	28	27	0.037
60.5–70.5	25.64%	106	103	0.087
70.5–80.5	38.91%	151	156	0.160
80.5–90.5	22.62%	97	90	0.544
90.5–100.5	5.04%	18	20	0.2
		400		1.028

H_0: Test scores have a normal distribution.

H_a: Test scores do not have a normal distribution.

(b) $\chi_0^2 = 9.488$; Reject H_0 if $\chi^2 > 9.488$.

(c) $\chi^2 \approx 1.028$

(d) Fail to reject H_0. There is not enough evidence at the 5% significance level to reject the claim that the distribution of test scores is normal.

10.2 INDEPENDENCE

10.2 Try It Yourself Solutions

1ab.

	Hotel	Leg Room	Rental Size	Other	Total
Business	36	108	14	22	180
Leisure	38	54	14	14	120
Total	74	162	28	36	300

c. $n = 300$

d.

	Hotel	Leg Room	Rental Size	Other
Business	44.4	97.2	16.8	21.6
Leisure	29.6	64.8	11.2	14.4

2a. H_0: Travel concern is independent of travel purpose.

   H_a: Travel concern is dependent on travel purpose. (claim)

b. $\alpha = 0.01$ **c.** $(r - 1)(c - 1) = 3$ **d.** $\chi_0^2 = 11.345$; Reject H_0 if $\chi^2 > 11.345$.

e. $\chi^2 = \sum \dfrac{(O - E)^2}{E} = \dfrac{(36 - 44.4)^2}{44.4} + \dfrac{(108 - 97.2)^2}{97.2} + \cdots + \dfrac{(14 - 14.4)^2}{14.4} \approx 8.158$

f. Fail to reject H_0.

g. There is not enough evidence to conclude that travel concern is dependent on travel purpose.

3a. H_0: The number of minutes adults spend online per day is independent on gender.

H_a: The number of minutes adults spend online per day is dependent on gender. (claim)

b. Enter the data.   **c.** $\chi_0^2 = 9.488$; Reject H_0 if $\chi^2 > 9.488$.

d. $\chi^2 \approx 65.619$ **e.** Reject H_0.

f. There is enough evidence to conclude that minutes spent online per day is dependent on gender.

10.2 EXERCISE SOLUTIONS

1. $E_{r,c} = \dfrac{(\text{Sum of row } r)(\text{Sum of column } c)}{\text{Sample size}}$

2. A large chi-square test statistic is evidence for rejecting the null hypothesis. So, the chi-square independence test is always a right-tailed test.

3. False. In order to use the χ^2 independence test, each expected frequency must be greater than or equal to 5.

4. False. A contingency table with two rows and five columns will have $(2 - 1)(5 - 1) = 4$ degrees of freedom.

5. False. If the two variables of a chi-square test for independence are dependent, then you can expect to find large differences between the observed frequencies and the expected frequencies.

6. True

7. (a)

Result	Athlete has		
	Stretched	Not stretched	
Injury	18	22	40
No injury	211	189	400
	229	211	440

(b)

Result	Athlete has	
	Stretched	Not stretched
Injury	20.818	19.182
No injury	208.182	191.818

8. (a)

Result	Treatment		
	Flu shot	No flu shot	
Got the flu	54	157	211
Did not get the flu	196	103	299
	250	260	510

(b)

Result	Treatment	
	Flu shot	No flu shot
Got the flu	103.431	107.569
Did not get the flu	146.569	152.431

9. (a)

Result	Treatment			
	Brand-name	**Generic**	**Placebo**	
Improvement	24	21	10	55
No change	12	13	45	70
	36	34	55	125

(b)

Result	Treatment		
	Brand-name	**Generic**	**Placebo**
Improvement	15.84	14.96	24.2
No change	20.16	19.04	30.8

10. (a)

Size	Rating			
	Excellent	**Fair**	**Poor**	
Seats 100 or fewer	182	203	165	550
Seats over 100	180	311	159	650
	362	514	324	1200

(b)

Size	Rating		
	Excellent	**Fair**	**Poor**
Seats 100 or fewer	165.917	235.583	148.500
Seats over 100	196.083	278.417	175.500

11. (a)

Gender	Type of Car				
	Compact	**Full-size**	**SUV**	**Truck/Van**	
Male	28	39	21	22	110
Female	24	32	20	14	90
	52	71	41	36	200

(b)

Gender	Type of Car			
	Compact	**Full-size**	**SUV**	**Truck/Van**
Male	28.60	39.05	22.55	19.80
Female	23.40	31.95	18.45	16.20

12. (a)

Result	Age				
	18–30	**31–42**	**43–61**	**62+**	
Willing to buy hybrid	72	66	73	69	280
Not willing to buy hybrid	14	21	19	21	75
	86	87	92	90	355

(b)

Result	Age			
	18–30	**31–42**	**43–61**	**62+**
Willing to buy hybrid	67.831	68.620	72.563	70.986
Not willing to buy hybrid	18.169	18.380	19.437	19.014

13. (a) H_0: Skill level in a subject is independent of location. (claim)

    H_a: Skill level in a subject is dependent on location.

    (b) d.f. $= (r - 1)(c - 1) = 2$

    $\chi_0^2 = 9.210$; Reject H_0 if $\chi^2 > 9.210$.

    (c) $\chi^2 \approx 0.297$

    (d) Fail to reject H_0. There is not enough evidence at the 1% significance level to reject the claim that skill level in a subject is independent of location.

14. (a) H_0: Attitudes about safety are independent of the type of school.

    H_a: Attitudes about safety are dependent on the type of school. (claim)

    (b) d.f. $= (r - 1)(c - 1) = 1$

    $\chi_0^2 = 6.635$; Reject H_0 if $\chi^2 > 6.635$.

    (c) $\chi^2 \approx 8.691$

    (d) Reject H_0. There is enough evidence at the 1% significance level to conclude that attitudes about safety are dependent on the type of school.

15. (a) H_0: Grades are independent of the institution.

    H_a: Grades are dependent on the institution. (claim)

    (b) d.f. $= (r - 1)(c - 1) = 8$

    $\chi_0^2 = 15.507$; Reject H_0 if $\chi^2 > 15.507$.

    (c) $\chi^2 \approx 48.488$

    (d) Reject H_0. There is enough evidence at the 5% significance level to conclude that grades are dependent on the institution.

16. (a) H_0: Adults' ratings are independent of the type of school.

    H_a: Adults' ratings are dependent on the type of school. (claim)

    (b) d.f. $= (r - 1)(c - 1) = 4$

    $\chi_0^2 = 7.815$; Reject H_0 if $\chi^2 > 7.815$.

    (c) $\chi^2 \approx 148.389$

    (d) Reject H_0. There is enough evidence at the 5% significance level to conclude that adults' ratings are dependent on the type of school.

17. (a) H_0: Results are independent of the type of treatment.

    H_a: Results are dependent on the type of treatment. (claim)

    (b) d.f. $= (r - 1)(c - 1) = 1$

    $\chi_0^2 = 2.706$; Reject H_0 if $\chi^2 > 2.706$.

    (c) $\chi^2 \approx 5.106$

    (d) Reject H_0. There is enough evidence at the 10% significance level to conclude that results are dependent on the type of treatment. I would recommend using the drug.

18. (a) H_0: Results are independent of the type of treatment.

    H_a: Results are dependent on the type of treatment. (claim)

    (b) d.f. $= (r - 1)(c - 1) = 1$

    $\chi_0^2 = 2.706$; Reject H_0 if $\chi^2 > 2.706$.

    (c) $\chi^2 \approx 1.032$

    (d) Fail to reject H_0. There is not enough evidence at the 10% significance level to conclude that results are dependent on the type of treatment. I would not recommend using the drug.

19. (a) H_0: Reasons are independent of the type of worker.

    H_a: Reasons are dependent on the type of worker. (claim)

    (b) d.f. $= (r - 1)(c - 1) = 2$

    $\chi_0^2 = 9.210$; Reject H_0 if $\chi^2 > 9.210$.

    (c) $\chi^2 \approx 7.326$

    (d) Fail to reject H_0. There is not enough evidence at the 1% significance level to conclude that the reason(s) for continuing education are dependent on the type of worker. Based on these results, marketing strategies should not differ between technical and non-technical audiences in regard to reason(s) for continuing education.

20. (a) H_0: Most important aspect of career development is independent of age.

    H_a: Most important aspect of career development is dependent on age. (claim)

    (b) d.f. $= (r - 1)(c - 1) = 4$

    $\chi_0^2 = 13.277$; Reject H_0 if $\chi^2 > 13.277$.

    (c) $\chi^2 \approx 5.757$

    (d) Fail to reject H_0. There is not enough evidence at the 1% significance level to conclude that the most important aspect of career development is dependent on age.

21. (a) H_0: Type of crash is independent of the type of vehicle.

    H_a: Type of crash is dependent on the type of vehicle. (claim)

    (b) d.f. $= (r - 1)(c - 1) = 2$

    $\chi_0^2 = 5.991$; Reject H_0 if $\chi^2 > 5.991$.

    (c) $\chi^2 \approx 108.913$

    (d) Reject H_0. There is enough evidence at the 5% significance level to conclude that type of crash is dependent on the type of vehicle.

22. (a) H_0: Age is independent of gender.

    H_a: Age is dependent on gender. (claim)

    (b) d.f. $= (r - 1)(c - 1) = 5$

    $\chi_0^2 = 11.071$; Reject H_0 if $\chi^2 > 11.071$.

    (c) $\chi^2 \approx 1.997$

    (d) Fail to reject H_0. There is not enough evidence at the 5% significance level to conclude that age is dependent on gender in such alcohol-related accidents.

23. (a) H_0: Subject is independent of coauthorship.

   H_a: Subject and coauthorship are dependent. (claim)

   (b) d.f. $= (r - 1)(c - 1) = 4$

   $\chi_0^2 = 7.779$; Reject H_0 if $\chi^2 > 7.779$.

   (c) $\chi^2 \approx 5.610$

   (d) Fail to reject H_0. There is not enough evidence at the 10% significance level to conclude that the subject matter and coauthorship are related.

24. (a) H_0: Access speed is independent of metropolitan status.

   H_1: Access speed is dependent on metropolitan status. (claim)

   (b) d.f. $= (r - 1)(c - 1) = 4$

   $\chi_0^2 = 13.277$; Reject H_0 if $\chi^2 > 13.277$.

   (c) $\chi^2 = 18.145$

   (d) Reject H_0. There is not enough evidence at the 1% significance level to conclude that access speed is dependent on metropolitan status.

25. H_0: The proportions are equal. (claim)

   H_a: At least one of the proportions is different from the others.

   d.f. $= (r - 1)(c - 1) = 7$

   $\chi_0^2 = 14.067$; Reject H_0 if $\chi^2 > 14.067$.

   $\chi^2 \approx 7.462$

   Fail to reject H_0. There is not enough evidence at the 5% significance level to reject the claim that the proportions are equal.

26. (a) H_0: The proportions are equal. (claim)

   H_1: At least one proportion is different from the others.

   (b) d.f. $= (r - 1)(c - 1) = 1$

   $\chi_0^2 = 2.706$; Reject H_0 if $\chi^2 > 2.706$.

   (c) $\chi^2 = 5.106$

   (d) Reject H_0. There is not enough evidence at the 10% significance level to reject the claim that the proportions are equal.

27.

Status	Educational Attainment			
	Not a high school graduate	High school graduate	Some college, no degree	Associate's, Bachelor's, or advanced degree
Employed	0.0610	0.1903	0.1129	0.2725
Unemployed	0.0058	0.0111	0.0053	0.0074
Not in labor force	0.0817	0.1204	0.0498	0.0817

28. 11.3%   **29.** 8.2%   **30.** 6.1%

31. 63.7%   **32.** 14.9%

33. Several of the expected frequencies are less than 5.

34.

Educational Attainment				
Status	Not a high school graduate	High school graduate	Some college, no degree	Associate's, Bachelor's, or advanced degree
Employed	0.0958	0.2989	0.1774	0.4280
Unemployed	0.1964	0.3750	0.1786	0.2500
Not in labor force	0.2448	0.3609	0.1494	0.2448

35. 29.9%   **36.** 14.9%

37.

Educational Attainment				
Status	Not a high school graduate	High school graduate	Some college, no degree	Associate's, Bachelor's, or advanced degree
Employed	0.4107	0.5914	0.6719	0.7537
Unemployed	0.0393	0.0346	0.0315	0.0205
Not in labor force	0.5500	0.3740	0.2965	0.2258

38. 22.6%   **39.** 3.9%

40.

41. As educational attainment increases, employment increases.

10.3 COMPARING TWO VARIANCES

10.3 Try It Yourself Solutions

1a. $\alpha = 0.01$ **b.** $F = 5.42$

2a. $\alpha = 0.01$ **b.** $F = 18.31$

3a. $H_0: \sigma_1^2 \leq \sigma_2^2$; $H_a: \sigma_1^2 > \sigma_2^2$ (claim)

b. $\alpha = 0.01$

c. $\text{d.f.}_N = n_1 - 1 = 24$
   $\text{d.f.}_D = n_2 - 1 = 19$

d. $F_0 = 2.92$; Reject H_0 if $F > 2.92$.

e. $F = \dfrac{s_1^2}{s_2^2} = \dfrac{180}{56} \approx 3.21$

f. Reject H_0.

g. There is enough evidence to support the claim.

4a. H_0: $\sigma_1 = \sigma_2$ (claim); H_a: $\sigma_1 \neq \sigma_2$

b. $\alpha = 0.01$

c. d.f.$_N$ = $n_1 - 1 = 15$
d.f.$_D$ = $n_2 - 1 = 21$

d. $F_0 = 3.43$; Reject H_0 if $F > 3.43$.

e. $F = \dfrac{s_1^2}{s_2^2} = \dfrac{(0.95)^2}{(0.78)^2} \approx 1.48$

f. Fail to reject H_0.

g. There is not enough evidence to reject the claim.

10.3 EXERCISE SOLUTIONS

1. Specify the level of significance α. Determine the degrees of freedom for the numerator and denominator. Use Table 7 in Appendix B to find the critical value F.

2. (1) The F-distribution is a family of curves determined by two types of degrees of freedom, d.f.$_N$ and d.f.$_D$.

   (2) F-distributions are positively skewed.

   (3) The area under the F-distribution curve is equal to 1.

   (4) F-values are always greater than or equal to zero.

   (5) For all F-distributions, the mean value of F is approximately equal to 1.

3. (1) The samples must be randomly selected, (2) The samples must be independent, and (3) Each population must have a normal distribution.

4. Determine the sample whose variance is greater. Use the size of this sample to find d.f.$_N$. Use the size of the other sample to find d.f.$_D$.

5. $F = 2.54$ 6. $F = 5.61$ 7. $F = 4.86$

8. $F = 2.63$ 9. $F = 2.06$ 10. $F = 14.62$

11. H_0: $\sigma_1^2 \leq \sigma_2^2$; H_a: $\sigma_1^2 > \sigma_2^2$ (claim)

    d.f.$_N$ = 4
    d.f.$_D$ = 5

    $F_0 = 3.52$; Reject H_0 if $F > 3.52$.

    $F = \dfrac{s_1^2}{s_2^2} = \dfrac{773}{765} \approx 1.010$

    Fail to reject H_0. There is not enough evidence to support the claim.

12. $H_0: \sigma_1^2 = \sigma_2^2$ (claim); $H_a: \sigma_1^2 \neq \sigma_2^2$

d.f.$_N$ = 6

d.f.$_D$ = 7

$F_0 = 5.12$; Reject H_0 if $F > 5.12$.

$F = \dfrac{s_1^2}{s_2^2} = \dfrac{310}{297} \approx 1.044$

Fail to reject H_0. There is not enough evidence to reject the claim.

13. $H_0: \sigma_1^2 \leq \sigma_2^2$ (claim); $H_a: \sigma_1^2 > \sigma_2^2$

d.f.$_N$ = 10

d.f.$_D$ = 9

$F_0 = 5.26$; Reject H_0 if $F > 5.26$.

$F = \dfrac{s_1^2}{s_2^2} = \dfrac{842}{836} \approx 1.007$

Fail to reject H_0. There is not enough evidence to reject the claim.

14. $H_0: \sigma_1^2 = \sigma_2^2$; $H_a: \sigma_1^2 \neq \sigma_2^2$ (claim)

d.f.$_N$ = 30

d.f.$_D$ = 27

$F_0 = 1.88$; Reject H_0 if $F > 1.88$.

$F = \dfrac{s_1^2}{s_2^2} = \dfrac{245}{112} \approx 2.188$

Reject H_0. There is enough evidence to support the claim.

15. $H_0: \sigma_1^2 = \sigma_2^2$ (claim); $H_a: \sigma_1^2 \neq \sigma_2^2$

d.f.$_N$ = 12

d.f.$_D$ = 19

$F_0 = 3.30$; Reject H_0 if $F > 3.30$.

$F = \dfrac{s_1^2}{s_2^2} = \dfrac{9.8}{2.5} \approx 3.920$

Reject H_0. There is enough evidence to reject the claim.

16. $H_0: \sigma_1^2 \leq \sigma_2^2$; $H_a: \sigma_1^2 > \sigma_2^2$ (claim)

d.f.$_N$ = 15

d.f.$_D$ = 11

$F_0 = 2.72$; Reject H_0 if $F > 2.72$.

$F = \dfrac{s_1^2}{s_2^2} = \dfrac{44.6}{39.3} \approx 1.135$

Fail to reject H_0. There is not enough evidence to support the claim.

17. Population 1: Company B

Population 2: Company A

(a) $H_0: \sigma_1^2 \leq \sigma_2^2; H_a: \sigma_1^2 > \sigma_2^2$ (claim)

(b) $\text{d.f.}_N = 24$

$\text{d.f.}_D = 19$

$F_0 = 2.11$; Reject H_0 if $F > 2.11$.

(c) $F = \dfrac{s_1^2}{s_2^2} = \dfrac{2.8}{2.6} \approx 1.08$

(d) Fail to reject H_0.

(e) There is not enough evidence at the 5% significance level to support Company A's claim that the variance of life of its appliances is less than the variance of life of Company B appliances.

18. (a) Population 1: Competitor

Population 2: Auto Manufacturer

$H_0: \sigma_1^2 \leq \sigma_2^2; H_a: \sigma_1^2 > \sigma_2^2$ (claim)

(b) $\text{d.f.}_N = 20$

$\text{d.f.}_D = 18$

$F_0 = 2.19$; Reject H_0 if $F > 2.19$.

(c) $F = \dfrac{s_1^2}{s_2^2} = \dfrac{22.5}{16.2} \approx 2.01$

(d) Fail to reject H_0.

(e) There is not enough evidence at the 5% significance level to conclude that the variance of the fuel consumption for the company's hybrid vehicles is less than that of the competitor's hybrid vehicles.

19. Population 1: District 1

Population 2: District 2

(a) $H_0: \sigma_1^2 = \sigma_2^2$ (claim); $H_a: \sigma_1^2 \neq \sigma_2^2$

(b) $\text{d.f.}_N = 11$

$\text{d.f.}_D = 13$

$F_0 = 2.635$; Reject H_0 if $F > 2.635$.

(c) $F = \dfrac{s_1^2}{s_2^2} = \dfrac{(36.8)^2}{(32.5)^2} \approx 1.282$

(d) Fail to reject H_0.

(e) There is not enough evidence at the 10% significance level to reject the claim that the standard deviation of science assessment test scores for eighth grade students is the same in Districts 1 and 2.

20. (a) Population 1: District 1

Population 2: District 2

$H_0: \sigma_1^2 = \sigma_2^2$ (claim); $H_a: \sigma_1^2 \neq \sigma_2^2$

(b) d.f.$_N$ = 9

d.f.$_D$ = 12

$F_0 = 5.20$; Reject H_0 if $F > 5.20$.

(c) $F = \dfrac{s_1^2}{s_2^2} = \dfrac{(33.9)^2}{(30.2)^2} \approx 1.26$

(d) Fail to reject H_0.

(e) There is not enough evidence at the 1% significance level to reject the claim that the standard deviation of U.S. history assessment test scores for eighth grade students is the same for Districts 1 and 2.

21. Population 1: Before new admissions procedure

Population 2: After new admissions procedure

(a) $H_0: \sigma_1^2 \leq \sigma_2^2$; $H_a: \sigma_1^2 > \sigma_2^2$ (claim)

(b) d.f.$_N$ = 24

d.f.$_D$ = 20

$F_0 = 1.77$; Reject H_0 if $F > 1.77$.

(c) $F = \dfrac{s_1^2}{s_2^2} = \dfrac{(0.7)^2}{(0.5)^2} \approx 1.96$

(d) Reject H_0.

(e) There is enough evidence at the 10% significance level to support the claim that the standard deviation of waiting times has decreased.

22. (a) Population 1: 2nd city

Population 2: 1st city

$H_0: \sigma_1^2 = \sigma_2^2$ (claim); $H_a: \sigma_1^2 \neq \sigma_2^2$

(b) d.f.$_N$ = 30

d.f.$_D$ = 27

$F_0 = 2.73$; Reject H_0 if $F > 2.73$.

(c) $F = \dfrac{s_1^2}{s_2^2} = \dfrac{(39.50)^2}{(23.75)^2} \approx 2.77$

(d) Reject H_0.

(e) There is enough evidence at the 1% significance level to reject the claim that the standard deviations of hotel room rates for the two cities are the same.

23. (a) Population 1: New York

Population 2: California

$H_0: \sigma_1^2 \le \sigma_2^2; H_a: \sigma_1^2 > \sigma_2^2$ (claim)

(b) d.f.$_N$ = 15

d.f.$_D$ = 16

$F_0 = 2.35$; Reject H_0 if $F > 2.35$.

(c) $F = \dfrac{s_1^2}{s_2^2} = \dfrac{(14,900)^2}{(9,600)^2} \approx 2.41$

(d) Reject H_0.

(e) There is enough evidence at the 5% significance level to conclude the standard deviation of annual salaries for actuaries is greater in New York than in California.

24. (a) Population 1: Florida

Population 2: Louisiana

$H_0: \sigma_1^2 \le \sigma_2^2; H_a: \sigma_1^2 > \sigma_2^2$ (claim)

(b) d.f.$_N$ = 27

d.f.$_D$ = 23

$F_0 = 1.985$; Reject H_0 if $F > 1.985$.

(c) $F = \dfrac{s_1^2}{s_2^2} = \dfrac{(10,100)^2}{(6,400)^2} \approx 2.490$

(d) Reject H_0.

(e) There is enough evidence at the 5% significance level to support the claim that the standard deviation of the annual salaries for public relations managers is greater in Florida than in Louisiana.

25. Right-tailed: $F_R = 14.73$

Left-tailed:

(1) d.f.$_N$ = 3 and d.f.$_D$ = 6

(2) $F = 6.60$

(3) Critical value is $\dfrac{1}{F} = \dfrac{1}{6.60} \approx 0.15$.

26. Right-tailed: $F = 2.33$

Left-tailed: (1) d.f.$_N$ = 15 and d.f.$_D$ = 20

(2) $F = 2.20$

(3) Critical value is $\dfrac{1}{F} = \dfrac{1}{2.20} \approx 0.45$.

27. $\dfrac{s_1^2}{s_2^2}F_L < \dfrac{\sigma_1^2}{\sigma_2^2} < \dfrac{s_1^2}{s_2^2}F_R \rightarrow \dfrac{10.89}{9.61}\,0.331 < \dfrac{\sigma_1^2}{\sigma_2^2} < \dfrac{10.89}{9.61}\,3.33 \rightarrow 0.375 < \dfrac{\sigma_1^2}{\sigma_2^2} < 3.774$

28. $\dfrac{s_1^2}{s_2^2}F_L < \dfrac{\sigma_1^2}{\sigma_2^2} < \dfrac{s_1^2}{s_2^2}F_R \rightarrow \dfrac{5.29}{3.61}\,0.331 < \dfrac{\sigma_1^2}{\sigma_2^2} < \dfrac{5.29}{3.61}\,3.33 \rightarrow 0.485 < \dfrac{\sigma_1^2}{\sigma_2^2} < 4.880$

10.4 ANALYSIS OF VARIANCE

10.4 Try It Yourself Solutions

1a. H_0: $\mu_1 = \mu_2 = \mu_3 = \mu_4$

   H_a: At least one mean is different from the others. (claim)

b. $\alpha = 0.05$

c. $\text{d.f.}_N = 3$

   $\text{d.f.}_D = 14$

d. $F_0 = 3.34$; Reject H_0 if $F > 3.34$.

e.

Variation	Sum of Squares	Degrees of Freedom	Mean Squares	F
Between	549.8	3	183.3	4.22
Within	608.0	14	43.4	

   $F \approx 4.22$

f. Reject H_0.

g. There is enough evidence to conclude that at least one mean is different from the others.

2a. Enter the data.

b. H_0: $\mu_1 = \mu_2 = \mu_3 = \mu_4$

   H_a: At least one mean is different from the others. (claim)

Variation	Sum of Squares	Degrees of Freedom	Mean Squares	F
Between	0.584	3	0.195	1.34
Within	4.360	30	0.145	

   $F = 1.34 \rightarrow P\text{-value} = 0.280$

c. $0.280 > 0.05$

d. Fail to reject H_0. There is not enough evidence to conclude that at least one mean is different from the others.

10.4 EXERCISE SOLUTIONS

1. H_0: $\mu_1 = \mu_2 = \ldots = \mu_k$

   H_a: At least one of the means is different from the others.

2. Each sample must be randomly selected from a normal, or approximately normal, population. The samples must be independent of each other. Each population must have the same variance.

3. MS_B measures the differences related to the treatment given to each sample.

   MS_W measures the differences related to entries within the same sample.

4. H_{0A}: There is no difference among the treatment means of Factor A.

H_{aA}: There is at least one difference among the treatment means of Factor A.

H_{0B}: There is no difference among the treatment means of Factor B.

H_{aB}: There is at least one difference among the treatment means of Factor B.

H_{0AB}: There is no interaction between Factor A and Factor B.

H_{aAB}: There is no interaction between Factor A and Factor B.

5. (a) $H_0: \mu_1 = \mu_2 = \mu_3$

H_a: At least one mean is different from the others. (claim)

(b) $\text{d.f.}_N = k - 1 = 2$

$\text{d.f.}_D = N - k = 26$

$F_0 = 3.37$; Reject H_0 if $F > 3.37$.

(c)

Variation	Sum of Squares	Degrees of Freedom	Mean Squares	F
Between	0.518	2	0.259	1.017
Within	6.629	26	0.255	

$F \approx 1.02$

(d) Fail to reject H_0. There is not enough evidence at the 5% significance level to conclude that the mean costs per ounce are different.

6. (a) $H_0: \mu_1 = \mu_2 = \mu_3$

H_a: At least one mean is different from the others. (claim)

(b) $\text{d.f.}_N = k - 1 = 2$

$\text{d.f.}_D = N - k = 14$

$F_0 = 3.74$; Reject H_0 if $F > 3.74$.

(c)

Variation	Sum of Squares	Degrees of Freedom	Mean Squares	F
Between	342.549	2	171.275	0.583
Within	4113.333	14	293.810	

$F \approx 0.58$

(d) Fail to reject H_0. There is not enough evidence at the 5% significance level to conclude that at least one of the mean battery prices is different from the others.

7. (a) $H_0: \mu_1 = \mu_2 = \mu_3$ (claim)

H_a: At least one mean is different from the others.

(b) $\text{d.f.}_N = k - 1 = 2$

$\text{d.f.}_D = N - k = 12$

$F_0 = 2.81$; Reject H_0 if $F > 2.81$.

(c)

Variation	Sum of Squares	Degrees of Freedom	Mean Squares	F
Between	302.6	2	151.3	1.77
Within	1024.2	12	85.4	

$F \approx 1.77$

(d) Fail to reject H_0. There is not enough evidence at the 10% significance level to reject the claim that the mean prices are all the same for the three types of treatment.

8. (a) $H_0: \mu_1 = \mu_2 = \mu_3 = \mu_4$ (claim)

H_a: At least one mean is different from the others.

(b) $\text{d.f.}_N = k - 1 = 3$

$\text{d.f.}_D = N - k = 23$

$F_0 = 2.34$; Reject H_0 if $F > 2.34$.

(c)

Variation	Sum of Squares	Degrees of Freedom	Mean Squares	F
Between	13,466	3	4488.667	1.337
Within	77,234	23	3358	

$F \approx 1.34$

(d) Fail to reject H_0. There is not enough evidence at the 10% significance level to reject the claim that the mean annual amounts are the same in all regions.

9. (a) $H_0: \mu_1 = \mu_2 = \mu_3 = \mu_4$ (claim)

H_a: At least one mean is different from the others.

(b) $\text{d.f.}_N = k - 1 = 3$

$\text{d.f.}_D = N - k = 29$

$F_0 = 4.54$; Reject H_0 if $F > 4.54$.

(c)

Variation	Sum of Squares	Degrees of Freedom	Mean Squares	F
Between	5.608	3	1.869	0.557
Within	97.302	29	3.355	

$F \approx 0.56$

(d) Fail to reject H_0. There is not enough evidence at the 1% significance level to reject the claim that the mean number of days patients spend in the hospital is the same for all four regions.

10. (a) $H_0: \mu_1 = \mu_2 = \mu_3 = \mu_4$

H_a: At least one mean is different from the others. (claim)

(b) $\text{d.f.}_N = k - 1 = 3$

$\text{d.f.}_D = N - k = 50$

$F_0 = 2.205$; Reject H_0 if $F > 2.205$.

(c)

Variation	Sum of Squares	Degrees of Freedom	Mean Squares	F
Between	216.277	3	72.092	3.686
Within	977.796	50	19.556	

$F \approx 3.686$

(d) Reject H_0. There is enough evidence at the 10% significance level to conclude that the mean square footage for at least one of the four regions is different from the others.

11. (a) $H_0: \mu_1 = \mu_2 = \mu_3 = \mu_4$ (claim)

H_a: At least one mean is different from the others.

(b) $\text{d.f.}_N = k - 1 = 3$

$\text{d.f.}_D = N - k = 40$

$F_0 = 2.23$; Reject H_0 if $F > 2.23$.

(c)

Variation	Sum of Squares	Degrees of Freedom	Mean Squares	F
Between	15,095.256	3	5031.752	2.757
Within	73,015.903	43	1825.398	

$F \approx 2.76$

(d) Reject H_0. There is enough evidence at the 10% significance level to reject the claim that the mean price is the same for all four cities.

12. (a) $H_0: \mu_1 = \mu_2 = \mu_3 = \mu_4$

H_a: At least one mean is different from the others. (claim)

(b) $\text{d.f.}_N = k - 1 = 3$

$\text{d.f.}_D = N - k = 16$

$F_0 = 3.24$; Reject H_0 if $F > 3.24$.

(c)

Variation	Sum of Squares	Degrees of Freedom	Mean Squares	F
Between	115,823,113.8	3	38,607,704.58	2.571
Within	240,248,350	16	15,015,521.88	

$F \approx 2.57$

(d) Fail to reject H_0. There is not enough evidence at the 5% significance level to conclude that the mean salary is different in at least one of the areas.

13. (a) $H_0: \mu_1 = \mu_2 = \mu_3 = \mu_4$

H_a: At least one mean is different from the others. (claim)

(b) $\text{d.f.}_N = k - 1 = 3$

$\text{d.f.}_D = N - k = 26$

$F_0 = 2.98$; Reject H_0 if $F > 2.98$

(c)

Variation	Sum of Squares	Degrees of Freedom	Mean Squares	F
Between	2,220,266,803	3	740,088,934	16.195
Within	1,188,146,984	16	45,697,960.9	

$F \approx 16.20$

(d) Reject H_0. There is enough evidence at the 10% significance level to conclude that at least one of the mean prices is different from the others.

14. (a) $H_0: \mu_1 = \mu_2 = \mu_3 = \mu_4$

H_a: At least one mean is different from the others. (claim)

(b) d.f.$_N$ = $k - 1 = 3$

d.f.$_D$ = $N - k = 35$

$F_0 = 2.255$; Reject H_0 if $F > 2.255$.

(c)

Variation	Sum of Squares	Degrees of Freedom	Mean Squares	F
Between	13,935.933	3	4645.311	2.823
Within	57,600.361	35	1645.725	

$F \approx 2.823$

(d) Reject H_0. There is enough evidence at the 1% significance level to conclude that the mean energy consumption for at least one region is different from the others.

15. (a) $H_0: \mu_1 = \mu_2 = \mu_3 = \mu_4$ (claim)

H_a: At least one mean is different from the others.

(b) d.f.$_N$ = $k - 1 = 3$

d.f.$_D$ = $N - k = 28$

$F_0 = 4.57$; Reject H_0 if $F > 4.57$.

(c)

Variation	Sum of Squares	Degrees of Freedom	Mean Squares	F
Between	771.25	3	257.083	0.459
Within	15,674.75	28	559.813	

$F \approx 0.46$

(d) Fail to reject H_0. There is not enough evidence at the 1% significance level to reject the claim that the mean numbers of female students are equal for all grades.

16. (a) $H_0: \mu_1 = \mu_2 = \mu_3 = \mu_4$ (claim)

H_a: At least one mean is different from the others.

(b) d.f.$_N$ = $k - 1 = 3$

d.f.$_D$ = $N - k = 30$

$F_0 = 4.51$; Reject H_0 if $F > 4.51$.

(c)

Variation	Sum of Squares	Degrees of Freedom	Mean Squares	F
Between	1,428,783.49	3	476,261.162	1.719
Within	8,309,413.01	30	276,980.434	

$F \approx 1.72$

(d) Fail to reject H_0. There is not enough evidence at the 1% level to reject the claim that the mean amounts spent are equal for all regions.

17. H_0: Advertising medium has no effect on mean ratings.
H_a: Advertising medium has an effect on mean ratings.

H_0: Length of ad has no effect on mean ratings.
H_a: Length of ad has an effect on mean ratings.

H_0: There is no interaction effect between advertising medium and length of ad on mean ratings.
H_a: There is an interaction effect between advertising medium and length of ad on mean ratings.

Source	d.f.	SS	MS	F	P
Ad medium	1	1.25	1.25	0.57	0.459
Length of ad	1	0.45	0.45	0.21	0.655
Interaction	1	0.45	0.45	0.21	0.655
Error	16	34.80	2.17		
Total	19	36.95			

None of the null hypotheses can be rejected at the 10% significance level.

18. H_0: Type of vehicle has no effect on the mean number of vehicles sold.
H_a: Type of vehicle has an effect on the mean number of vehicles sold.

H_0: Gender has no effect on the mean number of vehicles sold.
H_a: Gender has an effect on the mean number of vehicles sold.

H_0: There is no interaction effect between type of vehicle and gender on the mean number of vehicles sold.
H_a: There is an interaction effect between type of vehicle and gender on the mean number of vehicles sold.

Source	d.f.	SS	MS	F	P
Type of vehicle	2	84.08	42.04	34.01	0.000
Gender	1	0.38	0.38	0.30	0.589
Interaction	2	12.25	6.12	4.96	0.019
Error	18	22.25	1.24		
Total	23	118.96			

There appears to be an interaction effect between type of vehicle and gender on the mean number of vehicles sold. Also, there appears that type of vehicle has an effect on the mean number of vehicles sold.

19. H_0: Age has no effect on mean GPA.

H_a: Age has an effect on mean GPA.

H_0: Gender has no effect on mean GPA.

H_a: Gender has an effect on mean GPA.

H_0: There is no interaction effect between age and gender on mean GPA.

H_a: There is an interaction effect between age and gender on mean GPA.

Source	d.f.	SS	MS	F	P
Age	3	0.41	0.14	0.12	0.948
Gender	1	0.18	0.18	0.16	0.697
Interaction	3	0.29	0.10	0.08	0.968
Error	16	18.66	1.17		
Total	23	19.55			

None of the null hypotheses can be rejected at the 10% significance level.

20. H_0: Technicians have no effect on mean repair time.

H_a: Technicians have an effect on mean repair time.

H_0: Brand has no effect on mean repair time.

H_a: Brand has an effect on mean repair time.

H_0: There is no interaction effect between technicians and brand on mean repair time.

H_a: There is an interaction effect between technicians and brand on mean repair time.

Source	d.f.	SS	MS	F	P
Technicians	3	714	238	1.74	0.185
Brand	2	382	191	1.40	0.266
Interaction	6	2007	334	2.45	0.054
Error	24	3277	137		
Total	35	6381			

There appears to be an interaction effect between technicians and brand on the mean repair time at the 10% significance level.

21.

	Mean	Size
Pop 1	17.82	13
Pop 2	13.50	13
Pop 3	13.12	14
Pop 4	12.91	14

$SS_w = 977.796$

$\Sigma(n_i - 1) = N - k = 50$

$F_0 = 2.205 \rightarrow CV_{\text{Scheffé}} = 2.205(4 - 1) = 6.615$

$$\frac{(\bar{x}_1 - \bar{x}_2)^2}{\dfrac{SS_w}{\Sigma(n_i - 1)}\left[\dfrac{1}{n_1} + \dfrac{1}{n_2}\right]} \approx 6.212 \rightarrow \text{No difference}$$

$$\frac{(\bar{x}_1 - \bar{x}_3)^2}{\dfrac{SS_w}{\Sigma(n_i - 1)}\left[\dfrac{1}{n_1} + \dfrac{1}{n_3}\right]} \approx 7.620 \rightarrow \text{Significant difference}$$

$$\frac{(\bar{x}_1 - \bar{x}_4)^2}{\dfrac{SS_w}{\Sigma(n_i - 1)}\left[\dfrac{1}{n_1} + \dfrac{1}{n_4}\right]} \approx 8.330 \rightarrow \text{Significant difference}$$

$$\frac{(\bar{x}_2 - \bar{x}_3)^2}{\dfrac{SS_w}{\Sigma(n_i - 1)}\left[\dfrac{1}{n_2} + \dfrac{1}{n_3}\right]} \approx 0.049 \rightarrow \text{No difference}$$

$$\frac{(\bar{x}_2 - \bar{x}_4)^2}{\dfrac{SS_w}{\Sigma(n_i - 1)}\left[\dfrac{1}{n_2} + \dfrac{1}{n_4}\right]} \approx 0.121 \rightarrow \text{No difference}$$

$$\frac{(\bar{x}_3 - \bar{x}_4)^2}{\dfrac{SS_w}{\Sigma(n_i - 1)}\left[\dfrac{1}{n_3} + \dfrac{1}{n_4}\right]} \approx 0.016 \rightarrow \text{No difference}$$

22.

	Mean	Size
Pop 1	216.67	11
Pop 2	213.56	10
Pop 3	197.22	12
Pop 4	247.70	11

$SS_w = 73{,}015.903$

$\Sigma(n_i - 1) = N - k = 40$

$F_0 = 2.23 \rightarrow CV_{\text{Scheffé}} = 2.23(4 - 1) = 6.69$

$$\frac{(\bar{x}_1 - \bar{x}_2)^2}{\dfrac{SS_w}{\Sigma(n_i - 1)}\left[\dfrac{1}{n_1} + \dfrac{1}{n_2}\right]} \approx 0.03 \rightarrow \text{No difference}$$

$$\frac{(\bar{x}_1 - \bar{x}_3)^2}{\dfrac{SS_w}{\Sigma(n_i - 1)}\left[\dfrac{1}{n_1} + \dfrac{1}{n_3}\right]} \approx 1.19 \rightarrow \text{No difference}$$

$$\frac{(\bar{x}_1 - \bar{x}_4)^2}{\dfrac{SS_w}{\Sigma(n_i - 1)}\left[\dfrac{1}{n_1} + \dfrac{1}{n_4}\right]} \approx 2.90 \rightarrow \text{No difference}$$

$$\frac{(\bar{x}_2 - \bar{x}_3)^2}{\dfrac{SS_w}{\Sigma(n_i - 1)}\left[\dfrac{1}{n_2} + \dfrac{1}{n_3}\right]} \approx 0.80 \rightarrow \text{No difference}$$

$$\frac{(\bar{x}_2 - \bar{x}_4)^2}{\dfrac{SS_w}{\Sigma(n_i - 1)}\left[\dfrac{1}{n_2} + \dfrac{1}{n_4}\right]} \approx 3.34 \rightarrow \text{No difference}$$

$$\frac{(\bar{x}_3 - \bar{x}_4)^2}{\dfrac{SS_w}{\Sigma(n_i - 1)}\left[\dfrac{1}{n_3} + \dfrac{1}{n_4}\right]} \approx 8.01 \rightarrow \text{Significant difference}$$

23.

	Mean	Size
Pop 1	66,492	8
Pop 2	60,528	7
Pop 3	55,504	9
Pop 4	79,500	6

$SS_w \approx 1{,}188{,}146{,}984$

$\Sigma(n_i - 1) = N - k = 26$

$F_0 = 2.98 \rightarrow CV_{\text{Scheffé}} = 2.98(4 - 1) = 8.94$

$$\dfrac{(\overline{x}_1 - \overline{x}_2)^2}{\dfrac{SS_w}{\Sigma(n_i - 1)}\left[\dfrac{1}{n_1} + \dfrac{1}{n_2}\right]} \approx 2.91 \rightarrow \text{No difference}$$

$$\dfrac{(\overline{x}_1 - \overline{x}_3)^2}{\dfrac{SS_w}{\Sigma(n_i - 1)}\left[\dfrac{1}{n_1} + \dfrac{1}{n_3}\right]} \approx 11.19 \rightarrow \text{Significant difference}$$

$$\dfrac{(\overline{x}_1 - \overline{x}_4)^2}{\dfrac{SS_w}{\Sigma(n_i - 1)}\left[\dfrac{1}{n_1} + \dfrac{1}{n_4}\right]} \approx 12.70 \rightarrow \text{Significant difference}$$

$$\dfrac{(\overline{x}_2 - \overline{x}_3)^2}{\dfrac{SS_w}{\Sigma(n_i - 1)}\left[\dfrac{1}{n_2} + \dfrac{1}{n_3}\right]} \approx 2.17 \rightarrow \text{No difference}$$

$$\dfrac{(\overline{x}_2 - \overline{x}_4)^2}{\dfrac{SS_w}{\Sigma(n_i - 1)}\left[\dfrac{1}{n_2} + \dfrac{1}{n_4}\right]} \approx 25.45 \rightarrow \text{Significant difference}$$

$$\dfrac{(\overline{x}_3 - \overline{x}_4)^2}{\dfrac{SS_w}{\Sigma(n_i - 1)}\left[\dfrac{1}{n_3} + \dfrac{1}{n_4}\right]} \approx 45.36 \rightarrow \text{Significant difference}$$

24.

	Mean	Size
Pop 1	106.26	9
Pop 2	116.71	11
Pop 3	82.49	9
Pop 4	70.07	10

$\dfrac{SS_w}{\Sigma(n_i - 1)} \approx 57{,}600.361$

$F_0 = 2.255 \rightarrow CV_{\text{Scheffé}} = 2.255(4 - 1) = 6.765$

$$\dfrac{(\overline{x}_1 - \overline{x}_2)^2}{\dfrac{SS_w}{\Sigma(n_i - 1)}\left[\dfrac{1}{n_1} + \dfrac{1}{n_2}\right]} \approx 0.329 \rightarrow \text{No difference}$$

$$\dfrac{(\overline{x}_1 - \overline{x}_3)^2}{\dfrac{SS_w}{\Sigma(n_i - 1)}\left[\dfrac{1}{n_1} + \dfrac{1}{n_3}\right]} \approx 1.545 \rightarrow \text{No difference}$$

$$\frac{(\bar{x}_1 - \bar{x}_4)^2}{\dfrac{SS_w}{\Sigma(n_i - 1)}\left[\dfrac{1}{n_1} + \dfrac{1}{n_4}\right]} \approx 3.769 \rightarrow \text{No difference}$$

$$\frac{(\bar{x}_2 - \bar{x}_3)^2}{\dfrac{SS_w}{\Sigma(n_i - 1)}\left[\dfrac{1}{n_2} + \dfrac{1}{n_3}\right]} \approx 3.522 \rightarrow \text{No difference}$$

$$\frac{(\bar{x}_2 - \bar{x}_4)^2}{\dfrac{SS_w}{\Sigma(n_i - 1)}\left[\dfrac{1}{n_2} + \dfrac{1}{n_4}\right]} \approx 6.923 \rightarrow \text{Significant difference}$$

$$\frac{(\bar{x}_3 - \bar{x}_4)^2}{\dfrac{SS_w}{\Sigma(n_i - 1)}\left[\dfrac{1}{n_3} + \dfrac{1}{n_4}\right]} \approx 0.444 \rightarrow \text{No difference}$$

CHAPTER 10 REVIEW EXERCISE SOLUTIONS

1. Claimed distribution:

Category	Distribution
0	16%
1–3	46%
4–9	25%
10+	13%

H_0: Distribution of health care visits is as shown in table above.

H_a: Distribution of health care visits differs from the claimed distribution.

$\chi_0^2 = 7.815$

Category	Distribution	Observed	Expected	$\dfrac{(O - E)^2}{E}$
0	16%	99	117.44	2.895
1–3	46%	376	337.64	4.368
4–9	25%	167	183.50	1.484
10+	13%	92	95.42	0.123
		734		8.860

$\chi_0^2 = 8.860$

Reject H_0. There is enough evidence at the 5% significance level to reject the claimed or expected distribution.

2. Claimed distribution:

Advertisers	Distribution
Short-game	65%
Approach	22%
Driver	9%
Putting	4%

H_0: Distribution of advertisers is as shown in table above.

H_a: Distribution of advertisers differs from the claimed distribution.

$\chi_0^2 = 6.251$

Advertisers	Distribution	Observed	Expected	$\frac{(O-E)^2}{E}$
Short-game	65%	276	282.75	0.161
Approach	22%	99	95.70	0.114
Driver	9%	42	39.15	0.207
Putting	4%	18	17.40	0.021
		435		0.503

$\chi^2 = 0.503$

Fail to reject H_0. There is not enough evidence at the 10% significance level to reject the claimed or expected distribution.

3. Claimed distribution:

Days	Distribution
0	19%
1–5	30%
6–10	30%
11–15	12%
16–20	3%
21+	6%

H_0: Distribution of days is as shown in the table above.

H_a: Distribution of days differs from the claimed distribution.

$\chi_0^2 = 9.236$

Days	Distribution	Observed	Expected	$\frac{(O-E)^2}{E}$
0	19%	165	152	1.112
1–5	30%	237	240	0.038
6–10	30%	245	240	0.104
11–15	12%	88	96	0.667
16–20	3%	19	24	1.042
21+	6%	46	48	0.083
		800		3.045

$\chi^2 \approx 3.045$

Fail to reject H_0. There is not enough evidence at the 10% significance level to support the claim.

4. Claimed distribution:

Age	Distribution
21–29	20.5%
30–39	21.7%
40–49	18.1%
50–59	17.3%
60+	22.4%

H_0: Distribution of ages is as shown in table above.

H_a: Distribution of ages differs from the claimed distribution.

$\chi_0^2 = 13.277$

Age	Distribution	Observed	Expected	$\frac{(O-E)^2}{E}$
21–29	20.5%	45	205	124.878
30–39	21.7%	128	217	36.502
40–49	18.1%	244	181	21.928
50–59	17.3%	224	173	15.035
60+	22.4%	359	224	81.362
		1000		279.705

$\chi^2 \approx 279.705$

Reject H_0. There is enough evidence at the 1% significance level to support the claim that the distribution of ages of the jury differs from the age distribution of available jurors.

5. (a) Expected frequencies:

	HS–did not complete	HS complete	College 1–3 years	College 4+ years	Total
25–44	663.49	1440.47	1137.54	1237.5	4479
45+	856.51	1859.53	1468.46	1597.5	5782
Total	1520	3300	2606	2835	10,261

(b) H_0: Education is independent of age.

H_a: Education is dependent on age.

d.f. = 3

$\chi_0^2 = 6.251$

$\chi^2 = \sum \frac{(O-E)^2}{E} \approx 66.128$

Reject H_0.

(c) There is enough evidence at the 10% significance level to conclude that education level of people in the United States and their age are dependent.

6. (a) Expected frequencies:

	Types of vehicles owned				
Gender	Car	Truck	SUV	Van	Total
Male	94.25	82.65	50.75	4.35	232
Female	100.75	88.35	54.25	4.65	248
Total	195	171	105	9	480

(b) H_0: Type of vehicle is independent of gender.

H_a: Type of vehicle is dependent on gender.

$\chi_0^2 = 7.815$

$\chi^2 = \sum \frac{(O-E)^2}{E} = 8.403$

Reject H_0.

(c) There is enough evidence at the 5% significance level to conclude that type of vehicle owned is dependent on gender.

7. (a) Expected frequencies:

Gender	16–20	21–30	31–40	41–50	51–60	61+	Total
			Age Group				
Male	143.76	284.17	274.14	237.37	147.10	43.46	1130
Female	71.24	140.83	135.86	117.63	72.90	21.54	560
Total	215	425	410	355	220	65	1690

(b) H_0: Gender is independent of age.

H_a: Gender and age are dependent.

d.f. = 5

$\chi_0^2 = 11.071$

$\chi^2 = \sum \dfrac{(O - E)^2}{E} = 9.951$

Fail to reject H_0.

(c) There is not enough evidence at the 10% significance level to conclude that gender and age group are dependent.

8. (a) Expected frequencies:

Gender	12 A.M.–5:59 A.M.	6 A.M.–11:59 A.M.	12 P.M.–5:59 P.M.	6 P.M.–11:59 P.M.	Total
			Time of day		
Male	582.68	612.81	967.93	918.57	3082
Female	326.32	343.19	542.07	514.43	1726
Total	909	956	1510	1432	4808

(b) H_0: Type of vehicle is independent of gender.

H_a: Type of vehicle is dependent on gender.

$\chi_0^2 = 6.251$

$\chi^2 = 36.742$

Reject H_0.

(c) There is enough evidence at the 10% significance level to conclude that time and gender are dependent.

9. $F_0 \approx 2.295$ **10.** $F_0 = 4.71$ **11.** $F_0 = 2.39$ **12.** $F_0 = 2.01$

13. H_0: $\sigma_1^2 \le \sigma_2^2$ (claim); H_a: $\sigma_1^2 > \sigma_2^2$

d.f.$_N$ = 15

d.f.$_D$ = 20

$F_0 = 3.09$; Reject H_0 if $F > 3.09$.

$F = \dfrac{s_1^2}{s_2^2} = \dfrac{653}{270} \approx 2.419$

Fail to reject H_0. There is not enough evidence to reject the claim.

14. $H_0: \sigma_1^2 = \sigma_2^2$; $H_a: \sigma_1^2 \neq \sigma_2^2$ (claim)

  d.f.$_N$ = 5

  d.f.$_D$ = 10

  $F_0 = 3.33$; Reject H_0 if $F > 3.33$.

  $F = \dfrac{s_1^2}{s_2^2} = \dfrac{112{,}676}{49{,}572} \approx 2.273$

  Fail to reject H_0. There is not enough evidence at the 10% significance level to support the claim.

15. Population 1: Garfield County

  Population 2: Kay County

  $H_0: \sigma_1^2 \leq \sigma_2^2$; $H_a: \sigma_1^2 > \sigma_2^2$ (claim)

  d.f.$_N$ = 20

  d.f.$_D$ = 15

  $F_0 = 1.92$; Reject H_0 if $F > 1.92$.

  $F = \dfrac{s_1^2}{s_2^2} = \dfrac{(0.76)^2}{(0.58)^2} \approx 1.717$

  Fail to reject H_0. There is not enough evidence at the 10% significance level to support the claim that the variation in wheat production is greater in Garfield County than in Kay County.

16. Population 1: Nontempered

  Population 2: Tempered

  $H_0: \sigma_1^2 \leq \sigma_2^2$; $H_a: \sigma_1^2 > \sigma_2^2$ (claim)

  d.f.$_N$ = 8

  d.f.$_D$ = 8

  $F_0 = 3.44$; Reject H_0 if $F > 3.44$.

  $F = \dfrac{s_1^2}{s_2^2} = \dfrac{(25.4)^2}{(13.1)^2} \approx 3.759$

  Reject H_0. There is enough evidence at the 5% significance level to support the claim that the yield strength of the nontempered couplings is more variable than that of the tempered couplings.

17. Population 1: Male $\rightarrow s_1^2 = 18{,}486.26$

  Population 2: Female $\rightarrow s_2^2 = 12{,}102.78$

  $H_0: \sigma_1^2 = \sigma_2^2$; $H_a: \sigma_1^2 \neq \sigma_2^2$ (claim)

  d.f.$_N$ = 13

  d.f.$_D$ = 8

  $F_0 = 6.94$; Reject H_0 if $F > 6.94$.

  $F = \dfrac{s_1^2}{s_2^2} = \dfrac{18{,}486.26}{12{,}102.78} \approx 1.527$

  Fail to reject H_0. There is not enough evidence at the 1% significance level to support the claim that the test score variance for females is different than that for males.

18. Population 1: Current $\rightarrow s_1^2 = 0.00146$

Population 2: New $\rightarrow s_2^2 = 0.00050$

$H_0: \sigma_1^2 \leq \sigma_2^2; H_a: \sigma_1^2 > \sigma_2^2$ (claim)

d.f.$_N$ = 11

d.f.$_D$ = 11

$F_0 = 2.82$; Reject H_0 if $F > 2.82$.

$F = \dfrac{s_1^2}{s_2^2} = \dfrac{0.00146}{0.00050} \approx 2.92$

Reject H_0. There is enough evidence at the 5% significance level to support the claim that the new mold produces inserts that are less variable in diameter than the current mold.

19. $H_0: \mu_1 = \mu_2 = \mu_3 = \mu_4$

H_a: At least one mean is different from the others. (claim)

d.f.$_N$ = $k - 1 = 3$

d.f.$_D$ = $N - k = 28$

$F_0 = 2.29$; Reject H_0 if $F > 2.29$.

Variation	Sum of Squares	Degrees of Freedom	Mean Squares	F
Between	512.457	3	170.819	8.508
Within	562.162	28	20.077	

$F \approx 8.508$

Reject H_0. There is enough evidence at the 10% significance level to conclude that the mean residential energy expenditures are not the same for all four regions.

20. $H_0: \mu_1 = \mu_2 = \mu_3 = \mu_4$

H_a: At least one mean is different from the others. (claim)

d.f.$_N$ = $k - 1 = 3$

d.f.$_D$ = $N - k = 20$

$F_0 = 3.10$; Reject H_0 if $F > 3.10$.

Variation	Sum of Squares	Degrees of Freedom	Mean Squares	F
Between	461,446,125	3	153,815,375	1.087
Within	2,829,846,458	20	141,492,323	

$F \approx 1.087$

Fail to reject H_0. There is not enough evidence a the 5% significance level to conclude that the average annual incomes are not the same for the four regions.

CHAPTER 10 QUIZ SOLUTIONS

1. (a) Population 1: San Jose $\rightarrow s_1^2 \approx 430.084$

Population 2: Dallas $\rightarrow s_2^2 \approx 120.409$

$H_0: \sigma_1^2 = \sigma_2^2; H_a: \sigma_1^2 \neq \sigma_2^2$ (claim)

(b) $\alpha = 0.01$

(cd) d.f.$_N$ = 12

d.f.$_D$ = 15

$F_0 = 4.25$; Reject H_0 if $F > 4.25$.

(e) $F = \dfrac{s_1^2}{s_2^2} = \dfrac{430.084}{120.409} \approx 3.57$

(f) Fail to reject H_0.

(g) There is not enough evidence at the 1% significance level to conclude that the variances in annual wages for San Jose, CA and Dallas, TX are different.

2. (a) $H_0: \mu_1 = \mu_2 = \mu_3$ (claim)

H_a: At least one mean is different from the others.

(b) $\alpha = 0.10$

(cd) d.f.$_N$ = $k - 1 = 2$

d.f.$_D$ = $N - k = 40$

$F_0 = 2.44$; Reject H_0 if $F > 2.44$.

Variation	Sum of Squares	Degrees of Freedom	Mean Squares	F
Between	5021.896	2	2510.948	12.213
Within	8224.121	40	205.603	

(e) $F \approx 12.21$

(f) Reject H_0.

(g) There is enough evidence at the 10% significance level to reject the claim that the mean annual wages for the three cities are not all equal.

3. (a) Claimed distribution:

Education	25 & Over
Not a HS graduate	14.8%
HS graduate	32.2%
Some college, no degree	16.8%
Associate's degree	8.6%
Bachelor's degree	18.1%
Advanced degree	9.5%

H_0: Distribution of educational achievement for people in the United States ages 35–44 is as shown in table above.

H_a: Distribution of educational achievement for people in the United States ages 35–44 differs from the claimed distribution. (claim)

(b) $\alpha = 0.01$

(cd) $\chi_0^2 = 15.086$; Reject H_0 if $\chi^2 > 15.086$.

(e)

Education	25 & Over	Observed	Expected	$\frac{(O-E)^2}{E}$
Not a HS graduate	14.8%	35	44.548	2.046
HS graduate	32.2%	95	96.922	0.038
Some college, no degree	16.8%	51	50.568	0.004
Associate's degree	8.6%	30	25.886	0.654
Bachelor's degree	18.1%	61	54.481	0.780
Advanced degree	9.5%	29	28.595	0.006
		301		3.528

$\chi^2 = 3.528$

(f) Fail to reject H_0.

(g) There is not enough evidence at the 1% significance level to conclude that the distribution of educational achievement for people in the United States ages 35–44 differs from the claimed or expected distribution.

4. (a) Claimed distribution:

Education	25 & Over
Not a HS graduate	14.8%
HS graduate	32.2%
Some college, no degree	16.8%
Associate's degree	8.6%
Bachelor's degree	18.1%
Advanced degree	9.5%

H_0: Distribution of educational achievement for people in the United States ages 65–74 is as shown in table above.

H_a: Distribution of educational achievement for people in the United States ages 65–74 differs from the claimed distribution. (claim)

(b) $\alpha = 0.05$

(c) $\chi_0^2 = 11.071$

(d) Reject H_0 if $\chi^2 > 11.071$.

(e)

Education	25 & Over	Observed	Expected	$\frac{(O-E)^2}{E}$
Not a HS graduate	14.8%	91	60.088	15.903
HS graduate	32.2%	151	130.730	3.142
Some college, no degree	16.8%	58	68.208	1.528
Associate's degree	8.6%	23	34.916	4.067
Bachelor's degree	18.1%	50	73.486	7.506
Advanced degree	9.5%	33	38.570	0.804
		406		32.950

$\chi^2 = 32.950$

(f) Reject H_0.

(g) There is enough evidence at the 5% significance level to conclude that the distribution of educational achievement for people in the United States ages 65–74 differs from the claimed or expected distribution.

11.1 Try It Yourself Solutions

1a. H_0: median ≤ 2500; H_a: median > 2500 (claim)

b. $\alpha = 0.025$

c. $n = 22$

d. The critical value is 5.

e. $x = 10$

f. Fail to reject H_0.

g. There is not enough evidence to support the claim.

2a. H_0: median = \$134,500 (claim); H_a: median $\neq$ \$134,500

b. $\alpha = 0.10$

c. $n = 81$

d. The critical value is $z_0 = -1.645$.

e. $x = 30$

$$z = \frac{(x + 0.5) - 0.5(n)}{\frac{\sqrt{n}}{2}} = \frac{(30 + 0.5) - 0.5(81)}{\frac{\sqrt{81}}{2}} = \frac{-10}{4.5} = -2.22$$

f. Reject H_0.

g. There is enough evidence to reject the claim.

3a. H_0: The number of colds will not decrease.

   H_a: The number of colds will decrease. (claim)

b. $\alpha = 0.05$

c. $n = 11$

d. The critical value is 2.

e. $x = 2$

f. Reject H_0.

g. There is enough evidence to support the claim.

11.1 EXERCISE SOLUTIONS

1. A nonparametric test is a hypothesis test that does not require any specific conditions concerning the shape of populations or the value of any population parameters.

   A nonparametric test is usually easier to perform than its corresponding parametric test, but the nonparametric test is usually less efficient.

325

2. Identify the claim and state H_0 and H_a. Identify the level of significance and sample size. Find the critical value using Table 8 (if $n \le 25$) or Table 4 ($n > 25$). Calculate the test statistic. Make a decision and interpret in the context of the problem.

3. (a) H_0: median $\le$ \$300; H_a: median $>$ \$300 (claim)

   (b) Critical value is 1.

   (c) $x = 5$

   (d) Fail to reject H_0.

   (e) There is not enough evidence at the 1% significance level to support the claim that the median amount of new credit card charges for the previous month was more than \$300.

4. (a) H_0: median $= 83$ (claim); H_a: median $\ne 83$

   (b) The critical value is 1.

   (c) $x = 4$

   (d) Fail to reject H_0.

   (e) There is not enough evidence at the 1% significance level to reject the claim that the daily median temperature for the month of July in Pittsburgh is 83° Fahrenheit.

5. (a) H_0: median $\le$ \$210,000 (claim); H_a: median $>$ \$210,000

   (b) Critical value is 1.

   (c) $x = 3$

   (d) Fail to reject H_0.

   (e) There is not enough evidence at the 5% significance level to reject the claim that the median sales price of new privately owned one-family homes sold in the past year is \$210,000 or less.

6. (a) H_0: median $= 66$ (claim); H_a: median $\ne 66$

   (b) The critical value is 3.

   (c) $x = 0$

   (d) Reject H_0.

   (e) There is enough evidence at the 1% significance level to reject the claim that the daily median temperature for the month of January in San Diego is 66°F.

7. (a) H_0: median $\ge$ \$2200 (claim); H_a: median $<$ \$2200

   (b) Critical value is $z_0 = -2.05$.

   (c) $x = 44$
$$z = \frac{(x + 0.5) - 0.5(n)}{\frac{\sqrt{n}}{2}} = \frac{(44 + 0.5) - 0.5(104)}{\frac{\sqrt{104}}{2}} = \frac{-7.5}{5.099} \approx -1.47$$

   (d) Fail to reject H_0.

   (e) There is not enough evidence at the 2% significance level to reject the claim that the median amount of credit card debt for families holding such debts is at least \$2200.

8. (a) H_0: median $\geq$ \$50,000; H_a: median $<$ \$50,000 (claim)

   (b) The critical value is $z_0 = -1.96$.

   (c) $x = 24$

$$z = \frac{(x + 0.5) - 0.5(n)}{\frac{\sqrt{n}}{2}} = \frac{(24 + 0.5) - 0.5(70)}{\frac{\sqrt{70}}{2}} = \frac{-10.5}{4.183} \approx -2.51$$

   (d) Reject H_0.

   (e) There is enough evidence at the 2.5% significance level to support the claim that the median amount of financial debt for families holding such debts is less than \$50,000.

9. (a) H_0: median $\leq$ 30; H_a: median $>$ 30 (claim)

   (b) Critical value is 2.

   (c) $x = 4$

   (d) Fail to reject H_0.

   (e) There is not enough evidence at the 1% significance level to support the claim that the median age of recipients of engineering doctorates is greater than 30 years.

10. (a) H_0: median $\geq$ 32; H_a: median $<$ 32 (claim)

   (b) The critical value is 5.

   (c) $x = 5$

   (d) Reject H_0.

   (e) There is enough evidence at the 5% significance level to support the claim that the median age of recipients of biological science doctorates is less than 32 years.

11. (a) H_0: median $=$ 4 (claim); H_a: median $\neq$ 4

   (b) Critical value is $z_0 = -1.96$.

   (c) $x = 13$

$$z = \frac{(x + 0.5) - 0.5(n)}{\frac{\sqrt{n}}{2}} = \frac{(13 + 0.5) - 0.5(33)}{\frac{\sqrt{33}}{2}} = \frac{-3}{2.872} \approx -1.04$$

   (d) Fail to reject H_0.

   (e) There is not enough evidence at the 5% significance level to reject the claim that the median number of rooms in renter-occupied units is 4.

12. (a) H_0: median $=$ 1000 (claim); H_a: median $\neq$ 1000

   (b) The critical value is 5.

   (c) $x = 5$

   (d) Reject H_0.

   (e) There is enough evidence at the 10% significance level to reject the claim that the median square footage of renter-occupied units is 1000 square feet.

13. (a) H_0: median = \$12.16 (claim); H_a: median $\neq$ \$12.16

(b) Critical value is $z_0 = -2.575$.

(c) $x = 16$

$$z = \frac{(x + 0.5) - 0.5(n)}{\frac{\sqrt{n}}{2}} = \frac{(16 + 0.5) - 0.5(39)}{\frac{\sqrt{39}}{2}} = \frac{-3}{3.1225} \approx -0.961$$

(d) Fail to reject H_0.

(e) There is not enough evidence at the 1% significance level to reject the claim that the median hourly earnings of male workers paid hourly rates is \$12.16.

14. (a) H_0: median $\leq$ \$10.31 (claim); H_a: median > \$10.31

(b) The critical value is 5.

(c) $x = 9$

(d) Fail to reject H_0.

(e) There is not enough evidence at the 5% significance level to reject the claim that the median hourly earnings of female workers paid hourly rates is at most \$10.31.

15. (a) H_0: The lower back pain intensity scores have not decreased.

H_a: The lower back pain intensity scores have decreased. (claim)

(b) Critical value is 1.

(c) $x = 0$

(d) Reject H_0.

(e) There is enough evidence at the 5% significance level to conclude that the lower back pain intensity scores were reduced after accupunture.

16. (a) H_0: The lower back pain intensity scores have not decreased.

H_a: The lower back pain intensity scores have decreased. (claim)

(b) The critical value is 2.

(c) $x = 4$

(d) Fail to reject H_0.

(e) There is not enough evidence at the 5% significance level to support the claim that the lower back pain intensity scores have decreased.

17. (a) H_0: The SAT scores have not improved.

H_a: The SAT scores have improved. (claim)

(b) Critical value is 2.

(c) $x = 4$

(d) Fail to reject H_0.

(e) There is not enough evidence at the 5% significance level to support the claim that the verbal SAT scores improved.

18. (a) H_0: The SAT scores have not improved.

H_a: The SAT scores have improved. (claim)

(b) The critical value is 1.

(c) $x = 3$

(d) Fail to reject H_0.

(e) There is not enough evidence at the 1% significance level to support the claim that the verbal SAT scores have improved.

19. (a) H_0: The proportion of adults who prefer unplanned travel activities is equal to the proportion of adults who prefer planned travel activities. (claim)

H_a: The proportion of adults who prefer unplanned travel activities is not equal to the proportion of adults who prefer planned travel activities.

Critical value is 3.

$\alpha = 0.05$

$x = 5$

Fail to reject H_0.

(b) There is not enough evidence at the 5% significance level to reject the claim that the proportion of adults who prefer unplanned travel activities is equal to the proportions of adults who prefer planned travel activities.

20. (a) H_0: The proportion of adults who contact their parents by phone weekly is equal to the proportion of adults who contact their parents by phone daily. (claim)

H_a: The proportion of adults who contact their parents by phone weekly is not equal to the proportion of adults who contact their parents by phone daily.

The critical value is 6. ($\alpha = 0.05$).

$x = 9$

Fail to reject H_0.

(b) There is not enough evidence at the 5% significance level to reject the claim that the proportion of adults who contact their parents by phone weekly is equal to the proportion of adults who contact their parents by phone daily.

21. (a) H_0: median $\leq$ \$585 (claim); H_a: median $>$ \$585

(b) Critical value is $z_0 = 2.33$.

(c) $x = 29$

$$z = \frac{(x - 0.5) - 0.5(n)}{\frac{\sqrt{n}}{2}} = \frac{(29 - 0.5) - 0.5(47)}{\frac{\sqrt{47}}{2}} = \frac{5}{3.428} \approx 1.46$$

(d) Fail to reject H_0.

(e) There is not enough evidence at the 1% significance level to reject the claim that the median weekly earnings of female workers is less than or equal to \$585.

22. (a) H_0: median $\leq$ \$720; H_a: median $>$ \$720 (claim)

(b) The critical value is $z_0 = 2.33$.

(c) $x = 45$

$$z = \frac{(x - 0.5) - 0.5(n)}{\frac{\sqrt{n}}{2}} = \frac{(45 - 0.5) - 0.5(68)}{\frac{\sqrt{68}}{2}} = \frac{10.5}{4.123} \approx 2.55$$

(d) Reject H_0.

(e) There is enough evidence at the 1% significance level to support the claim that the median weekly earnings of male workers is greater than \$720.

23. (a) H_0: median $\leq$ 25.5; H_a: median $>$ 25.5 (claim)

(b) Critical value is $z_0 = 1.645$.

(c) $x = 38$

$$z = \frac{(x + 0.5) - 0.5(n)}{\frac{\sqrt{n}}{2}} = \frac{(38 + 0.5) - 0.5(60)}{\frac{\sqrt{60}}{2}} = \frac{8.5}{3.873} \approx 1.936$$

(d) Reject H_0.

(e) There is enough evidence at the 5% significance level to support the claim that the median age of first-time brides is greater than 25.5 years.

24. (a) H_0: median $\leq$ 27 (claim); H_a: median $>$ 27

(b) The critical value is $z_0 = 1.645$.

(c) $x = 33$

$$z = \frac{(x - 0.5) - 0.5(n)}{\frac{\sqrt{n}}{2}} = \frac{(33 - 0.5) - 0.5(56)}{\frac{\sqrt{56}}{2}} = \frac{4.5}{3.742} \approx 1.203$$

(d) Fail to reject H_0.

(e) There is not enough evidence at the 5% significance level to reject the claim that the median age of first-time grooms is less than or equal to 27.

11.2 THE WILCOXON TESTS

11.2 Try It Yourself Solutions

1a. H_0: The water repellent is not effective.

H_a: The water repellent is effective. (claim)

b. $\alpha = 0.01$

c. $n = 11$

d. Critical value is 5.

e.

No repellent	Repellent applied	Difference	Absolute value	Rank	Signed rank
8	15	−7	7	11	−11
7	12	−5	5	9	−9
7	11	−4	4	7.5	−7.5
4	6	−2	2	3.5	−3.5
6	6	0	0	−	−
10	8	2	2	3.5	3.5
9	8	1	1	1.5	1.5
5	6	−1	1	1.5	−1.5
9	12	−3	3	5.5	−5.5
11	8	3	3	5.5	5.5
8	14	−6	6	10	−10
4	8	−4	4	7.5	−7.5

Sum of negative ranks = −55.5
Sum of positive ranks = 10.5

$w_s = 10.5$

f. Fail to reject H_0.

g. There is not enough evidence at the 1% significance level to support the claim.

2a. H_0: There is no difference in the claims paid by the companies.
H_a: There is a difference in the claims paid by the companies. (claim)

b. $\alpha = 0.05$

c. The critical values are $z_0 = \pm 1.96$.

d. $n_1 = 12$ and $n_2 = 12$

e.

Ordered data	Sample	Rank	Ordered data	Sample	Rank
1.7	B	1	5.3	B	13
1.8	B	2	5.6	B	14
2.2	B	3	5.8	A	15
2.5	A	4	6.0	A	16
3.0	A	5.5	6.2	A	17
3.0	B	5.5	6.3	A	18
3.4	B	7	6.5	A	19
3.9	A	8	7.3	B	20
4.1	B	9	7.4	A	21
4.4	B	10	9.9	A	22
4.5	A	11	10.6	A	23
4.7	B	12	10.8	B	24

R = sum ranks of company B = 120.5

f. $\mu_R = \dfrac{n_1(n_1 + n_2 + 1)}{2} = \dfrac{12(12 + 12 + 1)}{2} = 150$

$\sigma_R = \sqrt{\dfrac{n_1 n_2(n_1 + n_2 + 1)}{12}} = \sqrt{\dfrac{(12)(12)(12 + 12 + 1)}{12}} \approx 17.321$

$z = \dfrac{R - \mu_R}{\sigma_R} = \dfrac{120.5 - 150}{17.321} \approx -1.703$

g. Fail to reject H_0.

h. There is not enough evidence to conclude that there is a difference in the claims paid by both companies.

11.2 EXERCISE SOLUTIONS

1. The Wilcoxon signed-rank test is used to determine whether two dependent samples were selected from populations having the same distribution. The Wilcoxon rank sum test is used to determine whether two independent samples were selected from populations having the same distribution.

2. The sample size of both samples must be at least 10.

3. (a) H_0: There is no reduction in diastolic blood pressure. (claim)

   H_a: There is a reduction in diastolic blood pressure.

   (b) Wilcoxon signed-rank test

   (c) Critical value is 10.

   (d) $w_s = 17$

   (e) Fail to reject H_0.

   (f) There is not enough evidence at the 1% significance level to reject the claim that there was no reduction in diastolic blood pressure.

4. (a) H_0: There is no difference in salaries. (claim)

   H_a: There is a difference in salaries.

   (b) Wilcoxon rank sum test

   (c) The critical value is $z_0 = \pm 1.645$.

   (d) $R = 127$

   $$\mu_R = \frac{n_1(n_1 + n_2 + 1)}{2} = \frac{10(10 + 10 + 1)}{2} = 105$$

   $$\sigma_R = \sqrt{\frac{n_1 n_2 (n_1 + n_2 + 1)}{12}} = \sqrt{\frac{(10)(10)(10 + 10 + 1)}{12}} = 13.229$$

   $$z = \frac{R - \mu_R}{\sigma_R} = \frac{127 - 105}{13.229} \approx 1.663$$

   (e) Reject H_0.

   (f) There is enough evidence at the 10% significance level to reject the claim that there is no difference in the salaries.

5. (a) H_0: There is no difference in the earnings.

   H_a: There is a difference in the earnings. (claim)

   (b) Wilcoxon rank sum test

   (c) The critical values are $z_0 = \pm 1.96$.

(d) $R = 58$

$$\mu_R = \frac{n_1(n_1 + n_2 + 1)}{2} = \frac{11(11 + 10 + 1)}{2} = 110$$

$$\sigma_R = \sqrt{\frac{n_1 n_2(n_1 + n_2 + 1)}{12}} = \sqrt{\frac{(11)(10)(11 + 10 + 1)}{12}} \approx 14.201$$

$$z = \frac{R - \mu_R}{\sigma_R} = \frac{58 - 110}{14.201} \approx -3.66$$

(e) Reject H_0.

(f) There is enough evidence at the 5% significance level to support the claim that there is a difference in the earnings.

6. (a) H_0: There is no difference in the number of months mothers breast-feed their babies. (claim)

H_a: There is a difference in the number of months mothers breast-feed their babies.

(b) Wilcoxon rank sum test

(c) The critical value is $z_0 = \pm 2.575$.

(d) $R = 104$

$$\mu_R = \frac{n_1(n_1 + n_2 + 1)}{2} = \frac{11(11 + 12 + 1)}{2} = 132$$

$$\sigma_R = \sqrt{\frac{n_1 n_2(n_1 + n_2 + 1)}{12}} = \sqrt{\frac{(11)(12)(11 + 12 + 1)}{12}} = 16.248$$

$$z = \frac{R - \mu_R}{\sigma_R} = \frac{104 - 132}{16.248} \approx -1.723$$

(e) Fail to reject H_0.

(f) There is not enough evidence at the 1% significance level to reject the claim that there is no difference in the number of months mothers breast-feed their babies.

7. (a) H_0: There is not a difference in salaries.

H_a: There is a difference in salaries. (claim)

(b) Wilcoxon rank sum test

(c) The critical values are $z_0 = \pm 1.96$.

(d) $R = 117$

$$\mu_R = \frac{n_1(n_1 + n_2 + 1)}{2} = \frac{12(12 + 12 + 1)}{2} = 150$$

$$\sigma_R = \sqrt{\frac{n_1 n_2(n_1 + n_2 + 1)}{12}} = \sqrt{\frac{(12)(12)(12 + 12 + 1)}{12}} \approx 17.321$$

$$z = \frac{R - \mu_R}{\sigma_R} = \frac{117 - 150}{17.321} \approx -1.91$$

(e) Fail to reject H_0.

(f) There is not enough evidence at the 5% significance level to support the claim that there is a difference in salaries.

8. (a) H_0: The new drug does not affect the number of headache hours.

   H_a: The new drug does affect the number of headache hours. (claim)

(b) Wilcoxon signed-rank test

(c) The critical value is 2.

(d)

Before	After	Difference	Absolute value	Rank	Signed rank
0.8	1.6	−0.8	0.8	3	−3
2.4	1.3	1.1	1.1	4	4
2.8	1.6	1.2	1.2	6	6
2.6	1.4	1.2	1.2	6	6
2.7	1.5	1.2	1.2	6	6
0.9	1.6	−0.7	0.7	2	−2
1.2	1.7	−0.5	0.5	1	−1

The sum of the negative ranks is $-3 + (-2) + (-1) = -6$.

The sum of the positive ranks is $4 + 6 + 6 + 6 = 22$.

Because $|-6| < |22|$, the test statistic is $w_s = 6$.

(e) Fail to reject H_0.

(f) There is not enough evidence at the 5% significance level to conclude that the new drug affects the number of headache hours.

9. H_0: The fuel additive does not improve gas mileage.

   H_a: The fuel additive does improve gas mileage. (claim)

   Critical value is $z_0 = 1.282$.

$$w_s = 43.5$$

$$z = \frac{w_s - \frac{n(n+1)}{4}}{\sqrt{\frac{n(n+1)(2n+1)}{24}}} = \frac{43.5 - \frac{32(32+1)}{4}}{\sqrt{\frac{32(32+1)[(2)32+1]}{24}}} = \frac{-220.5}{\sqrt{2860}} \approx -4.123$$

Note: $n = 32$ because one of the differences is zero and should be discarded.

Reject H_0. There is enough evidence at the 10% level to conclude that the gas mileage is improved.

10. H_0: The fuel additive does not improve gas mileage.

   H_a: The fuel additive does improve gas mileage. (claim)

   The critical value is $z_0 = -1.645$.

$$w_s = 0$$

$$z = \frac{w_s - \frac{n(n+1)}{4}}{\sqrt{\frac{n(n+1)(2n+1)}{24}}} = \frac{0 - 264}{\sqrt{2860}} = -4.937$$

Reject H_0. There is enough evidence at the 5% level to conclude that the fuel additive improves gas mileage.

11.3 THE KRUSKAL-WALLIS TEST

11.3 Try It Yourself Solutions

1a. H_0: There is no difference in the salaries in the three states.

H_a: There is a difference in the salaries in the three states. (claim)

b. $\alpha = 0.10$

c. d.f. $= k - 1 = 2$

d. Critical value is $\chi_0^2 = 4.605$; Reject H_0 if $\chi^2 > 4.605$.

e.

Ordered data	State	Rank	Ordered data	State	Rank
86.85	MI	1	94.72	NC	16
87.25	NC	2	94.75	NC	17
87.70	CO	3	95.10	MI	18
89.65	MI	4	95.36	NC	19
89.75	CO	5	96.02	NC	20
89.92	NC	6	96.24	CO	21
91.17	CO	7	96.31	MI	22
91.55	CO	8	97.35	CO	23
92.85	CO	9	98.21	MI	24
93.12	CO	10	98.34	NC	25
93.76	MI	11	98.99	CO	26
93.92	MI	12	100.27	NC	27
94.42	MI	13	105.77	NC	28
94.45	MI	14	106.78	MI	29
94.55	CO	15	110.99	NC	30

$R_1 = 127 \quad R_2 = 148 \quad R_3 = 190$

f. $H = \dfrac{12}{N(N+1)} \left(\dfrac{R_1^2}{n_1} + \dfrac{R_2^2}{n_2} + \dfrac{R_3^2}{n_3} \right) - 3(N+1)$

$= \dfrac{12}{30(30+1)} \left(\dfrac{(127)^2}{10} + \dfrac{(148)^2}{10} + \dfrac{(190)^2}{10} \right) - 3(30+1) = 2.655$

g. Fail to reject H_0.

h. There is not enough evidence to support the claim.

11.3 EXERCISE SOLUTIONS

1. Each sample must be randomly selected and the size of each sample must be at least 5.

2. The Kruskal-Wallis test is always a right-tailed test because the null hypothesis is only rejected when H is significantly large.

3. (a) H_0: There is no difference in the premiums.

H_a: There is a difference in the premiums. (claim)

(b) Critical value is 5.991.

(c) $H \approx 13.091$

(d) Reject H_0.

(e) There is enough evidence at the 5% significance level to support the claim that the distributions of the annual premiums in Arizona, Florida, and Louisiana are different.

4. (a) H_0: There is no difference in the premiums.

H_a: There is a difference in the premiums. (claim)

(b) The critical value is 5.991.

(c) $H \approx 9.841$

(d) Reject H_0.

(e) There is enough evidence at the 5% significance level to support the claim that the distributions of the annual premiums in the three states are different.

5. (a) H_0: There is no difference in the salaries.

H_a: There is a difference in the salaries. (claim)

(b) Critical value is 6.251.

(c) $H \approx 1.024$

(d) Fail to reject H_0.

(e) There is not enough evidence at the 10% significance level to support the claim that the distributions of the annual salaries in the four states are different.

6. (a) H_0: There is no difference in the salaries.

H_a: There is a difference in the salaries. (claim)

(b) The critical value is 6.251.

(c) $H \approx 15.713$

(d) Reject H_0.

(e) There is enough evidence at the 10% significance level to support the claim that the distributions of the annual salaries in the four states are different.

7. (a) H_0: There is no difference in the number of days spent in the hospital.

H_a: There is a difference in the number of days spent in the hospital. (claim)

The critical value is 11.345.

$H = 1.51$;

Fail to reject H_0.

(b)

Variation	Sum of squares	Degrees of freedom	Mean Squares	F
Between	9.17	3	3.06	0.52
Within	194.72	33	5.90	

For $\alpha = 0.01$, the critical value is about 4.45. Because $F = 0.52$ is less than the critical value, the decision is to fail to reject H_0. There is not enough evidence to support the claim.

(c) Both tests come to the same decision, which is that there is not enough evidence to support the claim that there is a difference in the number of days spent in the hospital.

8. (a) H_0: There is no difference in the mean energy consumptions.

H_a: There is a difference in the mean energy consumptions. (claim)

The critical value is 11.345.

$H \approx 16.26$

Reject H_0. There is enough evidence to support the claim.

(b)

Variation	Sum of Squares	Degrees of Freedom	Mean Squares	F
Between	32,116	3	10,705	8.18
Within	45,794	35	1308	

For $\alpha = 0.01$, the critical value is about 4.41. Because $F = 8.18$ is greater than the critical value, the decision is to reject H_0. There is enough evidence to support the claim.

(c) Both tests come to the same decision, which is that the mean energy consumptions are different.

11.4 RANK CORRELATION

11.4 Try It Yourself Solutions

1a. H_0: $\rho_s = 0$; H_a: $\rho_s \neq 0$

b. $\alpha = 0.05$

c. Critical value is 0.700.

d.

Oat	Rank	Wheat	Rank	d	d^2
1.10	1.5	2.65	3	−1.5	2.25
1.12	3	2.48	1	2	4
1.10	1.5	2.62	2	−0.5	0.25
1.59	6	2.78	4	2	4
1.81	8	3.56	8	0	0
1.48	4.5	3.40	5.5	−1	1
1.48	4.5	3.40	5.5	−1	1
1.63	7	4.25	7	0	0
1.85	9	4.25	9	0	0
					$\Sigma = 12.5$

$\Sigma d^2 = 12.5$

e. $r_s = 1 - \dfrac{6\Sigma d^2}{n(n^2 - 1)} = 0.896$

f. Reject H_0.

g. There is enough evidence to conclude that a significant correction exists.

11.4 EXERCISE SOLUTIONS

1. The Spearman rank correlation coefficient can (1) be used to describe the relationship between linear and nonlinear data, (2) be used for data at the ordinal level, and (3) is easier to calculate by hand than the Pearson coefficient.

2. The ranks of corresponding data pairs are identical.
The ranks of corresponding data pairs are in reverse order.
The ranks of corresponding data pairs have no relationship.

3. (a) $H_0: \rho_s = 0; H_a: \rho_s \neq 0$ (claim)

(b) Critical value is 0.929.

(c) $\Sigma d^2 = 8$
$$r_s = 1 - \frac{6\Sigma d^2}{n(n^2 - 1)} \approx 0.857$$

(d) Fail to reject H_0.

(e) There is not enough evidence at the 1% significance level to support the claim that there is a correlation between debt and income in the farming business.

4. (a) $H_0: \rho_s = 0; H_a: \rho_s \neq 0$ (claim)

(b) The critical value is 0.618.

(c) $\Sigma d^2 = 65.6$
$$r_s = 1 - \frac{6\Sigma d^2}{n(n^2 - 1)} \approx 0.702$$

(d) Reject H_0.

(e) There is enough evidence at the 5% significance level to support the claim that there is a correlation between the overall score and price.

5. (a) $H_0: \rho_s = 0; H_a: \rho_s \neq 0$ (claim)

(b) The critical value is 0.881.

(c) $\Sigma d^2 = 66.5$
$$r_s = 1 - \frac{6\Sigma d^2}{n(n^2 - 1)} \approx 0.208$$

(d) Fail to reject H_0.

(e) There is not enough evidence at the 1% significance level to support the claim that there is a correlation between the overall score and price.

6. (a) $H_0: \rho_s = 0; H_a: \rho_s \neq 0$ (claim)

(b) Critical value is 0.497.

(c) $\Sigma d^2 = 165.5$
$$r_s = 1 - \frac{6\Sigma d^2}{n(n^2 - 1)} \approx 0.421$$

(d) Fail to reject H_0.

(e) There is not enough evidence at the 10% significance level to support the claim that there is a correlation between overall score and price.

7. $H_0: \rho_s = 0$; $H_a: \rho_s \neq 0$ (claim)

Critical value is 0.700.

$\Sigma d^2 = 121.5$

$$r_s = 1 - \frac{6\Sigma d^2}{n(n^2 - 1)} \approx -0.013$$

Fail to reject H_0. There is not enough evidence at the 5% significance level to conclude that there is a correlation between science achievement scores and GNP.

8. $H_0: \rho_s = 0$; $H_a: \rho_s \neq 0$ (claim)

The critical value is 0.700.

$\Sigma d^2 = 120$

$$r_s = 1 - \frac{6\Sigma d^2}{n(n^2 - 1)} \approx 0.000$$

Fail to reject H_0. There is not enough evidence at the 5% significance level to conclude that there is a correlation between mathematics achievement scores and GNP.

9. $H_0: \rho_s = 0$; $H_a: \rho_s \neq 0$ (claim)

Critical value is 0.700.

$\Sigma d^2 = 9.5$

$$r_s = 1 - \frac{6\Sigma d^2}{n(n^2 - 1)} \approx 0.921$$

Reject H_0. There is enough evidence at the 5% significance level to conclude that there is a correlation between science and mathematics achievement scores.

10. There is not enough evidence at the 5% significance level to conclude that there is a significant correlation between eighth grade science achievement scores and the GNP of a country, or between eighth grade math achievement scores and the GNP of a country. However, there is enough evidence at the 5% significance level to conclude there is a significant correlation.

11. $H_0: \rho_s = 0$; $H_a: \rho_s \neq 0$ (claim)

The critical value is $= \frac{\pm z}{\sqrt{n-1}} = \frac{\pm 1.96}{\sqrt{33-1}} \approx \pm 0.346$.

$\Sigma d^2 = 6673$

$$r_s = 1 - \frac{6\Sigma d^2}{n(n^2 - 1)} \approx -0.115$$

Fail to reject H_0. There is not enough evidence to support the claim.

12. $H_0: \rho_s = 0$; $H_a: \rho_s \neq 0$ (claim)

Critical value $= \frac{\pm z}{\sqrt{n-1}} = \frac{\pm 1.96}{\sqrt{34-1}} \approx \pm 0.341$.

$\Sigma d^2 = 7310.5$

$$r_s = 1 - \frac{6\Sigma d^2}{n(n^2 - 1)} \approx -0.017$$

Fail to reject H_0. There is not enough evidence to support the claim.

11.5 Try It Yourself Solutions

1a. *P P P F P F P P P P F F P F P P F F F P P P F P P P*

b. 13 groups $\Rightarrow$ 13 runs

c. 3, 1, 1, 1, 4, 2, 1, 1, 2, 3, 3, 1, 3

2a. H_0: The sequence of genders is random.
H_a: The sequence of genders is not random. (claim)

b. $\alpha = 0.05$

c. *F F F M M F F M F M M F F F*
n_1 = number of *F*s = 9
n_2 = number of *M*s = 5
G = number of runs = 7

d. $cv = 3$ and 12

e. $G = 7$

f. Fail to reject H_0.

g. At the 5% significance level, there is not enough evidence to support the claim that the sequence of genders is not random.

3a. H_0: The sequence of weather conditions is random.
H_a: The sequence of weather conditions is not random. (claim)

b. $\alpha = 0.05$

c. n_1 = number of *N*s = 21
n_2 = number of *S*s = 10
G = number of runs = 17

d. $cv = \pm 1.96$

e. $\mu_G = \dfrac{2n_1n_2}{n_1 + n_2} + 1 = \dfrac{2(21)(10)}{21 + 10} + 1 \approx 14.5$

$\sigma_G = \sqrt{\dfrac{2n_1n_2(2n_1n_2 - n_1 - n_2)}{(n_1 + n_2)^2(n_1 + n_2 - 1)}} = \sqrt{\dfrac{2(21)(10)(2(21)(10) - 21 - 10)}{(21 + 10)^2(21 + 10 - 1)}} \approx 2.4$

$z = \dfrac{G - \mu_G}{\sigma_G} = \dfrac{17 - 14.5}{2.4} = 1.04$

f. Fail to reject H_0.

g. At the 5% significance level, there is not enough evidence to support the claim that the sequence of weather conditions each day is not random.

11.5 EXERCISE SOLUTIONS

1. Number of runs = 8
Run lengths = 1, 1, 1, 1, 3, 3, 1, 1

2. Number of runs = 9
Run lengths = 2, 2, 1, 1, 2, 2, 1, 1, 2

3. Number of runs = 9
Run lengths = 1, 1, 1, 1, 1, 6, 3, 2, 4

4. Number of runs = 10
Run lengths = 3, 3, 1, 2, 6, 1, 2, 1, 1, 2

5. n_1 = number of Ts = 6
n_2 = number of Fs = 6

6. n_1 = number of Us = 8
n_2 = number of Ds = 6

7. n_1 = number of Ms = 10
n_2 = number of Fs = 10

8. n_1 = number of As = 13
n_2 = number of Bs = 9

9. n_1 = number of Ts = 6
n_1 = number of Fs = 6

cv = 3 and 11

10. n_1 = number of Ms = 9
n_2 = number of Fs = 3

cv = 2 and 8

11. n_1 = number of Ns = 11
n_1 = number of Ss = 7

cv = 5 and 14

12. n_1 = number of Xs = 7
n_2 = number of Ys = 14

cv = 5 and 15

13. (a) H_0: The coin tosses were random.
H_a: The coin tosses were not random. (claim)

(b) n_1 = number of Hs = 7
n_2 = number of Ts = 9

cv = 4 and 14

(c) G = 9 runs

(d) Fail to reject H_0.

(e) At the 5% significance level, there is not enough evidence to support the claim that the coin tosses were not random.

14. (a) H_0: The selection of members was random.
H_a: The selection of members was not random. (claim)

(b) n_1 = number of Rs = 18
n_2 = number of Ds = 16
cv = 11 and 25

(c) G = 19 runs

(d) Fail to reject H_0.

(e) At the 5% significance level, there is not enough evidence to support the claim that the selection of members was not random.

15. (a) H_0: The sequence of digits was randomly generated.

H_a: The sequence of digits was not randomly generated. (claim)

(b) n_1 = number of Os = 16

n_2 = number of Es = 16

cv = 11 and 23

(c) G = 9 runs

(d) Reject H_0.

(e) At the 5% significance level, there is enough evidence to support the claim that the sequence of digits was not randomly generated.

16. (a) H_0: The microchips are random by gender. (claim)

H_a: The microchips are not random by gender.

(b) n_1 = number of Ms = 9

n_2 = number of Fs = 20

cv = 8 and 18

(c) G = 12 runs

(d) Fail to reject H_0.

(e) At the 5% significance level, there is not enough evidence to reject the claim that the microchips are random by gender.

17. (a) H_0: The sequence is random.

H_a: The sequence is not random. (claim)

(b) n_1 = number of Ns = 40

n_2 = number of Ps = 9

cv = ± 1.96

(c) G = 14 runs

$$\mu_G = \frac{2n_1 n_2}{n_1 + n_2} + 1 = \frac{2(40)(9)}{40 + 9} + 1 \approx 15.7$$

$$\sigma_G = \sqrt{\frac{2n_1 n_2(2n_1 n_2 - n_1 - n_2)}{(n_1 + n_2)^2(n_1 + n_2 - 1)}} = \sqrt{\frac{2(40)(9)(2(40)(9) - 40 - 9)}{(40 + 9)^2(40 + 9 - 1)}} = 2.05$$

$$z = \frac{G - \mu_G}{\sigma_G} = \frac{14 - 15.7}{2.05} = -0.83$$

(d) Fail to reject H_0.

(e) At the 5% significance level, there is not enough evidence to support the claim that the sequence is not random.

18. (a) H_0: The sequence of past winners is random.

H_a: The sequence of past winners is not random. (claim)

(b) n_1 = number of Ts = 42

n_2 = number of Ls = 16

$cv = \pm 1.96$

(c) $G = 32$ runs

$$\mu_G = \frac{2n_1 n_2}{n_1 + n_2} + 1 = \frac{2(42)(16)}{42 + 16} + 1 = 24.2$$

$$\sigma_G = \sqrt{\frac{2n_1 n_2 (2n_1 n_2 - n_1 - n_2)}{(n_1 + n_2)^2 (n_1 + n_2 - 1)}} = \sqrt{\frac{2(42)(16)(2(42)(16) - 42 - 16)}{(42 + 16)^2 (42 + 16 - 1)}} = 3.0$$

$$z = \frac{G - \mu_G}{\sigma_G} = \frac{32 - 24.2}{3.0} = 2.6$$

(d) Reject H_0.

(e) At the 5% significance level, there is enough evidence to support the claim that the sequence is not random.

19. H_0: Daily high temperatures occur randomly.

H_a: Daily high temperatures do not occur randomly. (claim)

median = 87

n_1 = number above median = 14

n_2 = number below median = 13

cv = 9 and 20

$G = 11$ runs

Fail to reject H_0.

At the 5% significance level, there is not enough evidence to support the claim that the daily high temperatures do not occur randomly.

20. H_0: The selection of students' GPAs was random.

H_a: The selection of students' GPAs was not random. (claim)

median = 1.8

n_1 = number above median = 12

n_2 = number below median = 3

cv = 2 and 8

$G = 3$ runs

Fail to reject H_0.

At the 5% significance level, there is not enough evidence to support the claim that the selection of students' GPAs was not random.

21. Answers will vary.

CHAPTER 11 REVIEW EXERCISE SOLUTIONS

1. (a) H_0: median = \$24,300 (claim); H_a: median ≠ \$24,300

   (b) Critical value is 2.

   (c) $x = 7$

   (d) Fail to reject H_0.

   (e) There is not enough evidence at the 1% significance level to reject the claim that the median value of stock among families that own stock is \$24,300.

2. (a) H_0: median ≤ \$2000; H_a: median > \$2000 (claim)

   (b) The critical value is 1.

   (c) $x = 6$

   (d) Fail to reject H_0.

   (e) There is not enough evidence at the 1% significance level to support the claim that the median credit card debt is more than \$2000.

3. (a) H_0: median ≤ 6 (claim); H_a: median > 6

   (b) Critical value is $z_0 \approx -1.28$.

   (c) $x = 44$

   $$z = \frac{(x - 0.5) - 0.5(n)}{\frac{\sqrt{n}}{2}} = \frac{(44 - 0.5) - 0.5(70)}{\frac{\sqrt{70}}{2}} = \frac{8.5}{4.1833} \approx 2.03$$

   (d) Reject H_0.

   (e) There is enough evidence at the 10% significance level to reject the claim that the median turnover time is no more than 6 hours.

4. (a) H_0: There was no reduction in diastolic blood pressure. (claim)

   H_a: There was a reduction in diastolic blood pressure.

   (b) The critical value is 1.

   (c) $x = 4$

   (d) Fail to reject H_0.

   (e) There is not enough evidence at the 5% significance level to reject the claim that there was no reduction in diastolic blood pressure.

5. (a) H_0: There is no reduction in diastolic blood pressure. (claim)

   H_a: There is a reduction in diastolic blood pressure.

   (b) Critical value is 2.

   (c) $x = 3$

   (d) Fail to reject H_0.

   (e) There is not enough evidence at the 5% significance level to reject the claim that there was no reduction diastolic blood pressure.

6. (a) H_0: median = \$40,200 (claim); H_a: median ≠ \$40,200

(b) The critical values are $z_0 = \pm1.96$.

(c) $x = 21$

$$z = \frac{(x + 0.5) - 0.5(n)}{\dfrac{\sqrt{n}}{2}} = \frac{(21 + 0.5) - 0.5(54)}{\dfrac{\sqrt{54}}{2}} = \frac{-5.5}{3.674} \approx -1.50$$

(d) Fail to reject H_0.

(e) There is not enough evidence at the 5% significance level to reject the claim that the median starting salary is \$40,200.

7. (a) Independent; Wilcoxon Rank Sum Test

(b) H_0: There is no difference in the amount of time that it takes to earn a doctorate.

H_a: There is a difference in the amount of time that it takes to earn a doctorate. (claim)

(c) Critical values are $z_0 = \pm2.575$.

(d) $R = 173.5$

$$\mu_R = \frac{n_1(n_1 + n_2 + 1)}{2} = \frac{12(12 + 12 + 1)}{2} = 150$$

$$\sigma_R = \sqrt{\frac{n_1 n_2(n_1 + n_2 + 1)}{12}}\ \sqrt{\frac{(12)(12)(12 + 12 + 1)}{12}} \approx 17.321$$

$$z = \frac{R - \mu_R}{\sigma_R} = \frac{173.5 - 150}{17.321} \approx 1.357$$

(e) Fail to reject H_0.

(f) There is not enough evidence at the 1% significance level to support the claim that there is a difference in the amount of time that it takes to earn a doctorate.

8. (a) Dependent; Wilcoxon Signed-Rank Test

(b) H_0: The new drug does not affect the number of headache hours experienced.

H_a: The new drug does affect the number of headache hours experienced. (claim)

(c) The critical value is 4.

(d) $w_s = 1.5$

(e) Reject H_0.

(f) There is enough evidence at the 5% significance level to support the claim that the new drug does affect the number of headache hours experienced.

9. (a) H_0: There is no difference in salaries between the fields of study.

H_a: There is a difference in salaries between the fields of study. (claim)

(b) Critical value is 5.991.

(c) $H \approx 21.695$

(d) Reject H_0.

(e) There is enough evidence at the 5% significance level to conclude that there is a difference in salaries between the fields of study.

10. (a) H_0: There is no difference in the amount of time to earn a doctorate between the fields of study.

H_a: There is a difference in the amount of time to earn a doctorate between the fields of study. (claim)

(b) The critical value is 5.991.

(c) $H \approx 3.282$

(d) Fail to reject H_0.

(e) There is not enough evidence at the 5% significance level to support the claim that there is a difference in the amount of time to earn a doctorate between the fields of study.

11. (a) H_0: $\rho_s = 0$; H_a: $\rho_s \neq 0$ (claim)

(b) Critical value is 0.881.

(c) $\Sigma d^2 = 57$

$$r_s = 1 - \frac{6\Sigma d^2}{n(n^2 - 1)} \approx 0.321$$

(d) Fail to reject H_0.

(e) There is not enough evidence at the 1% significance level to support the claim that there is a correlation between overall score and price.

12. (a) H_0: $\rho_s = 0$; H_a: $\rho_s \neq 0$ (claim)

(b) The critical value is 0.700.

(c) $\Sigma d^2 = 94.5$

$$r_s = 1 - \frac{6\Sigma d^2}{n(n^2 - 1)} \approx 0.213$$

(d) Fail to reject H_0.

(e) There is not enough evidence at the 5% significance level to support the claim that there is a correlation between overall score and price.

13. (a) H_0: The traffic stops were random by gender.

H_a: The traffic stops were not random by gender. (claim)

(b) n_1 = number of Fs = 12

n_2 = number of Ms = 13

cv = 8 and 19

(c) G = 14 runs

(d) Fail to reject H_0.

(e) There is not enough evidence at the 5% significance level to support the claim that the traffic stops were not random.

14. (a) H_0: The departure status of buses is random.

 H_a: The departure status of buses is not random. (claim)

  (b) n_1 = number of Ts = 11

 n_2 = number of Ls = 7

 cv = 5 and 14

  (c) G = 5 runs

  (d) Reject H_0.

  (e) There is enough evidence at the 5% significance level to support the claim that the departure status of buses is not random.

CHAPTER 11 QUIZ SOLUTIONS

1. (a) H_0: There is no difference in the salaries between genders.

 H_a: There is a difference in the salaries between genders. (claim)

  (b) Wilcoxon Rank Sum Test

  (c) Critical values are $z_0 = \pm 1.645$.

  (d) $R = 67.5$

$$\mu_R = \frac{n_1(n_1 + n_2 + 1)}{2} = \frac{10(10 + 10 + 1)}{2} = 105$$

$$\sigma_R = \sqrt{\frac{n_1 n_2(n_1 + n_2 + 1)}{12}} = \sqrt{\frac{(10)(10)(10 + 10 + 1)}{12}} \approx 13.229$$

$$z = \frac{R - \mu_R}{\sigma_R} = \frac{67.5 - 105}{13.229} \approx -2.835$$

  (e) Reject H_0.

  (f) There is enough evidence at the 10% significance level to support the claim that there is a difference in the salaries between genders.

2. (a) H_0: median = 50 (claim); H_a: median $\neq$ 50

  (b) Sign Test   (c) Critical value is 5.

  (d) $x = 9$ (e) Fail to reject H_0.

  (f) There is not enough evidence at the 5% significance level to reject the claim that the median number of annual volunteer hours is 50.

3. (a) H_0: There is no difference in rent between regions.

 H_a: There is a difference in rent between regions. (claim)

  (b) Kruskal-Wallis Test   (c) Critical value is 7.815.

  (d) $H \approx 11.826$ (e) Reject H_0.

  (f) There is enough evidence at the 5% significance level to conclude that there is a difference in rent between regions.

4. (a) H_0: The days with rain are random.

H_a: The days with rain are not random. (claim)

(b) The Runs test

(c) n_1 = number of Ns = 15

n_2 = number of Rs = 15

$cv = 10$ and 22

(d) $G = 16$ runs

(e) Fail to reject H_0.

(f) There is not enough evidence at the 5% significance level to conclude that days with rain are not random.

CUMULATIVE REVIEW FOR CHAPTERS 9–11

1. (a)

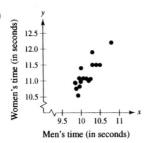

Men's time (in seconds)

$r = 0.828$

There is a strong positive linear correlation.

(b) H_0: $\rho = 0$

H_a: $\rho \neq 0$ (claim)

$t = 5.911$

$p = 0.0000591$

Reject H_0. There is enough evidence at the 5% significance level to conclude that there is a significant linear correlation.

(c) $\hat{y} = 1.423x - 3.204$

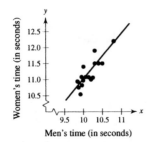

Men's time (in seconds)

(d) $\hat{y} = 1.423(9.90) - 3.204 = 10.884$ seconds

2. H_0: Median (Union) = Median (Non union)

H_1: Median (Union) ≠ Median (Non union) (claim)

$cv = \pm 1.96$

$R = 145.0$

$$\mu_R = \frac{n_1(n_1 + n_2 + 1)}{2} = \frac{10(10 + 10 + 1)}{2} = 105$$

$$\sigma_R = \sqrt{\frac{n_1 n_2(n_1 + n_2 + 1)}{12}} = \sqrt{\frac{10(10)(10 + 10 + 1)}{12}} = 13,229$$

$$z = \frac{R - \mu_R}{\sigma_R} = \frac{145.0 - 105}{13,229} = 4.375$$

Reject H_0. There is enough evidence at the 5% significance level to support the claim.

3. H_0: median $= 48$ (claim)

H_1: median $\neq 48$

$cv = 3$

$x = 7$

Fail to reject H_0. There is not enough evidence at the 5% significance level to reject the claim.

4. H_0: $\mu_1 = \mu_2 = \mu_3 = \mu_4$ (claim)

H_1: At least one μ if different.

$cv = 2.29$

$F = 2.476$

Reject H_0. There is enough evidence at the 10% significance level to reject the claim.

5. H_0: $\sigma_1^2 = \sigma_2^2$ (claim)

H_1: $\sigma_1^2 \neq \sigma_2^2$

$cv = 2.01$

$$F = \frac{s_1^2}{s_2^2} = \frac{34.6}{33.2} = 1.042$$

Fail to reject H_0. There is not enough evidence at the 10% significance level to reject the claim.

6. H_0: The medians are all equal.

H_1: The medians are not all equal. (claim)

$cv = 11.345$

$H = 14.78$

Reject H_0. There is enough evidence at the 1% significance level to support the claim.

7.

Physicians	Distribution
1	0.36
2–4	0.32
5–9	0.20
10+	0.12

H_0: The distribution is as claimed. (claim)

H_1: The distribution is not as claimed.

Physician	Distribution	Observed	Expected	$\dfrac{(O - E)^2}{E}$
1	0.36	116	104.04	1.375
2–4	0.32	84	92.48	0.778
5–9	0.20	66	57.80	1.163
10+	0.12	23	34.68	3.934
		289		$x^2 = 7.250$

$cv = 7.815$

$x^2 = 7.250$

Fail to reject H_0. There is not enough evidence at the 5% significance level to reject the claim.

8. (a) $r^2 = 0.733$

Metacarpal bone length explains 73.3% of the variability in height. About 26.7% of the variation is unexplained.

(b) $s_e = 4.255$

(c) $\hat{y} = 94.428 + 1.700(50) = 179.428$

$$E = t_c S_e \sqrt{1 + \frac{1}{n} + \frac{n(x_0 - \bar{x})^2}{n\Sigma x^2 - (\Sigma x)^2}} = 2.365(4.255)\sqrt{1 + \frac{1}{9} + \frac{9(50 - 45,444)^2}{9(18,707) - (409)^2}}$$

$$= 2.365(4.255)\sqrt{1.284} = 11.402$$

$\hat{y} \pm E \Longrightarrow (168.026, 190.83)$

You can be 95% confident that the height will be between 168.026 centimeters and 190.83 centimeters when the metacarpal bone length is 50 centimeters.

9.

Score	Rank	Price	Rank	d	d^2
85	8	81	7	1.0	1.00
83	6.5	78	6	0.5	0.25
83	6.5	56	1	5.5	30.25
81	5	77	5	0	0.00
77	4	62	3.5	0.5	0.25
72	3	85	8	−5.0	25.00
70	2	62	3.5	−1.5	2.25
66	1	61	2	−1.0	1.00
					$\Sigma d^2 = 60$

H_0: $\rho_s = 0$

H_1: $\rho_s \neq 0$ (claim)

$cv = 0.643$

$$r_s = 1 - \frac{6\Sigma d^2}{n(n^2 - 1)} = 1 - \frac{60(60)}{8(8^2 - 1)} = 0.286$$

Fail to reject H_0. There is not enough evidence at the 10% significance level to support the claim.

10. (a) $\hat{y} = 91.113 - 0.014(4325) + 0.018(1900) = 64.763$ bushels

(b) $\hat{y} = 91.113 - 0.014(4900) + 0.018(2163) = 61.447$ bushels

Alternative Presentation of the Standard Normal Distribution

Try It Yourself Solutions

1 (1) 0.4857

   (2) $z = \pm 2.17$

2a.

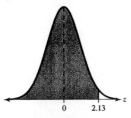

b. 0.4834

c. Area $= 0.5 + 0.4834 = 0.9834$

3a.

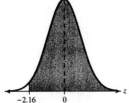

b. 0.4846

c. Area $= 0.5 + 0.4846 = 0.9846$

4a.

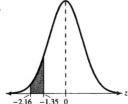

b. $z = -2.16$: Area $= 0.4846$

   $z = -1.35$: Area $= 0.4115$

c. Area $= 0.4846 - 0.4115 = 0.0731$

Normal Probability Plots and Their Graphs

Try It Yourself Solutions

1a.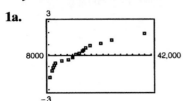

The points do not appear to be approximately linear.

b. 39,860 is a possible outlier because it is far removed from the other entries in the data set.

c. Because the points do not appear to be approximately linear and there is an outlier, you can conclude that the sample data do not come from a population that has a normal distribution.

ACTIVITY 1.3

1. {1, 17, 17, 17, 14, 10, 15, 6} Answers will vary.

   This list is a random sample taken with replacement because 17 appears 3 times.

2. Min = 1

   Max = 731

   Number of samples = 8

   {565, 718, 305, 75, 364, 221, 230, 231} Answers will vary.

   The random number generator is easier to use than a random number table.

ACTIVITY 2.3

1. When the mean is equal to the median, the shape of the distribution will be symmetric.

   When a few points are added below 10, the mean shifts downward.

   As points are added below 10, both the mean and median will shift downward.

2. No, the mean and median cannot be any of the points that were plotted.

ACTIVITY 2.4

1. {18, 10, 15, 20, 16, 14, 19, 17, 13, 15}

   $\bar{x} = 15.7$

   $s = 2.98$

   By adding $x = 15$, $\bar{x} = 15.6$ and $s = 2.84$, the mean moves slightly closer to 15 while the standard deviation decreases.

   By adding $x = 20$, $\bar{x} = 16$ and $s = 2.98$, the mean and standard deviation increase.

2. {30, 30, 30, 30, 40, 40, 40, 40}

   $s = 5.35$

   {35, 35, 35, 35, 35, 35, 35, 35}

   $s = 0$

   When the values in a data set are all equal, the standard deviation will be zero.

ACTIVITY 3.1

1. Answers will vary.

2. P(market goes up on day 36) = 0.5

1. $P(3 \text{ or } 4) = \frac{1}{6} + \frac{1}{6} = \frac{1}{3}$

2. Answers will vary.

3. The green line will increase to $\frac{1}{2}$.

1. $\{7, 8, 7, 7, 7, 7, 8, 8, 9, 10\}$ Answers will vary.

   (a) $P(x = 5) = \frac{0}{10}$

   (b) $P(\text{at least } 8) = \frac{5}{10}$

   (c) $P(\text{at most } 7) = \frac{5}{10}$

2. $\{3, 1, 2, 2, 1, 2, 2, 2, 5, 3\}$ Answers will vary.

   (a) $P(x = 4) = \frac{0}{10}$

   (b) $P(\text{at least } 5) = \frac{1}{10}$

   (c) $P(\text{less than } 4) = \frac{9}{10}$

3. Answers will vary.

   $P(x = 5) \approx 0.103$

   When using $N = 100$, the estimated probability of exactly five is closer to 0.103.

1. The mean of the sampling distribution of a uniform, bell-shaped, and skewed distribution will be approximately 25.

2. The estimated standard deviation of the sampling distribution will be approximately $\frac{\sigma}{\sqrt{50}}$.

1. Approximately 95% of the z and t confidence intervals will contain the mean of 25.

2. Approximately 95% of the z and t confidence intervals will contain the mean of 7.26.

   Because $n = 24$, we should use a t-CI.

1. Approximately 95% and 99% of the CI's will contain 0.6.

2. Approximately 95% and 99% of the CI's will contain 0.4.

ACTIVITY 7.2

1. At $\alpha = 0.05$, approximately 50 of the 1000 null hypotheses will be rejected.

   At $\alpha = 0.01$, approximately 10 of the 1000 null hypotheses will be rejected.

2. If the null hypothesis is rejected at the 0.01 level, the p-value will be smaller than 0.01. So, it would also be rejected at the 0.05 level.

   Suppose a null hypothesis is rejected at the 0.05 level. It will not necessarily be rejected at the 0.01 level because $0.01 < p\text{-value} < 0.05$.

3. H_0: $\mu \geq 27$

   The proportion of rejected null hypotheses will be larger than 0.05 and 0.01 because the true mean is 25 (i.e. $\mu < 27$).

ACTIVITY 7.4

1. Approximately 50 null hypotheses will be rejected at the 0.05 level while approximately 10 null hypotheses will be rejected at the 0.01 level.

2. H_0: $p \geq 0.4$

   The proportion of null hypotheses rejected will be less than 0.05 and 0.01 because the true $p = 0.6$.

ACTIVITY 9.1

1. If the scatter plot is linear with a positive slope, $r \approx 1$. If the scatter plot is linear with a negative slope $r \approx -1$.

2. Create a data set that is linear with a positive slope.

3. Create a data set that is nonlinear.

4. $r \approx -0.9$ has a negative slope while using $r \approx 0.9$ will have a positive slope.

ACTIVITY 9.2

1. Answers will vary.

2. The regression line is influenced by the new point.

3. The regression line is not as influenced by the new point due to more points being used to calculate the regression line.

4. As the sample size increases, the slope changes less.

CHAPTER 1

CASE STUDY: *RATING TELEVISION SHOWS IN THE UNITED STATES*

1. Yes. A rating of 8.4 is equivalent to 9,357,600 households which is twice the number of households at a rating of 4.2.

2. $\frac{10,000}{1,114,000} \approx 0.00897 \Rightarrow 0.897\%$

3. Program Name and Network

4. Rank and Rank Last Week
   Data can be arranged in increasing or decreasing order.

5. Day, Time
   Data can be chronologically ordered.
   Hours or minutes

6. Rating, Share, and Audience

7. Shows are ranked by rating. Share is not arranged in decreasing order.

8. A decision of whether a program should be cancelled or not can be made based on the Nielsen ratings.

CHAPTER 2

CASE STUDY: *EARNINGS OF ATHLETES*

1. NFL has $3,041,000,000 in total revenue.

2. MLB: $\mu = \$3,152,305.83$ MLS: $\mu = \$265,576.32$

   NBA: $\mu = \$4,414,977.48$ NFL: $\mu = \$1,620,138.52$

   NHL: $\mu = \$1,960,454.32$ NASCAR: $\mu = \$2,500,000.00$

   PGA: $\mu = \$1,167,300.38$

3. NBA has a mean player revenue of $4,414,977.

4. MLB: $s = \$4,380,876.18$ MLS: $s = \$124,022.66$

   NBA: $s = \$4,533,634.75$ NFL: $s = \$2,220,852.88$

   NHL: $s = \$1,588,168.87$ NASCAR: $S = \$2,228,796.19$

   PGA: $1,271,590.31$

5. NBA

6. By examining the frequency distributions, the NBA is more bell-shaped.

CHAPTER 3

CASE STUDY: *PROBABILITY AND PARKING LOT STRATEGIES*

1. No, each parking space is not equally likely to be empty. Spaces near the entrances are more likely to be filled.

2. Pick-a-Lane strategy appears to take less time (in both time to find a parking space and total time in parking lot).

3. Cycling strategy appears to give a shorter average walking distance to the store.

4. In order to assume that each space is equally likely to be empty, drivers must be able to see which spaces are available as soon as they enter a row.

5. $P(\text{1st row}) = \frac{1}{7}$ since there are seven rows.

6. $P(\text{in one of 20 closest spaces}) = \frac{20}{72}$

7. Answers will vary. Using the Pick-a-Lane strategy, the probability that you can find a parking space in the most desirable category is low. Using the Cycling strategy, the probability that you can find a parking space in the most desirable category depends on how long you cycle. The longer you cycle, the probability that you can find a parking space in the most desirable category will increase.

CHAPTER 4

CASE STUDY: *BINOMIAL DISTRIBUTION OF AIRPLANE ACCIDENTS*

1. $P(\text{crash in 2006}) = \dfrac{2}{11,000,000} = 0.00000018$

2. (a) $P(4) = 0.192$

   (b) $P(10) = 0.009$

   (c) $P(1 \leq x \leq 5) = P(1) + P(2) + P(3) + P(4) + P(5)$
   $$= 0.054 + 0.119 + 0.174 + 0.192 + 0.169 = 0.708$$

3. $n = 11,000,000$, $p = 0.0000008$

   (a) $P(4) \approx 0.0376641$

   (b) $P(10) \approx 0.1156838$

   (c) $P(1 \leq x \leq 5) \approx 0.12823581$

x	$P(x)$
0	0.0001507
1	0.0013264
2	0.0058364
3	0.0171200
4	0.0376641
5	0.0662889
6	0.0972237
7	0.1222241
8	0.1344465
9	0.1314588
10	0.1156838
11	0.0925470
12	0.0678678

4. No, because each flight is not an independent trial. There are many flights that use the same plane.

5. Assume $p = 0.0000004$. The number of flights would be 64,000 years $\cdot$ 365 days per year $= 23,360,000$ flights. The average number of fatal accidents for that number of flights would be 9.344. So, the claim cannot be justified.

CHAPTER 5

CASE STUDY: *BIRTH WEIGHTS IN AMERICA*

1. (a) 40 weeks ($\mu = 7.69$)

(b) 32 to 35 weeks ($\mu = 5.71$)

(c) 41 weeks ($\mu = 7.79$)

2. (a) $P(x < 5.5) = P(z < 3.02) = 0.9987 \rightarrow 99.87\%$

(b) $P(x < 5.5) = P(z < -0.14) = 0.4432 \rightarrow 44.32\%$

(c) $P(x < 5.5) = P(z < -1.66) = 0.0484 \rightarrow 4.84\%$

(d) $P(x < 5.5) = P(z < -1.90) = 0.0287 \rightarrow 2.87\%$

3. (a) top 10% $\rightarrow$ 90th percentile $\rightarrow z \approx 1.285$

$x = \mu + z\sigma = 7.31 + (1.285)(1.09) \approx 8.711$

Birth weights must be at least 8.711 pounds to be in the top 10%.

(b) top 10% $\rightarrow$ 90th percentile $\rightarrow z \approx 1.285$

$x = \mu + z\sigma = 7.61 + (1.285)(1.12) \approx 9.036$

Birth weights must be at least 9.036 pounds to be in the top 10%.

4. (a) $P(6 < x < 9) = P(0.20 < z < 2.24) = 0.9875 - 0.5793 = 0.4082$

(b) $P(6 < x < 9) = P(-1.20 < z < 1.55) = 0.9394 - 0.1151 = 0.8243$

(c) $P(6 < x < 9) = P(-1.45 < z < 1.25) = 0.8944 - 0.0735 = 0.8209$

5. (a) $P(x < 3.3) = P(z < 1.18) = 0.8810$

(b) $P(x < 3.3) = P(z < -1.64) = 0.0505$

(c) $P(x < 3.3) = P(z < -3.68) = 0.001$ (using technology)

CHAPTER 6

CASE STUDY: *SHOULDER HEIGHTS OF APPALACHIAN BLACK BEARS*

1. (a) $\bar{x} = 79.7$

(b) $\bar{x} = 75.7$

2. (a) $s = 12.1$

(b) $s = 7.6$

3. (a) $\bar{x} \pm z_c \dfrac{s}{\sqrt{n}} \Rightarrow 79.7 \pm 1.96 \dfrac{12.1}{\sqrt{40}} = (76.0, 83.5)$

(b) $\bar{x} \pm t_c \dfrac{s}{\sqrt{n}} \Rightarrow 75.7 \pm 2.048 \dfrac{7.6}{\sqrt{28}} = (72.7, 78.6)$

4. $\bar{x} \pm z_c \dfrac{s}{\sqrt{n}} \Rightarrow 78.1 \pm 1.96 \dfrac{10.6}{\sqrt{68}} = (75.5, 80.6)$

5. (a) $n = \left(\dfrac{z_c \sigma}{E}\right)^2 = \left(\dfrac{2.575 \cdot 12.4}{0.5}\right)^2 = 4078.10 \Rightarrow 4079$ bears

(b) $n = \left(\dfrac{z_c \sigma}{E}\right)^2 = \left(\dfrac{2.575 \cdot 7.8}{0.5}\right)^2 = 1613.63 \Rightarrow 1614$ bears

CHAPTER 7

CASE STUDY: *HUMAN BODY TEMPERATURE: WHAT'S NORMAL?*

1. (a) (see part d)

(b) $z_0 = \pm 1.96$ (see part d)

(c) Rejection regions: $z < -1.96$ and $z > 1.96$ (see part d)

(d) $\bar{x} \approx 98.25, \quad s \approx 0.73$

$$z = \frac{\bar{x} - \mu}{\dfrac{s}{\sqrt{n}}} = \frac{98.25 - 98.6}{\dfrac{0.73}{\sqrt{130}}} = \frac{-0.35}{0.0640} \approx -5.469$$

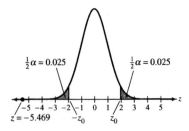

(e) Reject H_0.

(f) There is sufficient evidence at the 5% level to reject the claim that the mean body temperature of adult humans is 98.6° F.

2. No, $\alpha = 0.01 \rightarrow z_0 = \pm 2.575 \rightarrow$ Reject H_0

3. H_0: $\mu_m = 98.6$ and H_a: $\mu_m \neq 98.6$

$\alpha = 0.01 \rightarrow z_0 = \pm 2.575$

$\bar{x} \approx 98.105,\quad s \approx 0.699$

$z = \dfrac{\bar{x} - \mu}{\frac{s}{\sqrt{n}}} = \dfrac{98.105 - 98.6}{\frac{0.699}{\sqrt{65}}} = \dfrac{-0.495}{0.0867} \approx -5.709$

Reject H_0. There is sufficient evidence to reject the claim.

4. H_0: $\mu_w = 98.6$ and H_a: $\mu_w \neq 98.6$

$\alpha = 0.01 \rightarrow z_0 = \pm 2.575$

$\bar{x} \approx 98.394,\quad s \approx 0.743$

$z = \dfrac{\bar{x} - \mu}{\frac{s}{\sqrt{n}}} = \dfrac{98.394 - 98.6}{\frac{0.743}{\sqrt{65}}} = \dfrac{-0.206}{0.0922} \approx -2.234$

Fail to reject H_0. There is insufficient evidence to reject the claim.

5. $\bar{x} \approx 98.25,\quad s \approx 0.73$

$\bar{x} \pm z_c \dfrac{s}{\sqrt{n}} = 98.25 \pm 2.575 \dfrac{0.73}{\sqrt{130}}\ 98.25 \pm 0.165 \approx (98.05, 98.415)$

6. Wunderlich may have sampled more women than men, thus causing an overestimated body temperature.

CHAPTER 8

CASE STUDY: *DASH DIET AND BLOOD PRESSURE*

1. H_0: $\mu_1 = \mu_2$ vs H_1: $\mu_1 \neq \mu_2$ (claim)

$z = \pm 1.96$ Reject H_0 if $z < -1.96$ or $z > 1.96$

$z = \dfrac{(\bar{x}_1 - \bar{x}_2) - (0)}{\sqrt{\frac{s_1^2}{n_1} + \frac{s_2^2}{n_2}}} = \dfrac{10.36 - 5.73}{\sqrt{\frac{(15.79)^2}{77} + \frac{(12.50)^2}{79}}} = \dfrac{4.63}{\sqrt{5.216}}\ 2.03$

Reject H_0. There is enough evidence to support the claim.

2. H_0: $\mu_3 = \mu_4$ vs H_2: $\mu_3 \neq \mu_4$ (claim)

$z_0 = \pm 1.96$ Reject H_0 if $z < -1.96$ or $z > 1.96$.

$z = \dfrac{(\bar{x}_3 - \bar{x}_4) - (0)}{\sqrt{\frac{s_3^2}{n_3} + \frac{s_4^2}{n_4}}} = \dfrac{4.85 - 3.53}{\sqrt{\frac{(13.73)^2}{76} + \frac{(12)^2}{79}}} = \dfrac{1.32}{\sqrt{4.303}} \approx 0.64$

Fail to reject H_0. There is not enough evidence to support the claim.

3. H_0: $\mu_2 = \mu_4$ vs H_1: $\mu_2 \neq \mu_4$ (claim)

$z = \pm 1.96$ Reject H_0 if $z < -1.96$ or $z > 1.96$.

$$z = \frac{(\bar{x}_2 - \bar{x}_4) - (0)}{\sqrt{\dfrac{s_2^2}{n_2} + \dfrac{s_4^2}{n_4}}} = \frac{5.73 - 3.53}{\sqrt{\dfrac{(12.50)^2}{79} + \dfrac{(12.00)^2}{79}}} = \frac{2.2}{\sqrt{3.801}} = 1.13$$

Fail to reject H_0. There is not enough evidence to support the claim.

4. H_0: $\mu_1 = \mu_3$ vs H_1: $\mu_1 \neq \mu_3$ (claim)

$z_0 = \pm 1.96$ Reject H_0 if $z < -1.96$ or $z > 1.96$.

$$z = \frac{(\bar{x}_1 - \bar{x}_3) - (0)}{\sqrt{\dfrac{s_1^2}{n_1} + \dfrac{s_3^2}{n_3}}} = \frac{10.36 - 4.85}{\sqrt{\dfrac{(15.79)^2}{77} + \dfrac{(13.73)^2}{76}}} = \frac{5.51}{\sqrt{5.718}} = 2.30$$

Reject H_0. There is enough evidence to support the claim.

5. (a) H_0: $\mu_1 = \mu_5$ vs H_1: $\mu_1 \neq \mu_5$ (claim)

$z = \pm 2.576$ Reject H_0 if $z < -2.576$ or $z > 2.576$.

$$z = \frac{(\bar{x}_1 - \bar{x}_2) - (0)}{\sqrt{\dfrac{s_1^2}{n_1} + \dfrac{s_5^2}{n_5}}} = \frac{10.36 - 7.05}{\sqrt{\dfrac{(15.79)^2}{77} + \dfrac{(7.28)^2}{42}}} = \frac{3.31}{\sqrt{4.500}} = 1.56$$

Fail to reject H_0. There is not enough evidence to support the claim.

(b) H_0: $\mu_4 = \mu_5$ vs H_1: $\mu_4 \neq \mu_5$ (claim)

$z = \pm 2.576$ Reject H_0 if $z < -2.576$ or $z > 2.576$.

$$z = \frac{(\bar{x}_4 - \bar{x}_2) - (0)}{\sqrt{\dfrac{s_4^2}{n_4} + \dfrac{s_5^2}{n_5}}} = \frac{3.53 - 7.05}{\sqrt{\dfrac{(12)^2}{79} + \dfrac{(7.28)^2}{42}}} = \frac{-3.52}{\sqrt{3.085}} = -2.00$$

Fail to reject H_0. There is not enough evidence to support the claim.

6. There was enough evidence to support the claim that the mean weight losses of the Atkins diet and the LEARN diet were different.

There was enough evidence to support the claim that the mean weight losses of the Atkins diet and the Ornish diet were different.

CHAPTER 9

CASE STUDY: *CORRELATION OF BODY MEASUREMENTS*

1. Answers will vary.

2.

	r
a	0.698
b	0.746
c	0.351
d	0.953
e	0.205
f	0.580
g	0.844
h	0.798
i	0.116
j	0.954
k	0.710
l	0.710

3. d, g, and j have strong correlations ($r > 0.8$).

   (d) $\hat{y} = 1.129x - 8.817$

   (g) $\hat{y} = 1.717x + 23.144$

   (j) $\hat{y} = 1.279x + 10.451$

4. (a) $\hat{y} = 0.086(180) + 21.923 = 37.403$

   (b) $\hat{y} = 1.026(100) - 15.323 = 87.277$

5. (weight, chest) $\rightarrow r = 0.888$
(weight, hip) $r = 0.930$
(neck, wrist) $r = 0.849$
(chest, hip) $r = 0.953$
(chest, thigh) $r = 0.936$
(chest, knee) $r = 0.855$
(chest, ankle) $r = 0.908$
(abdom, thigh) $r = 0.863$
(abdom, knee) $r = 0.903$
(hip, thigh) $r = 0.894$
(hip, ankle) $r = 0.908$
(thigh, knee) $r = 0.954$
(thigh, ankle) $r = 0.874$
(ankle, knee) $r = 0.857$
(forearm, wrist) $r = 0.854$

CHAPTER 10

CASE STUDY: *TRAFFIC SAFETY FACTS*

1. $43{,}443(0.24) \approx 10{,}426.32 \approx 10{,}427$

2. $\dfrac{3478}{14312} \approx 0.243 > 0.24 \rightarrow$ Central United States and

$\dfrac{1655}{6751} \approx 0.245 > 0.24 \rightarrow$ Western United States

3. The number of fatalities in the Eastern, Central, and Western United States for the 25–34 age group exceeded the expected number of fatalities.

4. H_0: Region is independent of age.

H_a: Region is dependent on age.

d.f. $= (r - 1)(c - 1) = 14$

$\chi_0^2 = 23.685 \rightarrow$ Reject H_0 if $\chi^2 > 23.685$.

$\chi^2 \approx 71.504$

Reject H_0. There is enough evidence at the 5% significance level to conclude that region is dependent on age.

5. H_0: Eastern United States has same distribution as the entire United States.
H_a: Eastern United States has a different distribution than the entire United States.

$\chi_0^2 = 14.067 \rightarrow$ Reject H_0 if $\chi^2 > 14.067$.

$\chi^2 \approx 14.694$

Reject H_0. There is enough evidence at the 5% significance level to conclude that the eastern United States has a different distribution than the entire United States.

6. H_0: Central United States has same distribution as the entire United States.

H_a: Central United States has a different distribution than the entire United States.

$\chi_0^2 = 14.067 \rightarrow$ Reject H_0 if $\chi^2 > 14.067$.

$\chi^2 \approx 38.911$

Reject H_0. There is enough evidence at the 5% significance level to conclude that the central United States has a different distribution than the entire United States.

7. H_0: Western United States has same distribution as the entire United States.

H_a: Western United States has a different distribution than the entire United States.

$\chi_0^2 = 14.067 \rightarrow$ Reject H_0 if $\chi^2 > 14.067$.

$\chi^2 \approx 58.980$

Reject H_0. There is enough evidence at the 5% significance level to conclude that the western United States has a different distribution than the entire United States.

8. Alcohol, vehicle speed, driving conditions, etc.

CHAPTER 11

CASE STUDY: *EARNINGS BY COLLEGE DEGREE*

1. (a), (b), (c), (d)

The doctorate groups appear to make more money than the bachelor groups.

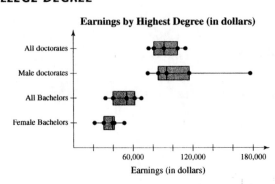

Earnings by Highest Degree (in dollars)

Earnings (in dollars)

2. H_0: median $\leq$ \$92,000 (claim); H_a: median $>$ \$92,000

The critical value is 1.

$x = 5$

Fail to reject H_0. There is not enough evidence to reject the claim.

3. H_0: median = \$102,500 (claim); H_a: median $\neq$ \$102,500

The critical value is 1.

$x = 3$

Fail to reject H_0. There is not enough evidence to reject the claim.

4. H_0: median $\geq$ \$56,000 (claim); H_a: median $<$ \$56,000

The critical value is 1.

$x = 4$

Fail to reject H_0. There is not enough evidence to reject the claim.

5. H_0: median = \$42,000; H_a: median $\neq$ \$42,000 (claim)

The critical value is 1.

$x = 1$

Reject H_0. There is enough evidence to support the claim.

6. H_0: There is no difference in median earnings. (claim)

H_a: There is a difference in median earnings.

The critical values are $z_0 = \pm 2.575$.

$R = 98$

$$\mu_R = \frac{n_1(n_1 + n_2 + 1)}{2} = \frac{10(10 + 10 + 1)}{2} = 105$$

$$\sigma_R = \sqrt{\frac{n_1 n_2(n_1 + n_2 + 1)}{12}} = \sqrt{\frac{(10)(10)(10 + 10 + 1)}{12}} \approx 13.2$$

$$z = \frac{R - \mu_R}{\sigma_R} = \frac{98 - 105}{13.2} \approx -0.530$$

Fail to reject H_0. There is not enough evidence to reject the claim.

7. H_0: There is no difference in median earnings.

H_a: There is a difference in median earnings. (claim)

The critical values are $z_0 = \pm 2.575$.

$R = 140$

$$\mu_R = \frac{n_1(n_1 + n_2 + 1)}{2} = \frac{10(10 + 10 + 1)}{2} = 105$$

$$\sigma_R = \sqrt{\frac{n_1 n_2(n_1 + n_2 + 1)}{12}} = \sqrt{\frac{(10)(10)(10 + 10 + 1)}{12}} \approx 13.2$$

$$z = \frac{R - \mu_R}{\sigma_R} = \frac{140 - 105}{13.2} \approx 2.652$$

Reject H_0. There is enough evidence to support the claim.

CHAPTER 1

USES AND ABUSES

1. Answers will vary.

2. Answers will vary.

CHAPTER 2

USES AND ABUSES

1. Answers will vary.

2. Answers will vary.

CHAPTER 3

USES AND ABUSES

1. (a) P(winning Tues and Wed) = P(winning Tues) $\cdot$ P(winning Wed)

$$= \left(\frac{1}{1000}\right) \cdot \left(\frac{1}{1000}\right)$$

$$= 0.000001$$

(b) P(winning Wed given won Tues) = P(winning Wed)

$$= \frac{1}{1000}$$

$$= 0.001$$

(c) P(winning Wed given didn't win Tues) = P(winning Wed)

$$= \frac{1}{1000}$$

$$= 0.001$$

2. Answers will vary.

P(pickup or SUV) $\leq$ 0.55 because P(pickup) = 0.25 and P(SUV) = 0.30, but a person may own both a pick-up and an SUV (ie not mutually exclusive events). So, P(pickup and SUV) $\geq$ 0 and this probability would have to be subtracted from 0.55.

If the events were mutually exclusive, the value of the probability would be 0.55. Otherwise, the probability will be less than 0.55 (not 0.60).

USES AND ABUSES

1. 40

$P(40) = 0.081$

2. $P(35 \le x \le 45) = P(35) + P(36) + \cdots + P(45)$

$= 0.7386$

3. The probability of finding 36 adults out of 100 who prefer Brand A is 0.059. So the manufacturer's claim is hard to believe.

4. The probability of finding 25 adults out of 100 who prefer Brand A is 0.000627. So the manufacturer's claim is not be believable.

CHAPTER 5

USES AND ABUSES

1. $\mu = 100$

$\sigma = 15$

(a) $z = \dfrac{\bar{x} - \mu}{\sigma / \sqrt{n}} = \dfrac{115 - 100}{15 / \sqrt{3}} = 1.73$

$\bar{x} = 115$ is not an unusual sample mean.

(b) $z = \dfrac{\bar{x} - \mu}{\sigma / \sqrt{n}} = \dfrac{105 - 100}{15 / \sqrt{20}} = 1.49$

$\bar{x} = 105$ is not an unusual sample mean.

2. The problem does not state the population of ages is normally distributed.

3. Answers will vary.

CHAPTER 6

USES AND ABUSES

1. Answers will vary.

2. Answers will vary.

CHAPTER 7

USES AND ABUSES

1. Randomly sample car dealerships and gather data necessary to answer question. (Answers will vary.)

2. H_0: $p = 0.57$
   We cannot prove that $p = 0.57$. We can only show that there is insufficient evidence in our sample to reject H_0: $p = 0.57$. (Answers will vary.)

3. If we gather sufficient evidence to reject the null hypothesis when it is really true, then a Type I error would occur. (Answers will vary.)

4. If we fail to gather sufficient evidence to reject the null hypothesis when it is really false, then a Type II error would occur. (Answers will vary.)

CHAPTER 8

USES AND ABUSES

1. Age and health. (Answers will vary.)

2. Blind: The patients do not know which group (medicine or placebo) they belong to.

   Double Blind: Both the researcher and patient do not know which group (medicine or placebo) that the patient belongs to.

   (Answers will vary.)

CHAPTER 9

USES AND ABUSES

1. Answers will vary.

2. Answers will vary. One example would be temperature and output of sulfuric acid. When sulfuric acid is manufactured, the amount of acid that is produced depends on the temperature at which the industrial process is run. As the temperature increases, so does the output of acid, up to a point. Once the temperature passes that point, the output begins to decrease.

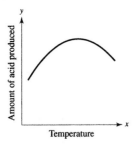

CHAPTER 10

USES AND ABUSES

1. Answers will vary. An example would be a biologist who is analyzing effectiveness of three types of pesticides. The biologists sprays one pesticide on five different acres during week one. At the end of week one, he calculates the mean number of insects in the acres. He sprays the second pesticide on the same five acres during week two. At the end of week two, he calculates the mean number of insects in the acres. He follows a similar procedure with the third pesticide during week three. He wants to determine whether there is a difference in the mean number of insects per acre. Since the same five acres are treated each time, it is unclear as to the effectiveness of each of the pesticides.

2. Answers will vary. An example would be another biologist who is analyzing the effectiveness of three types of pesticides. He has fifteen acres to test. He randomly assigns each pesticide to five acres and treats all of the acres for three weeks. At the end of three weeks, the biologist calculates the mean number of insects for each of the pesticides. He wants to determine whether there is a difference in the mean number of insects per acre.

   Rejecting the null hypothesis would mean that at least one of the means differs from the others.

   In this example, rejecting the null hypothesis means that there is at least one pesticide whose mean number of insects per acre differs from the other pesticides.

CHAPTER 11

USES AND ABUSES

1. (Answers will vary.)

   H_0: median $\geq$ 10

   H_0: median $<$ 10

   Using $n = 5$ and $\alpha = 0.05$, the $cv = 0$ ($\alpha = 0.031$). Thus, every item in the sample would have to be less than 10 in order to reject the H_0. (i.e. 100% of the sample would need to be less than 10.)

   However, using $n = 20$ and $\alpha = 0.05$, the $cv = 6$ ($\alpha = 0.057$). Now only 14 items (or 70%) of the sample would need to be less than 10 in order to reject H_0.

2. | *Nonparametric Test* | *Parametric Test* |
   |---|---|
   | (a) Sign Test | z-test |
   | (b) Paired Sample Sign Test | t-test |
   | (c) Wilcoxon Signed-Rank Test | t-test |
   | (d) Wilcoxon Rank Sum Test | z-test or t-test |
   | (e) Kruskal-Wallis Test | one-way Anova |
   | (f) Spearmans Rank Correlation Coefficient | Pearson correlation coefficient |

CHAPTER 1

REAL STATISTICS–REAL DECISIONS

1. (a) If the survey identifies the type of reader that is responding, then stratified sampling ensures that representatives from each group of readers (engineers, manufacturers, researchers, and developers) are included in the sample.

   (b) Yes

   (c) After dividing the population of readers into their subgroups (strata), a simple random sample of readers of the same size is taken from each subgroup.

   (d) You may take too large a percentage of your sample from a subgroup of the population that is relatively small.

2. (a) Once the results have been compiled you have quantitative data, because "Percent Responding" is a variable consisting of numerical measurements.

   (b) "Percent Responding" is a ratio level of measurement since the data can be ordered and you can calculate meaningful differences between data entries.

   (c) Sample

   (d) Statistics

3. (a) Using Internet surveys introduces bias into the sampling process. The data was not collected randomly and may not adequately represent the population.

   (b) Stratified sampling would have been more appropriate assuming the population of readers consists of various subgroups that should be represented in the sample.

CHAPTER 2

REAL STATISTICS–REAL DECISIONS

1. (a) Examine data from the four cities and make comparisons using various statistical methodologies. For example, compare the average price of automobile insurance for each city.

   (b) Calculate: Mean, range, and population standard deviation for each city.

2. (a) Construct a Pareto chart because the data in use are quantitative and a Pareto chart positions data in order of decreasing height, with the tallest bar positioned at the left.

   (b)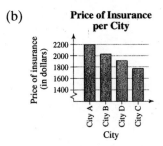

   (c) Yes

3. (a) Find the mean, range, and population standard deviation for each city.

(b) City A: $\bar{x}$ = \$2191.00 $\sigma \approx$ \$351.86   range = \$1015.00

City B: $\bar{x}$ = \$2029.20 $\sigma \approx$ \$437.54   range = \$1336.00

City C: $\bar{x}$ = \$1772.00 $\sigma \approx$ \$418.52   range = \$1347.00

City D: $\bar{x}$ = \$1909.30 $\sigma \approx$ \$361.14   range = \$1125.00

(c) The mean from Cities B, C, and D appear to be similar. However, City A appears to have the highest price of insurance. These measurements support the conclusion in Exercise 2.

4. (a) You would tell your readers that on average, the price of automobile insurance is higher in this city than in other cities.

(b) Location of city, weather, population.

CHAPTER 3

REAL STATISTICS–REAL DECISIONS

1. (a) Answers will vary. Investigate the probability of not matching any of the 5 white balls selected.

(b) You could use the Multiplication Rule, the Fundamental Counting Principle, and Combinations.

2. If you played only the red ball, the probability of matching it is $\frac{1}{42}$. However, because you must pick 5 white balls, you must get the white balls wrong. So, using the Multiplication Rule, we get

$$P\left(\begin{array}{c}\text{matching only the red ball and not} \\ \text{matching any of the 5 white balls}\end{array}\right) = P(\text{matching red ball}) \cdot P(\text{not matching any white balls})$$

$$= \frac{1}{42} \cdot \frac{50}{55} \cdot \frac{49}{54} \cdot \frac{48}{53} \cdot \frac{47}{52} \cdot \frac{46}{51}$$

$$\approx 0.015$$

$$\approx \frac{3}{200}.$$

3. Using combinations, calculate the number of ways to win, and divide by the number of outcomes.

CHAPTER 4

REAL STATISTICS–REAL DECISIONS

1. (a) Answers will vary. For example, calculate the probability of obtaining zero clinical pregnancies out of 10 randomly selected ART cycles.

   (b) Binomial
   Discrete, we are counting the number of successes (clinical pregnancies).

2. Using a binomial distribution where $n = 10$, $p = 0.337$, we find $P(0) = 0.0164$.

   It is not impossible to obtain zero clinical pregnancies out of 10 ART cycles. However, it is rather unlikely to occur.

3. (a) Using $n = 10$, $p = 0.230$, $P(8) = 0.0002$

   Suspicious, because the probability is very small.

   (b) Using $n = 10$, $p = 0.192$, $P(0) = 0.1186$

   Not suspicious, because the probability is not that small.

CHAPTER 5

REAL STATISTICS–REAL DECISIONS

1. (a) $z = \dfrac{x - \mu}{\sigma} = \dfrac{11.55 - 11.56}{0.05} = -0.2$

   $P(\text{not detecting shift}) = P(x < 11.55) = P(z < -0.2) = 0.4207$

   (b) $n = 15 \qquad p = 0.4207 \qquad q = 0.5793$

   $np = (15)(0.4207) \approx 6.3105 > 5 \quad$ and $\quad nq = (15)(0.5793) \approx 8.6895 > 5$

   $\sigma = \sqrt{npq} = \sqrt{(15)(0.4207)(0.5793)} = \sqrt{3.656} \approx 1.9121 \quad \mu = 6.3105$

   $x = 0.5 \leftarrow$ correction for continuity

   $z = \dfrac{0.5 - 6.3105}{1.9121} \approx \dfrac{-5.8105}{1.9121} \approx -3.04$

   $P(x \geq 0.5) = P(z \geq -3.04) \approx 1 - 0.0012 = 0.9988$

2. (a) $z = \dfrac{x - \mu}{\dfrac{\sigma}{\sqrt{n}}} = \dfrac{11.55 - 11.56}{\dfrac{0.05}{\sqrt{5}}} = \dfrac{-0.01}{0.0224} = -0.45$

   $P(\text{not detecting shift}) = P(\bar{x} < 11.55) = P(z < -0.45) = 0.3264$

(b) $n = 3$ $p = 0.3264$ $q = 0.6736$

$np = (3)(0.3264) \approx 0.9792 < 5$ and $nq = (3)(0.6736) \approx 2.0208 < 5$

$P(\text{at least } 1) = P(1) + P(2) + P(3)$

$= 0.4443 + 0.2153 + 0.0348$

$= 0.6944$

(c) The mean is more sensitive to change.

3. Answers will vary.

REAL STATISTICS–REAL DECISIONS

1. (a) No, there has not been a change in the mean concentration levels because the confidence interval for year 1 overlaps with the confidence interval for year 2.

   (b) Yes, there has been a change in the mean concentration levels because the confidence interval for year 2 does not overlap with the confidence interval for year 3.

   (c) Yes, there has been a change in the mean concentration levels because the confidence interval for year 1 does not overlap with the confidence interval for year 3.

2. Due to the fact the CI from year 1 does not overlap the CI from year 3, it is very likely that the efforts to reduce the concentration of formaldehyde in the air are significant over the 3-year period.

3. (a) Used the sampling distribution of the sample means due to the fact that the "mean concentration" was used. The point estimate is the most unbiased estimate of the population mean.

   (b) No, it is more likely the sample standard deviation of the current year's sample was used. Typically σ is unknown.

REAL STATISTICS–REAL DECISIONS

1. (a) Take a random sample to get a diverse group of people.
      Stratified random sampling

   (b) Random sample

   (c) Answers will vary.

2. $H_0: p \geq 0.40$; $H_a: p < 0.40$ (claim)

$z_0 = -1.282$

$\hat{p} = \dfrac{x}{n} = \dfrac{15}{32} \approx 0.469$

$z = \dfrac{\hat{p} - p}{\sqrt{\dfrac{pz}{n}}} = \dfrac{0.469 - 0.40}{\sqrt{\dfrac{(0.40)(0.60)}{32}}} = \dfrac{0.069}{0.087} = 0.794$

Fail to reject H_0. There is not enough evidence to support the claim.

3. $H_0: \mu \geq 60$ (claim); $H_a: \mu < 60$

$\bar{x} = 53.067 \quad s = 19.830 \quad n = 15$

$t_0 = -1.345$

$t = \dfrac{\bar{x} - \mu}{\dfrac{s}{\sqrt{n}}} = \dfrac{53.067 - 60}{\dfrac{19.83}{\sqrt{15}}} = \dfrac{-6.933}{5.120} = -1.354$

Reject H_0. There is enough evidence to reject the claim.

4. There is not enough evidence to conclude the proportions of Americans who think Social Security will have money to provide their benefits is less than 40%. There is enough evidence to conclude the mean age of those individuals responding "yes" is less than 60. (Answers will vary.)

CHAPTER 8

REAL STATISTICS–REAL DECISIONS

1. (a) Take a simple random sample of records today and 10 years ago from a random sample of hospitals.

Cluster sampling

(b) Answers will vary.

(c) Answers will vary.

2. t-test; Independent

3. $H_0: \mu_1 = \mu_2$; $H_a: \mu_1 \neq \mu_2$ (claim)

d.f. $= \min\{n_1 - 1, n_2 - 1\} = 19$

$t_0 = \pm 2.093$

$t = \dfrac{(\bar{x}_1 - \bar{x}_2) - (\mu_1 - \mu_2)}{\sqrt{\dfrac{s_1^2}{n_1} + \dfrac{s_2^2}{n_2}}} = \dfrac{(3.7 - 3.2) - (0)}{\sqrt{\dfrac{(1.5)^2}{20} + \dfrac{(0.8)^2}{25}}} = \dfrac{0.5}{\sqrt{0.1381}} \approx 1.345$

Fail to reject H_0. There is not enough evidence to support the claim.

CHAPTER 9

REAL STATISTICS–REAL DECISIONS

1. (a)

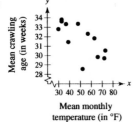

Negative linear correlation

(b)

Mean monthly temperature, x	Mean crawling age, y	xy	x^2	y^2
33	33.64	1110.12	1089.00	1131.65
30	32.82	984.60	900.00	1077.15
33	33.83	1116.39	1089.00	1144.47
37	33.35	1233.95	1369.00	1112.22
48	33.38	1602.24	2304.00	1114.22
57	32.32	1842.24	3249.00	1044.58
66	29.84	1969.44	4356.00	890.43
73	30.52	2227.96	5329.00	931.47
72	29.70	2138.40	5184.00	882.09
63	31.84	2005.92	3969.00	1013.79
52	28.58	1486.16	2704.00	816.82
39	31.44	1226.16	1521.00	988.47
$\Sigma x = 603$	$\Sigma y = 381.26$	$\Sigma xy = 18{,}943.58$	$\Sigma x^2 = 33{,}063$	$\Sigma y^2 = 12{,}147.36$

$$r = \frac{n(\Sigma xy) - (\Sigma x)(\Sigma y)}{\sqrt{n\Sigma x^2 - (\Sigma x)^2}\ \sqrt{n\Sigma y^2 - (\Sigma y)^2}} = \frac{12(18{,}943.58) - (603)(381.26)}{\sqrt{12(33{,}063) - (603)^2}\ \sqrt{12(12{,}147.36) - (381.26)^2}} = -0.700$$

(c) H_0: $\rho = 0$; H_a: $\rho \neq 0$

$n = 12 \Rightarrow cv = 0.576$

$|r| = 0.700 > 0.576 \Rightarrow$ Reject H_0. There is a significant linear correlation.

(d) $m = \dfrac{n(\Sigma xy) - (\Sigma x)(\Sigma y)}{n(\Sigma x^2) - (\Sigma x)^2} = \dfrac{12(18{,}943.58) - (603)(381.26)}{12(33{,}063) - (603)^2} = -0.078$

$b = \bar{y} - m\bar{x} = \left(\dfrac{381.26}{12}\right) - (-0.078)\left(\dfrac{603}{12}\right) = 35.678$

$\hat{y} = -0.078x + 35.678$

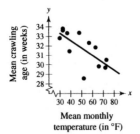

The regression line appears to be a good fit.

(e) Because the regression line is a good fit, it is reasonable to use temperature to predict mean crawling age.

(f) $r^2 = 0.490$

$s_e = 1.319$

49% of the variation in y is explained by the regression line. The standard error of estimate of mean crawling age for a specific mean monthly temperature is 1.319 weeks.

2. Agree. Due to high correlation, parents can "consider" the temperature when determining when their baby will crawl. However, the parents should not "conclude" that temperature is the other variable related to mean crawling age.

CHAPTER 10

REAL STATISTICS–REAL DECISIONS

1.

Ages	Distribution	Observed	Expected	$\frac{(O - E)^2}{E}$
Under 20	1%	30	10	40.000
20–29	13%	200	130	37.692
30–39	16%	300	160	122.500
40–49	19%	270	190	33.684
50–59	16%	150	160	0.625
60–69	13%	40	130	62.308
70+	22%	10	220	200.455
				$\chi^2 = 497.26$

H_0: Distribution of ages is as shown in the table above. (claim)

H_a: Distribution of ages differs from the claimed distribution.

d.f. $= n - 1 = 6$

$\chi_0^2 = 16.812$

Reject H_0. There is enough evidence at the 1% level to conclude the distribution of ages of telemarketing fraud victims differs from the survey.

2. (a)

Type of Fraud	Age								Total
	Under 20	20–29	30–39	40–49	50–59	60–69	70–79	80+	
Sweepstakes	10 (15)	60 (120)	70 (165)	130 (185)	90 (135)	160 (115)	280 (155)	200 (110)	1000
Credit cards	20 (15)	180 (120)	260 (165)	240 (185)	180 (135)	70 (115)	30 (155)	20 (110)	1000
Total	30	240	330	370	270	230	310	220	2000

(b) H_0: The type of fraud is independent of the victim's age.

H_a: The type of fraud is dependent on the victim's age.

d.f. $= (r - 1)(c - 1) = (1)(7) = 7$

$\chi_0^2 = 18.475$

$$\chi^2 = \Sigma \frac{(O - E)^2}{E} \approx 619.533$$

Reject H_0. There is enough evidence at the 1% significance level to conclude that the victim's ages and type of fraud are dependent.

CHAPTER 11

REAL STATISTICS–REAL DECISIONS

1. (a) random sample

   (b) Answers will vary.

   (c) Answers will vary.

2. (a) Answers will vary. For example, ask "Is the data a random sample?" and "Is the data normally distributed?"

   (b) Sign test; You need to use the nonparametric test because nothing is known about the shape of the population.

   (c) H_0: median ≥ 4.0; H_0: median < 4.0 (claim)

   (d) $\alpha = 0.05$

   $n = 19$

   $cv = 5$

   $x = 7$

   Fail to reject H_0. There is not enough evidence at the 5% significance level to support the claim that the median tenure is less than 4.0 years.

3. (a) Because the data from each sample appears non-normal, the Wilcoxon Rank Sum test should be used.

   (b) H_0: The median tenure for male workers is less than or equal to the median tenure for female workers.

   H_a: The median tenure for male workers is greater than the median tenure for female workers. (claim)

(c) $R = 164.5$

$$\mu_R = \frac{n_1(n_1 + n_2 + 1)}{2} = \frac{12(12 + 14 + 1)}{2} = 162$$

$$\sigma_R = \sqrt{\frac{n_1 n_2(n_1 + n_2 + 1)}{12}} = \sqrt{\frac{(12)(14)(12 + 14 + 1)}{12}} \approx 19.442$$

$$z = \frac{R - \mu_R}{\sigma_R} = \frac{164.5 - 162}{19.442} \approx 0.129$$

Using $\alpha = 0.05$, $z_0 = \pm 1.645$

Fail to reject H_0. There is not enough evidence at the 5% significance level to conclude that the median tenure for male workers is greater than the median tenure for female workers.

CHAPTER 1

TECHNOLOGY: *USING TECHNOLOGY IN STATISTICS*

1. From a list of numbers ranging from 1 to 86, randomly select eight of them. Answers will vary.

2. From a list of numbers ranging from 1 to 300, randomly select 25 of them. Answers will vary.

3. From a list of numbers ranging from 0 to 9, randomly select five of them. Repeat this two times. Average these three numbers and compare with the population average of 4.5. Answers will vary.

4. The average of the numbers 0 to 40 is 20. From a list of numbers ranging from 0 to 40, randomly select seven of them. Repeat this three times. Average these three numbers and compare with the population average of 20. Answers will vary.

5. Answers will vary.

6. No, we would anticipate 10-1's, 10-2's, 10-3's, 10-4's, 10-5's, and 10-6's. An inference that we might draw from the results is that the die is not a fair die.

7. Answers will vary.

8. No, we would anticipate 50 heads and 50 tails. An inference that we might draw from the results is that the coin is not fair.

9. The analyst could survey 10 counties by assigning each county a number from 1 to 47 and using a random number generator to find 10 numbers that correspond to certain counties.

CHAPTER 2

TECHNOLOGY: *MONTHLY MILK PRODUCTION*

1. $\bar{x} = 2270.5$

2. $s = 653.2$

3.

Lower limit	Upper limit	Frequency
1147	1646	7
1647	2146	15
2147	2646	13
2647	3146	11
3147	3646	3
3647	4146	0
4147	4646	1

4.

Monthly Milk Production

Monthly milk production (in pounds)

The distribution does not appear to be bell-shaped.

5. 74% of the entries were within one standard deviation of the mean (1617.3, 2923.7).
98% of the entries were within two standard deviations of the mean (964.1, 3576.9).

6. $\bar{x} = 2316.5$

7. $s \approx 641.75$

8. Answers will vary.

CHAPTER 3

TECHNOLOGY: *SIMULATION: COMPOSING MOZART VARIATIONS WITH DICE*

1. $2 + 11 = 13$ phrases were written.

2. $(11)^7 \cdot 2 \cdot (11)^7 \cdot 2 = 1.518999334333 \times 10^{15}$

3. (a) $\dfrac{1}{11} \approx 0.091$

(b) Results will vary.

4. (a) $P(\text{option 6, 7, or 8 for 1st bar}) = \dfrac{3}{11} = 0.273$

$P(\text{option 6, 7, or 8 for all 14 bars}) = \left(\dfrac{3}{11}\right)^{14} \approx 0.000000012595$

(b) Results will vary.

5. (a) $P(1) = \dfrac{1}{36}$ $\quad P(2) = \dfrac{2}{36}$ $\quad P(3) = \dfrac{3}{36}$ $\quad P(4) = \dfrac{4}{36}$

$P(5) = \dfrac{5}{36}$ $\quad P(6) = \dfrac{6}{36}$ $\quad P(7) = \dfrac{5}{36}$ $\quad P(8) = \dfrac{4}{36}$

$P(9) = \dfrac{3}{36}$ $\quad P(10) = \dfrac{2}{36}$ $\quad P(11) = \dfrac{1}{36}$

(b) Results will vary.

6. (a) $P(\text{option 6, 7, or 8 for 1st bar}) = \dfrac{15}{36} \approx 0.417$

$P(\text{option 6, 7, or 8 for all 14 bars}) = \left(\dfrac{15}{36}\right)^{14} \approx 0.00000475$

(b) Results will vary.

CHAPTER 4

TECHNOLOGY: *USING POISSON DISTRIBUTIONS AS QUEUING MODELS*

1.

x	P(x)
0	0.0183
1	0.0733
2	0.1465
3	0.1954
4	0.1954
5	0.1563
6	0.1042
7	0.0595
8	0.0298
9	0.0132
10	0.0053
11	0.0019
12	0.0006
13	0.0002
14	0.0001
15	0.0000
16	0.0000
17	0.0000
18	0.0000
19	0.0000
20	0.0000

2. (a) (See part b) 1, 2, 4, 7

(b)

Minute	Customers Entering Store	Total Customers in Store	Customers Serviced	Customers Remaining
1	3	3	3	0
2	3	3	3	0
3	3	3	3	0
4	3	3	3	0
5	5	5	4	1
6	5	6	4	2
7	6	8	4	4
8	7	11	4	7
9	3	10	4	6
10	6	12	4	8
11	3	11	4	7
12	5	12	4	8
13	6	14	4	10
14	3	13	4	9
15	4	13	4	9
16	6	15	4	11
17	2	13	4	9
18	2	11	4	7
19	4	11	4	7
20	1	8	4	4

3. Answers will vary.

4. 20 customers (an additional 1 customer would be forced to wait in line/minute)

5. Answers will vary.

6. $P(10) = 0.0181$

7. (a) It makes no difference if the customers were arriving during the 1st minute or the 3rd minute. The mean number of arrivals will still be 4 customers/minute.

$$P(3 \leq x \leq 5) = P(3) + P(4) + P(5) \approx 0.1954 + 0.1954 + 0.1563 = 0.5471$$

(b) $P(x > 4) = 1 - P(x \leq 4) = 1 - [P(0) + P(1) + \cdots + P(4)] = 1 - [0.6289] = 0.3711$

(c) $P(x > 4$ during **each** of the first four minutes$) = (0.3711)^4 \approx 0.019$

CHAPTER 5

TECHNOLOGY: *AGE DISTRIBUTION IN THE UNITED STATES*

1. $\mu \approx 36.59$

2. $\bar{x}$ of the 36 sample means ≈ 36.209
This agrees with the Central Limit Theorem.

3. No, the distribution of ages appears to be positively skewed.

4.
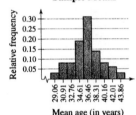
Distribution of 36 Sample Means

The distribution is approximately bell-shaped and symmetrical. This agrees with the Central Limit Theorem.

5. $\sigma \approx 22.499$

6. σ of the 36 sample means $= 3.552$.

$$\frac{\sigma}{\sqrt{n}} = \frac{22.499}{\sqrt{36}} \approx 3.750$$

This agrees with the Central Limit Theorem.

CHAPTER 6

TECHNOLOGY: *MOST ADMIRED POLLS*

1. $\hat{p} = 0.13$

$$\hat{p} \pm z_c \sqrt{\frac{\hat{p}\hat{q}}{n}} = 0.013 \pm 1.96 \sqrt{\frac{0.013 \cdot 0.87}{1010}} = 0.13 \pm 0.021 \approx (0.109, 0.151)$$

2. No, the proportion is 13% plus or minus 2.1%.

3. $\hat{p} = 0.09$

$$\hat{p} \pm z_c \sqrt{\frac{\hat{p}\hat{q}}{n}} = 0.09 \pm 1.96 \sqrt{\frac{0.09 \cdot 0.91}{1010}} = 0.09 \pm 0.018 \approx (0.072, 0.108)$$

4. Answers will vary.

5. No, because the 95% CI does not contain 10%.

CHAPTER 7

TECHNOLOGY: *THE CASE OF THE VANISHING WOMEN*

1. Reject H_0 with a P-value less than 0.01.

2. Type I error

3. Sampling process was non-random.

4. (a) H_0: $p = 0.2914$ (claim); H_a: $p \neq 0.2914$

   (b) Test and Confidence Interval for One Proportion

   Test of $p = 0.2914$ vs p not $= 0.2914$

Exact Sample	X	N	Sample p	99.0% CI	Z-Value	P-Value
1	9	100	0.090000	(0.016284, 0.163716)	−4.43	0.000

   (c) Reject H_0.

   (d) It was highly unlikely that random selection produced a sample of size 100 that contained only 9 women.

CHAPTER 8

TECHNOLOGY: *TAILS OVER HEADS*

1. Test and Confidence Interval for One Proportion

   Test of $p = 0.5$ vs. p not $= 0.5$

Exact Sample	X	N	Sample p	95.0% CI	P-Value
1	5772	11902	0.484961	(0.47598, 0.49394)	0.001

   Reject H_0: $p = 0.5$

2. Yes, obtaining 5772 heads is a very uncommon occurrence (see sampling distribution). The coins might not be fair.

3. $\frac{1}{500} \cdot 100 = 0.2\%$

4. $z = 8.802 \rightarrow$ Fail to reject H_0: $\mu_1 = \mu_2$ There is sufficient evidence at the 5% significance level to suppport the claim that there is a difference in the mint dates in Philadelphia and Denver. (Answers will vary.)

5. $z = 1.010 \rightarrow$ Fail to reject H_0: $\mu_1 = \mu_2$ There is insufficient evidence at the 5% significance level to support the claim that there is a difference in the value of the coins in Philadelphia and Denver. (Answers will vary.)

<hr/>

CHAPTER 9

TECHNOLOGY: *NUTRIENTS IN BREAKFAST CEREALS*

1.

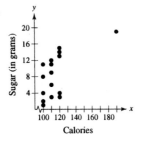

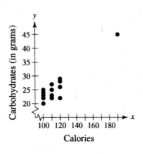

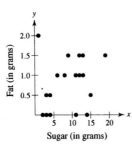

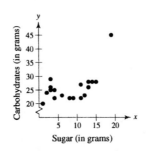

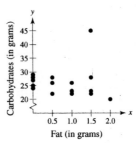

2. (calories, sugar), (calories, carbohydrates), (sugar, carbohydrates)

3.

(calories, sugar):	$r \approx 0.566$
(calories, fat):	$r \approx 0.194$
(calories, carbohydrates):	$r \approx 0.927$
(sugar, fat):	$r \approx 0.311$
(sugar, carbohydrates):	$r \approx 0.544$
(fat, carbohydrates):	$r \approx -0.021$

Largest r: (calories, carbohydrates)

4. (a) $\hat{y} = -8.823 + 0.151x$

(b) $\hat{y} = -3.417 + 0.252x$

5. (a) $\hat{y} = -8.823 + 0.151(120) = 9.297$ grams

(b) $\hat{y} = -3.417 + 0.252(120) = 26.823$ grams

6. (a) $C = 22.055 - 0.043S + 6.644F + 3.455R$

(b) $C = 29.427 + 0.329S + 3.240R$

7. $C = 29.427 + 0.329(10) + 3.240(25) = 113.717$ calories

CHAPTER 10

TECHNOLOGY: *TEACHER SALARIES*

1. Yes, since the salaries are from different states, it is safe to assume that the samples are independent.

2. Using a technology tool, it appears that all three samples were taken from approximately normal populations.

3. H_0: $\sigma_1^2 = \sigma_2^2$; H_a: $\sigma_1^2 \neq \sigma_2^2$ ($F_0 = 2.86$)

(CA, OH) $\rightarrow F \approx 1.048 \rightarrow$ Fail to reject H_0.

(CA, WY) $\rightarrow F \approx 1.114 \rightarrow$ Fail to reject H_0.

(OH, WY) $\rightarrow F \approx 1.167 \rightarrow$ Fail to reject H_0.

4. The three conditions for a one-way ANOVA test are satisfied.

H_0: $\mu_1 = \mu_2 = \mu_3$ (claim)

H_a: At least one mean is different from the others.

Variation	Sum of Squares	Degrees of Freedom	Mean Square	F
Between	2,345,804,805	2	1,172,902,402	19,275
Within	2,738,312,005	45	60,851,377.9	

$F \approx 19.275$

Reject H_0. There is not enough evidence at the 5% significance level to reject the claim that the mean salaries are the same.

5. The samples are independent.

H_0: $\sigma_1^2 = \sigma_2^2$; H_a: $\sigma_1^2 \neq \sigma_2^2$ ($F_0 = 2.86$)

(AK, NV) $\rightarrow F \approx 1.223 \rightarrow$ Fail to reject H_0.

(AK, NY) $\rightarrow F \approx 3.891 \rightarrow$ Reject H_0.

(NV, NY) $\rightarrow F \approx 4.760 \rightarrow$ Reject H_0.

The sample from the New York shows some signs of being drawn from a non-normal population.

The three conditions for one-way ANOVA are not satisfied.

388 | TECHNOLOGY

CHAPTER 11

TECHNOLOGY: *U.S. INCOME AND ECONOMIC RESEARCH*

1.

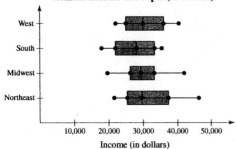

Annual Income of People (in dollars)

The median annual incomes do not appear to differ between regions.

2. H_0: median $\le$ \$25,000

   H_a: median $>$ \$25,000 (claim)

   The critical value is 2.

   $x = 1$

   Reject H_0. There is enough evidence to support the claim.

3. H_0: There is no difference in incomes in the Northeast and South. (claim)

   H_a: There is a difference in incomes in the Northeast and South.

   The critical values are $z_0 = \pm 1.96$.

   $R = 164.5$

   $$\mu_R = \frac{n_1(n_1 + n_2 + 1)}{2} = \frac{12(12 + 12 + 1)}{2} = 150$$

   $$\sigma_R = \sqrt{\frac{n_1 n_2(n_1 + n_2 + 1)}{12}} = \sqrt{\frac{(12)(12)(12 + 12 + 1)}{12}} \approx 17.3$$

   $$z = \frac{R - \mu_R}{\sigma_R} = \frac{164.5 - 150}{17.3} \approx 0.838$$

   Fail to reject H_0. There is not enough evidence to reject the claim.

4. H_0: There is no difference in the incomes for all four regions. (claim)

   H_a: There is a difference in the incomes for all four regions.

   $H = 1.03$

   Fail to reject H_0. There is not enough evidence to reject the claim.

© 2009 Pearson Education, Inc., Upper Saddle River, NJ. All rights reserved. This material is protected under all copyright laws as they currently exist. No portion of this material may be reproduced, in any form or by any means, without permission in writing from the publisher.

5. H_0: There is no difference in the incomes for all four regions. (claim)

H_a: There is a difference in the incomes for all four regions.

Analysis of Variance

Source	Sum of Squares	Degrees of Freedom	Mean Square	F	P
Factor	95,270,310	3	31,756,770	0.73	0.537
Error	1.903×10^9	44	43,247,860		
Total	1.998×10^9	47			

$F \approx 0.74 \rightarrow P\text{-value} = 0.537$

Fail to reject H_0. There is not enough evidence to reject the claim.

6. *(Box-and-whisker plot)*

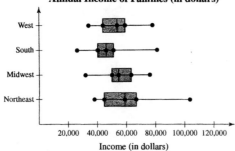

Annual Income of Families (in dollars)

The median family incomes appear to be higher in the Northeast and lower in the South.

(Wilcoxon rank sum test)

H_0: There is no difference in incomes in the Northeast and South. (claim)

H_a: There is a difference in incomes in the Northeast and South.

The critical values are $z_0 = \pm 1.96$.

$R = 281$

$$\mu_R = \frac{n_1(n_1 + n_2 + 1)}{2} = \frac{15(15 + 15 + 1)}{2} = 232.5$$

$$\sigma_R = \sqrt{\frac{n_1 n_2(n_1 + n_2 + 1)}{12}} = \sqrt{\frac{(15)(15)(15 + 15 + 1)}{12}} \approx 24.1$$

$$z = \frac{R - \mu_R}{\sigma_R} = \frac{281.0 - 232.5}{24.1} \approx 2.012$$

Reject H_0. There is enough evidence to reject the claim.

(Kruskal-Wallis test)

H_0: There is no difference in incomes for all four regions. (claim)

H_a: There is a difference in incomes for all four regions.

$H = 6.09$

Fail to reject H_0. There is not enough evidence to reject the claim.

(ANOVA test)

H_0: There is no difference in incomes for all four regions. (claim)

H_a: There is a difference in incomes for all four regions.

Analysis of Variance

Source	Sum of Squares	Degrees of Freedom	Mean Square	F	P
Factor	1.283×10^9	3	427,704,638	2.04	0.118
Error	1.172×10^{10}	56	209,309,693		
Total	1.300×10^{10}	59			

$F \approx 1.93 \rightarrow p\text{-value} = 0.135$

Fail to reject H_0. There is not enough evidence to reject the claim.